# MACHINE DESIGN CALCULATIONS REFERENCE GUIDE

# The McGraw-Hill Engineering Reference Guide Series

This series makes available to professionals and students a wide variety of engineering information and data available in McGraw-Hill's library of highly acclaimed books and publications. The books in the series are drawn directly from this vast resource of titles. Each one is either a condensation of a single title or a collection of sections culled from several titles. The Project Editors responsible for the books in the series are highly respected professionals in the engineering areas covered. Each Editor selected only the most relevant and current information available in the McGraw-Hill library, adding further details and commentary where necessary.

Hicks · CIVIL ENGINEERING CALCULATIONS REFERENCE GUIDE

Hicks · MACHINE DESIGN CALCULATIONS REFERENCE GUIDE

Hicks · PLUMBING DESIGN AND INSTALLATION REFERENCE GUIDE

Hicks · POWER GENERATION CALCULATIONS REFERENCE GUIDE

Hicks · POWER PLANT EVALUATION AND DESIGN REFERENCE GUIDE

Johnson & Jasik · ANTENNA APPLICATIONS REFERENCE GUIDE

Markus and Weston · CLASSIC CIRCUITS REFERENCE GUIDE

Merritt · CIVIL ENGINEERING REFERENCE GUIDE

Woodson · HUMAN FACTORS REFERENCE GUIDE FOR ELECTRONICS AND COMPUTER PROFESSIONALS

Woodson · HUMAN FACTORS REFERENCE GUIDE FOR PROCESS PLANTS

# MACHINE DESIGN CALCULATIONS REFERENCE GUIDE

**TYLER G. HICKS, P.E.,** EDITOR
International Engineering Associates

## Contributors

**GERALD M. EISENBERG**
Project Engineer Administrator
American Society of Mechanical Engineers

**STEPHEN M. EBER, P.E.**
Ebasco Services, Inc.

**RAYMOND J. ROARK**
Professor, University of Wisconsin

**S.W. SPIELVOGEL**
Piping Engineering Consultant

**RUFUS OLDENBURGER**
Professor, Purdue University

**McGRAW-HILL BOOK COMPANY**

New York St. Louis San Francisco Auckland Bogotá Hamburg
Johannesburg London Madrid Mexico Milan Montreal New Delhi
Panama Paris São Paulo Singapore Sydney Tokyo Toronto

**Library of Congress Cataloging-in-Publication Data**

Machine design calculations reference guide.

(The McGraw-Hill engineering reference guide series)
"Condensation of the Standard handbook of engineering calculations, 2nd edition"—Pref.
Includes index.
1. Machinery—Design—Handbooks, manuals, etc.
I. Hicks, Tyler Gregory. II. Eisenberg,
Gerald M. III. Standard handbook of engineering
calculations. IV. Series.
TJ230.M225 1987 621.8 86-33796
ISBN 0-07-028799-6

MACHINE DESIGN CALCULATIONS REFERENCE GUIDE



1234567890 DOC/DOC 893210987

ISBN 0-07-028799-6



Printed and bound by R.R. Donnelley and Sons Company

# CONTENTS

# PREFACE

This reference guide is a concise coverage of the key areas of machine design and metalworking calculations for engineers, designers, and shop personnel. The guide is a condensation of the *Standard Handbook of Engineering Calculations*, 2nd Edition.

Fully metricated, the guide contains hundreds of step-by-step calculation procedures for making quick and accurate analyses of many common and uncommon design situations. Specific topics covered include: flywheels, shafts, bearings, gears, power transmission, springs of various types, clutches, shock mounts, force and shrink fits, bolt selection and sizing, hydraulic power analysis, pneumatics, cutting speeds, time and power to drill, bore, mill, and ream; threading and tapping time; grinding, broaching, sawing, and metal plating; learning-curve analysis, hardness determination, economic cutting speed and production rate; optimum lot size in manufacturing; bending, dimpling, and drawing metal parts; breakeven considerations in manufacturing, plus much more.

Most procedures given have related calculations—that is, other items which can be computed using the same general technique. This expands the coverage of the guide enormously. The result is that engineers, designers, and shop personnel have a powerful guide for quick and accurate solution of hundreds of calculation procedures, along with worked-out real-life applications.

Both USCS and SI units are used throughout the guide. This permits easier and faster use of the guide in both the United States and overseas, where SI is widely used. Thus, a designer in the United States can easily do work for overseas applications. And an overseas designer can easily do work for applications in the United States.

The guide is thoroughly up-to-date in its coverage, with calculations devoted to the design of robots, noncircular shafts, external-spring clutches, bolted-joint analysis, and similar topics. Such coverage helps users of the guide to cope with new design jobs they meet in their daily work.

Users who will find the guide helpful include engineers in many fields—mechanical, electrical, chemical, nuclear, and aircraft; designers, drafters, machinists, and metalworking personnel. The step-by-step worked-out calculation procedures will allow all users to save time and energy. Further, the algorithms given are ideal for use in micro, mini, and mainframe computers to effect a greater time saving for users.

The editor thanks the many contributors whose work is cited in the guide. And if readers find any errors or deficiencies in the guide, the editor asks that they be pointed out to him. He will be grateful to every reader detecting such flaws who writes him in care of the publisher.

TYLER G. HICKS, P.E.

# Machine Design and Analysis

REFERENCES: Deutschman—*Machine Design: Theory and Practice*, Macmillan; Johnson—*Mechanical Design Synthesis*, Kreiger; Stephenson and Collander—*Engineering Design*, Wiley-Interscience; Creamer—*Machine Design*, Addison-Wesley; Artobolevskii—*Mechanisms in Modern Engineering Design*, MIR Publishers (Moscow); Sandor and Erdman—*Advanced Mechanism Design: Analysis and Synthesis*, Prentice-Hall; Dhillon—*Reliability Engineering in Systems Design and Operation*, VNR; Chakraborty and Dhande—*Kinematics and Geometry of Planar and Spatial Cam Mechanisms*, Wiley; Lynwander—*Gear Drive Systems: Design and Application*, Dekker; Schwartz—*Composite Materials Handbook*, McGraw-Hill; Taraman—*CAD/CAM: Meeting Today's Productivity Challenge*, SME; Shtipelman—*Design and Manufacture of Hypoid Gears*, Wiley; Roark—*Formulas for Stress and Strain*, McGraw-Hill; Church—*Mechanical Vibrations*, Wiley; *Machinery's Handbook*, Industrial Press; Johnson—*Optimum Design of Mechanical Elements*, Wiley; Slaymaker—*Mechanical Design Analysis*, Wiley; Chironis—*Spring Design and Application*, McGraw-Hill; Spotts—*Design of Machine Elements*, Prentice-Hall; AGMA *Standards Books*, American Gear Manufacturers Association; Doughtie and Vallance—*Design of Machine Members*, McGraw-Hill; Buckingham—*Manual of Gear Design*, Industrial Press; Fuller—*Theory and Practice of Lubrication for Engineers*, Wiley; Dudley—*Gear Handbook*, McGraw-Hill; Churchman—*Prediction and Optimal Decision*, Wiley; Crandall—*Engineering Analysis*, McGraw-Hill; Ver Planck and Téare—*Engineering Analysis*, Wiley; Wahl—*Mechanical Springs*, McGraw-Hill; Haberman—*Engineering Systems Analysis*, Merrill; Shigley—*Mechanical Engineering Design*, McGraw-Hill; Ryder—*Creative Engineering Analysis*, Prentice-Hall; Baumeister and Marks—*Standard Handbook for Mechanical Engineers*, McGraw-Hill; Church—*Kinematics of Machines*, Wiley; Carmichael—*Kent's Mechanical Engineers' Handbook*, Wiley; Faires—*Design of Machine Elements*, Macmillan; Black—*Machine Design*, McGraw-Hill; Maleev—*Machine Design*, International; Bradford and Eaton—*Machine Design*, Wiley; Dudley—*Practical Gear Design*, McGraw-Hill; Shigley—*Simulation of Mechanical Systems*, McGraw-Hill.

## ENERGY STORED IN A ROTATING FLYWHEEL

A 48-in (121.9-cm) diameter spoked steel flywheel having a 12-in wide × 10-in (30.5-cm × 25.4-cm) deep rim rotates at 200 r/min. How long a cut can be stamped in a 1-in (2.5-cm) thick aluminum plate if the stamping energy is obtained from this flywheel? The ultimate shearing strength of the aluminum is 40,000 lb/in$^2$ (275,789.9 kPa).

### Calculation Procedure:

#### 1. *Determine the kinetic energy of the flywheel*

In routine design calculations, the weight of a spoked or disk flywheel is assumed to be concentrated in the rim of the flywheel. The weight of the spokes or disk is neglected. In computing the kinetic energy of the flywheel, the weight of a rectangular, square, or circular rim is assumed to be concentrated at the horizontal centerline. Thus, for this rectangular rim, the weight is concentrated at a radius of $48/2 - 10/2 = 19$ in (48.3 cm) from the centerline of the shaft to which the flywheel is attached.

Then the kinetic energy $K = Wv^2/(2g)$, where $K$ = kinetic energy of the rotating shaft, ft·lb; $W$ = flywheel weight of flywheel rim, lb; $v$ = velocity of flywheel at the horizontal centerline of the rim, ft/s. The velocity of a rotating rim is $v = 2\pi RD/60$, where $\pi = 3.1416$; $R$ = rotational speed, r/min; $D$ = distance of the rim horizontal centerline from the center of rotation, ft. For this flywheel, $v = 2\pi(200)(19/12)/60 = 33.2$ ft/s (10.1 m/s).

The rim of the flywheel has a volume of (rim height, in)(rim width, in)(rim circumference measured at the horizontal centerline, in), or $(10)(12)(2\pi)(19) = 14{,}350$ in$^3$ (235,154.4 cm$^3$). Since machine steel weighs 0.28 lb/in$^3$ (7.75 g/cm$^3$), the weight of the flywheel rim is $(14{,}350)(0.28) = 4010$ lb (1818.9 kg). Then $K = (4010)(33.2)^2/[2(32.2)] = 68{,}700$ ft·lb (93,144.7 N·m).

#### 2. *Compute the dimensions of the hole that can be stamped*

A stamping operation is a shearing process. The area sheared is the product of the plate thickness and the length of the cut. Each square inch of the sheared area offers a resistance equal to the ultimate shearing strength of the material punched.

During stamping, the force exerted by the stamp varies from a maximum $F$ lb at the point of contact to 0 lb when the stamp emerges from the metal. Thus, the average force during stamping is $(F + 0)/2 = F/2$. The work done is the product of $F/2$ and the distance through which this force moves, or the plate thickness $t$ in. Therefore, the maximum length that can be stamped is that which occurs when the full kinetic energy of the flywheel is converted to stamping work.

With a 1-in (2.5-cm) thick aluminum plate, the work done is $W$ ft·lb = (force, lb)(distance, ft). The work done when all the flywheel kinetic energy is used is $W = K$. Substituting the kinetic energy from step 1 gives $W = K = 68{,}700$ ft·lb (93,144.7 N·m) $= (F/2)(1/12)$; and solving for the force yields $F = 1{,}650{,}000$ lb (7,339,566.3 N).

The force $F$ also equals the product of the plate area sheared and the ultimate shearing strength of the material stamped. Thus, $F = lts_u$, where $l$ = length of cut, in; $t$ = plate thickness, in; $s_u$ = ultimate shearing strength of the material. Substituting the known values and solving for $l$, we get $l = 1{,}650{,}000/[(1)(40{,}000)] = 41.25$ in (104.8 cm).

**Related Calculations:** The length of cut computed above can be distributed in any form—square, rectangular, circular, or irregular. This method is suitable for computing the energy stored in a flywheel used for any purpose. Use the general procedure in step 2 for computing the principal dimension in blanking, punching, piercing, trimming, bending, forming, drawing, or coining.

## SHAFT TORQUE, HORSEPOWER, AND DRIVER EFFICIENCY

A 4-in (10.2-cm) diameter shaft is driven at 3600 r/min by a 400-hp (298.3-kW) motor. The shaft drives a 48-in (121.9-cm) diameter chain sprocket having an output efficiency of 85 percent. Determine the torque in the shaft, the output force on the sprocket, and the power delivered by the sprocket.

### Calculation Procedure:

#### 1. *Compute the torque developed in the shaft*

For any shaft driven by any driver, the torque developed is $T$ lb·in $= 63{,}000hp/R$, where $hp$ = horsepower delivered to, or by, the shaft; $R$ = shaft rotative speed, r/min. Thus, the torque developed by this shaft is $T = (63{,}000)(400)/3600 = 7000$ lb·in (790.9 N·m).

### 2. *Compute the sprocket output force*

The force developed at the output surface, tooth, or other part of a rotating member is given by $F = T/r$, where $F$ = force developed, lb; $r$ = radius arm of the force, in. In this drive the radius is 48/2 = 24 in (61 cm). Hence, $F$ = 7000/24 = 291 lb (1294.4 N).

### 3. *Compute the power delivered by the sprocket*

The work input to this shaft is 400 hp (298.3 kW). But the work output is less than the input because the efficiency is less than 100 percent. Since efficiency = work output, hp/work input, hp, the work output, hp = (work input, hp)(efficiency), or output hp = (400)(0.85) = 340 hp (253.5 kW).

**Related Calculations:** Use this procedure for any shaft driven by any driver—electric motor, steam turbine, internal-combustion engine, gas turbine, belt, chain, sprocket, etc. When computing the radius of toothed or geared members, use the pitch-circle or pitch-line radius.

## PULLEY AND GEAR LOADS ON SHAFTS

A 500-r/min shaft is fitted with a 30-in (76.2-cm) diameter pulley weighing 250 lb (113.4 kg). This pulley delivers 35 hp (26.1 kW) to a load. The shaft is also fitted with a 24-in (61.0-cm) pitch-diameter gear weighing 200 lb (90.7 kg). This gear delivers 25 hp (18.6 kW) to a load. Determine the concentrated loads produced on the shaft by the pulley and the gear.

### Calculation Procedure:

### 1. *Determine the pulley concentrated load*

The largest concentrated load caused by the pulley occurs when the belt load acts vertically downward. Then the total pulley concentrated load is the sum of the belt load and pulley weight.

For a pulley in which the tension of the tight side of the belt is twice the tension in the slack side of the belt, the maximum belt load is $F_p = 3T/r$, where $F_p$ = tension force, lb, produced by the belt load; $T$ = torque acting on the pulley, lb·in; $r$ = pulley radius, in. The torque acting on a pulley is found from $T = 63{,}000hp/R$, where $hp$ = horsepower delivered by pulley; $R$ = revolutions per minute (rpm) of shaft.

For this pulley, $T$ = 63,000(35)/500 = 4410 lb·in (498.3 N·m). Hence, the total pulley concentrated load = 882 + 250 = 1132 lb (5035.1 N).

### 2. *Determine the gear concentrated load*

With a gear, the turning force acts only on the teeth engaged with the meshing gear. Hence, there is no slack force as in a belt. Therefore, $F_g = T/r$, where $F_g$ = gear tooth-thrust force, lb; $r$ = gear pitch radius, in: other symbols as before. The torque acting on the gear is found in the same way as for the pulley.

Thus, $T$ = 63,000(25)/500 = 3145 lb·in (355.3 N·m). Then $F_g$ = 3145/12 = 263 lb (1169.9 N). Hence, the total gear concentrated load is 263 + 200 = 463 lb (2059.5 N).

**Related Calculations:** Use this procedure to determine the concentrated load produced by any type of gear (spur, herringbone, worm, etc.), pulley (flat, V, or chain belt), sprocket, or their driving member. When the power transmission belt or chain leaves the belt or sprocket at an angle other than the vertical, take the vertical component of the pulley force and add it to the pulley weight to determine the concentrated load.

## SHAFT REACTIONS AND BENDING MOMENTS

A 30-ft (9.1-m) long steel shaft weighing 150 lb/ft (223.2 kg/m) of length has a 500-lb (2224.1-N) concentrated gear load 10 ft (3.0 m) from the left end of the shaft and a 2000-lb (8896.4-N) concentrated pulley load 15 ft (4.6 m) from the right end of the shaft. Determine the end reactions and the maximum bending moment in this shaft.

## Calculation Procedure:

### 1. *Draw a sketch of the shaft*

Figure 1*a* shows a sketch of the shaft. Label the left- and right-hand reactions $L_R$ and $R_R$, respectively.

### 2. *Compute the shaft end reactions*

Take moments about $R_R$ to determine the magnitude of $L_R$. Since the shaft has a uniform weight per foot of length, assume that the total weight of the shaft is concentrated at its midpoint. Then $30L_R - 500(20) - 50(30)(15) - 2000(15) = 0$; $L_R = 3583.33$ lb (15,939.4 N). Take moments about $L_R$ to determine $R_R$. Or, $30R_R - 500(10) - 150(30)(15) - 2000(15) = 0$; $R_R = 3416.67$ lb (15,198.1 N). Alternatively, the first reaction found could be subtracted from the sum of the vertical loads, or $500 + 30 \times 150 + 2000 - 3583.33 = 3416.67$ lb (15,198.1 N). However, taking moments about each support permits checking the results, because the sum of the reactions should equal the sum of the vertical loads, including the weight of the shaft.

### 3. *Compute the maximum bending moment*

The maximum bending moment in a shaft occurs where the shear is zero. Find the vertical shear at each point of applied load or reaction by taking the algebraic sum of the vertical forces to the left and right of the load. Use a plus sign for upward forces and a minus sign for downward forces. Designate each shear force by $V$ with a subscript number showing its location, in feet (meters) along the shaft from the left end. Use $L$ and $R$ to indicate whether the shear is to the left or right of the load. The shear at the left-hand reaction is $V_{LR} = +3583.33$ lb (+15,939.5 N); $V_{10L} = 3583.33 - 10 \times 150 = 2083.33$ lb (9267.1 N), where the product $10 \times 150 =$ the weight of the shaft from the point $V_{LR}$ to the 500-lb (2224.1-N) load. At this load, $V_{10R} = 2083.33 - 500 = 1583.33$ lb (7043.0 N). To the right of the 500-lb (2224.1-N) load, at the 2000-lb (8896.4-N) load, $V_{20L} = 1583.33 - 5 \times 150 = 833.33$ lb (3706.9 N). To the right of the 2000-lb (8896.4-N) load, $V_{20R} = 833.33 - 2000 = -1166.67$ lb (−5189.6 N). At the left of $V_R$, $V_{30L} = -1166.67 - 15 \times 150 = -3416.67$ lb (−15,198.1 N). At the right hand end of the shaft $V_{30R} = -3416.67 + 3416.67 = 0$.

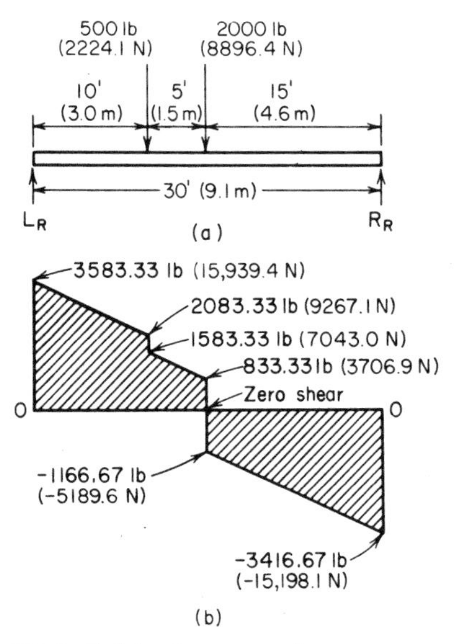


**FIG. 1** Shaft bending-moment diagram.

Draw the shear diagram (Fig. 1*b*). This diagram shows that zero shear occurs at a point 15 ft (4.6 m) from the left-hand reaction. Hence, the maximum bending moment $M_m$ on this shaft is $M_m = 3583.33(15) - 500(5) - 150(15)(7.5) = 34{,}340$ lb·ft (46,558.8 N·m).

**Related Calculations:** Use this procedure for shafts of any metal—steel, bronze, aluminum, plastic, etc.—if the shaft is of uniform cross section. For nonuniform shafts, use the procedures discussed later in this section.

## SOLID AND HOLLOW SHAFTS IN TORSION

A solid steel shaft will transmit 500 hp (372.8 kW) at 3600 r/min. What diameter shaft is required if the allowable stress in the shaft is 12,500 lb/in$^2$ (86,187.5 kPa)? What diameter hollow shaft is needed to transmit the same power if the inside diameter of the shaft is 1.0 in (2.5 cm)?

### Calculation Procedure:

**1. *Compute the torque in the solid shaft***

For any solid shaft, the torque $T$, lb·in $= 63{,}000hp/R$, where $R =$ shaft rpm. Thus, $T = 63{,}000(500)/3600 = 8750$ lb·in (988.6 N·m).

**2. *Compute the required shaft diameter***

For any solid shaft, the required diameter $d$, in $= 1.72\ (T/s)^{1/3}$, where $s =$ allowable stress in shaft, lb/in². Thus, for this shaft, $d = 1.72(8750/12{,}500)^{1/3} = 1.526$ in (3.9 cm).

**3. *Analyze the hollow shaft***

The usual practice is to size hollow shafts such that the ratio $q$ of the inside diameter $d_i$ into the outside diameter $d_o$ in is 1:2 to 1:3 or some intermediate value. With a $q$ in this range the shaft will have sufficient thickness to prevent failure in service.

Assume $q = d_i/d_o = 1/2$. Then with $d_i = 1.0$ in (2.5 cm), $d_o = d_i/q$, or $d_o = 1.0/0.5 = 2.0$ in (5.1 cm). With $q = 1/3$, $d_o = 1.0/0.33 = 3.0$ in (7.6 cm).

**4. *Compute the stress in each hollow shaft***

For the hollow shaft $s = 5.1T/d_o^3(1 - q^4)$, where the symbols are as defined above. Thus, for the 2-in (5.1-cm) outside-diameter shaft, $s = 5.1(8750)/[8(1 - 0.0625)] = 5950$ lb/in² (41,023.8 kPa).

By inspection, the stress in the 3-in (7.6-cm) outside-diameter shaft will be lower because the torque is constant. Thus, $s = 5.1\ (8750)/[27(1 - 0.0123)] = 1672$ lb/in² (11,528.0 kPa).

**5. *Choose the outside diameter of the hollow shaft***

Use a trial-and-error procedure to choose the hollow shaft's outside diameter. Since the stress in the 2-in (5.1-cm) outside-diameter shaft, 5950 lb/in² (41,023.8 kPa), is less than half the allowable stress of 12,500 lb/in² (86,187.5 kPa), select a smaller outside diameter and compute the stress while holding the inside diameter constant.

Thus, with a 1.5-in (3.8-cm) shaft and the same inside diameter, $s = 5.1(8750)/[3.38(1 - 0.197)] = 16{,}430$ lb/in² (113,284.9 kPa). This exceeds the allowable stress.

Try a larger outside diameter, 1.75 in (4.4 cm), to find the effect on the stress. Or $s = 5.1\ (8750)/[5.35(1 - 0.107)] = 9350$ lb/in² (64,468.3 kPa). This is lower than the allowable stress.

Since a 1.5-in (3.8-cm) shaft has a 16,430-lb/in² (113,284.9-kPa) stress and a 1.75-in (4.4-cm) shaft has a 9350-lb/in² (64,468.3-kPa) stress, a shaft of intermediate size will have a stress approaching 12,500 lb/in² (86,187.5 kPa). Trying 1.625 in (4.1 cm) gives $s = 5.1(8750)/[4.4(1 - 0.143)] = 11{,}820$ lb/in² (81,489.9 kPa). This is within 680 lb/in² (4688.6 kPa) of the allowable stress and is close enough for usual design calculations.

**Related Calculations:** Use this procedure to find the diameter of any solid or hollow shaft acted on only by torsional stress. Where bending and torsion occur, use the next calculation procedure. Find the allowable torsional stress for various materials in Baumeister and Marks—*Standard Handbook for Mechanical Engineers.*

## SOLID SHAFTS IN BENDING AND TORSION

A 30-ft (9.1-m) long solid shaft weighing 150 lb/ft (223.2 kg/m) is fitted with a pulley and a gear as shown in Fig. 2. The gear delivers 100 hp (74.6 kW) to the shaft while driving the shaft at 500 r/min. Determine the required diameter of the shaft if the allowable stress is 10,000 lb/in² (68,947.6 kPa).

### Calculation Procedure:

**1. *Compute the pulley and gear concentrated loads***

Using the method of the previous calculation procedure, we get $T = 63{,}000hp/R = 63{,}000(100)/500 = 12{,}600$ lb·in (1423.6 N·m). Assuming that the maximum tension of the tight side of the belt is twice the tension of the slack side, we see the maximum belt load is $R_P = 3T/r = 3(12{,}600)/24 = 1575$ lb (7005.9 N). Hence, the total pulley concentrated load = belt load + pulley weight $= 1575 + 750 = 2325$ lb (10,342.1 N).

The gear concentrated load is found from $F_g = T/r$, where the torque is the same as computed for the pulley, or $F_g = 12{,}600/9 = 1400$ lb (6227.5 N). Hence, the total gear concentrated load is 1400 + 75 = 1475 lb (6561.1 N).

Draw a sketch of the shaft showing the two concentrated loads in position (Fig. 2).

## 2. *Compute the end reactions of the shaft*

Take moments about $R_R$ to determine $L_R$, using the method of the previous calculation procedures. Thus, $L_R(30) - 2325(25) - 1475(8) - 150(30)(15) = 0$; $L_R = 4580$ lb (20,372.9 N). Taking moments about $L_R$ to determine $R_R$ yields $R_R(30) - 1475(22) - 2325(5) - 150(30)(15) = 0$; $R_R = 3720$ lb (16,547.4 N). Check by taking the sum of the upward forces: 4580 + 3720 = 8300 lb (36,920.2 N) = sum of the downward forces, or 2325 + 1475 + 4500 = 8300 lb (36,920.2 N).

## 3. *Compute the vertical shear acting on the shaft*

Using the method of the previous calculation procedures, we find $V_{LR} = 4580$ lb (20,372.9 N); $V_{5L} = 4580 - 5(150) = 3830$ lb (17,036.7 N); $V_{5R} = 3830 - 2325 = 1505$ lb (6694.6 N); $V_{22L} = 1505 - 17(150) = -1045$ lb (−4648.4 N); $V_{22R} = -1045 - 1475 = -2520$ lb (−11,209.5 N); $V_{30L} = -2520 - 8(150) = -3720$ lb (−16,547.4 N); $V_{30R} = -3720 + 3720 = 0$.

## 4. *Find the maximum bending moment on the shaft*

Draw the shear diagram shown in Fig. 2. Determine the point of zero shear by scaling it from the shear diagram or setting up an equation thus: positive shear $- x(150 \text{ lb/ft}) = 0$, where the positive shear is the last recorded plus value, $V_{5R}$ in this shaft, and $x$ = distance from $V_{5R}$ where the shear is zero. Substituting values gives $1505 - 150x = 0$; $x = 10.03$ ft (3.1 m). Then $M_m = 4580(15.03) - 2325(10.03) - (150)(5 + 10.03)[(5 + 10.03)/2] = 28{,}575$ lb (127,108.3 N).

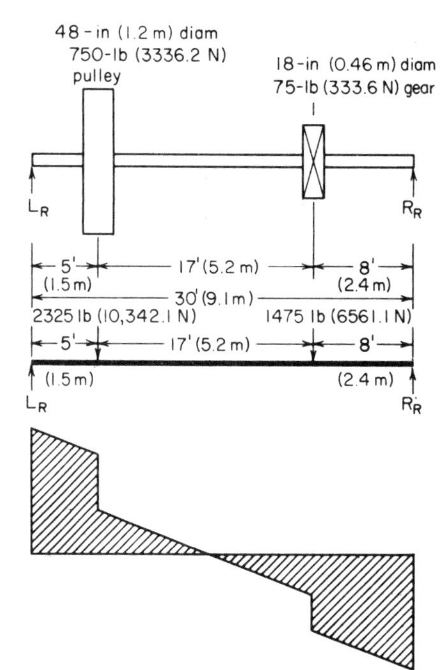


**FIG. 2** Solid-shaft bending moments.

## 5. *Determine the required shaft diameter*

Use the method of maximum shear theory to size the shaft. Determine the equivalent torque $T_e$ from $T_e = (M_m^2 + T^2)^{0.5}$, where $M_m$ is the maximum bending moment, lb·ft, acting on the shaft and $T$ is the maximum torque acting on the shaft. For this shaft, $T_e = [28{,}575^2 + (12{,}600/12)^2]^{0.5} = 28{,}600$ lb·ft (38,776.4 N·m), where the torque in pound-inches is divided by 12 to convert it to pound-feet. To convert $T_e$ to $T_{e'}$ lb·in, multiply by 12.

Once the equivalent torque is known, the shaft diameter $d$ in is computed from $d = 1.72(T_{e'}/s)^{1/3}$, where $s$ = allowable stress in the shaft. For this shaft, $d = 1.72(28{,}500)(12)/(10{,}000)^{1/3} = 5.59$ in (14.2 cm). Use a 6.0-in (15.2-cm) diameter shaft.

**Related Calculations:** Use this procedure for any solid shaft of uniform cross section made of metal—steel, aluminum, bronze, brass, etc. The equation used in step 4 to determine the location of zero shear is based on a strength-of-materials principle: When zero shear occurs between two concentrated loads, find its location by dividing the last *positive* shear by the uniform load. If desired, the maximum principal stress theory can be used to combine the bending and torsional stresses in a shaft. The results obtained approximate those of the maximum shear theory.

## EQUIVALENT BENDING MOMENT AND IDEAL TORQUE FOR A SHAFT

A 2-in (5.1-cm) diameter solid steel shaft has a maximum bending moment of 6000 lb·in (677.9 N·m) and an applied torque of 3000 lb·in (339.0 N·m). Is this shaft safe if the maximum allowable bending stress is 10,000 lb/in$^2$ (68,947.6 kPa)? What is the ideal torque for this shaft?

### Calculation Procedure:

**1.** ***Compute the equivalent bending moment***

The equivalent bending moment $M_e$ lb·in for a solid shaft is $M_e = 0.5[M + (M^2 + T^2)^{0.5}]$, where $M$ = maximum bending moment acting on the shaft, lb·in; $T$ = maximum torque acting on the shaft, lb·in. For this shaft, $M_e = 0.5[6000 + (6000^2 + 3000^2)^{0.5}] = 6355$ lb·in (718.0 N·m).

**2.** ***Compute the stress in the shaft***

Use the flexure relation $s = Mc/I$, where $s$ = stress developed in the shaft, lb/in$^2$; $M = M_e$ for a shaft; $I$ = section moment of inertia of the shaft about the neutral axis; in$^4$; $c$ = distance from shaft neutral axis to outside fibers, in. For a circular shaft, $I = \pi d^4/64 = \pi(2)^4/64 = 0.785$ in$^4$ (32.7 cm$^4$); $c = d/2 = 2/2 = 1.0$. Then $s = Mc/I = (6355)(1.0)/0.785 = 8100$ lb/in$^2$ (55,849.5 kPa). Thus, the actual bending stress is 1900 lb/in$^2$ (13,100.5 kPa) less than the maximum allowable bending stress. Therefore the shaft is safe. Alternatively, compute the maximum equivalent bending moment from $M_e = sI/c = (10{,}000)(0.785)/1.0 = 7850$ lb·in (886.9 N·m). This is 7850 − 6355 = 1495 lb·in (168.9 N·m) greater than the actual equivalent bending moment. Hence, the shaft is safe.

**3.** ***Compute the ideal torque for the shaft***

The ideal torque $T_i$ lb·in for a shaft is $T_i = M + (M^2 + T^2)^{0.5}$, where $M$ and $T$ are the bending and torsional moments, respectively, acting on the shaft, lb·in. For this shaft, $T_i = 6000 + (6000^2 + 3000^2)^{0.5} = 12{,}710$ lb·in (1436.0 N·m).

**Related Calculations:** Use this procedure for any shaft of uniform cross section made of metal—steel, aluminum, bronze, brass, etc.

## TORSIONAL DEFLECTION OF SOLID AND HOLLOW SHAFTS

What diameter solid steel shaft should be used for a 500-hp (372.8-kW) 250-r/min application if the allowable torsional deflection is 1°, the maximum allowable stress is 10,000 lb/in$^2$ (68,947.6 kPa), and the modulus of rigidity is $13 \times 10^6$ lb/in$^2$ ($89.6 \times 10^6$ kPa)? What diameter hollow steel shaft should be used if the ratio of the inside diameter to the outside diameter is 1:3, the allowable deflection is 1°, the allowable stress is 10,000 lb/in$^2$ (68,947.6 kPa), and the modulus of rigidity is $13 \times 10^6$ lb/in$^2$ ($89.6 \times 10^6$ kPa)? What shaft has the greatest weight?

### Calculation Procedure:

**1.** ***Determine the torque acting on the shaft***

For any shaft, $T = 63{,}000hp/R$; or for this shaft, $T = 63{,}000(500)/250 = 126{,}000$ lb·in (14,236.1 N·m).

**2.** ***Compute the required diameter of the solid shaft***

For a solid metal shaft, $d = (584Tl/G\alpha)^{1/3}$, where $l$ = shaft length expressed as a number of shaft diameters, in; $G$ = modulus of rigidity, lb/in$^2$; $\alpha$ = angle of torsion deflection, deg.

Usual specifications for noncritical applications of shafts require that the torsional deflection not exceed 1° in a shaft having a length equal to 20 diameters. Using this length gives $d = [584 \times 126{,}000 \times 20/(13 \times 10^6 \times 1.0)]^{1/3} = 4.84$ in (12.3 cm). Use a 5-in (12.7-cm) diameter shaft.

**3.** ***Compute the outside diameter of the hollow shaft***

Assume that the shaft has a length equal to 20 diameters. Then for a hollow shaft $d = [584Tl/G\alpha(1 - q^4)]^{1/3}$, where $q = d_i/d_o$; $d_i$ = inside diameter of the shaft, in; $d_o$ = outside diameter of the shaft, in. For this shaft, $d = \{584 \times 126{,}000 \times 20/(13 \times 10^6 \times 1.0)[1 - (1/3)^4]\}^{1/3} =$

4.86 in (12.3 cm). Use a 5-in (12.7-cm) outside-diameter shaft. The inside diameter would be 5.0/3 = 1.667 in (4.2 cm).

**4.** ***Compare the weight of the shafts***

Steel weighs approximately 480 lb/ft$^3$ (7688.9 kg/m$^3$). To find the weight of each shaft, compute its volume in cubic feet and multiply it by 480. Thus, for the 5-in (12.7-cm) diameter solid shaft, weight = $(\pi\, 5^2/4)(5 \times 20)(480)/1728$ = 540 lb (244.9 kg). The 5-in (12.7-cm) outside-diameter hollow shaft weighs $(\pi\, 5^2/4 - \pi 1.667^2/4)(5 \times 20)(480)/1728$ = 242 lb (109.8 kg). Thus, the hollow shaft weighs less than half the solid shaft. However, it would probably be more expensive to manufacture because drilling the central hole could be costly.

**Related Calculations:** Use this procedure to determine the steady-load torsional deflection of any shaft of uniform cross section made of any metal—steel, bronze, brass, aluminum, Monel, etc. The assumed torsional deflection of 1° for a shaft that is 20 times as long as the shaft diameter is typical for routine applications. Special shafts may be designed for considerably less torsional deflection.

## DEFLECTION OF A SHAFT CARRYING CONCENTRATED AND UNIFORM LOADS

A 2-in (5.1-cm) diameter steel shaft is 6 ft (1.8 m) long between bearing centers and turns at 500 r/min. The shaft carries a 600-lb (2668.9-N) concentrated gear load 3 ft (0.9 m) from the left-hand center. Determine the deflection of the shaft if the modulus of elasticity $E$ of the steel is $30 \times 10^6$ lb/in$^2$ ($206.8 \times 10^9$ Pa). What would the shaft deflection be if the load were 2 ft (0.6 m) for the left-hand bearing? The shaft weighs 10 lb/ft (14.9 kg/m).

### Calculation Procedure:

**1.** ***Compute the deflection caused by the concentrated load***

When a beam carries both a concentrated and a uniformly distributed load, compute the deflection for each load separately and find the sum. This sum is the total deflection caused by the two loads.

For a beam carrying a concentrated load, the deflection $\Delta$ in = $Wl^3/48EI$, where $W$ = concentrated load, lb; $l$ = length of beam, in; $E$ = modulus of elasticity, lb/in$^2$; $I$ = moment of inertia of shaft cross section, in$^4$. For a circular shaft, $I = \pi d^4/64 = \pi(2)^4/64 = 0.7854$ in$^4$ (32.7 cm$^4$). Then $\Delta = 600(72)^3/[48(30)(10^6)(0.7854)] = 0.198$ in (5.03 mm). The deflection per foot of shaft length is $\Delta_f = 0.198/6 = 0.033$ in/ft (2.75 mm/m) for the concentrated load.

**2.** ***Compute the deflection due to shaft weight***

For a shaft of uniform weight, $\Delta = 5wl^3/384EI$, where $w$ = total distributed load = weight of shaft, lb. Thus, $\Delta = 5(60)(72)^3/[384(30 \times 10^6)(0.7854)] = 0.0129$ in (0.328 mm). The deflection per foot of shaft length is $\Delta_f = 0.0129/6 = 0.00214$ in/ft (0.178 mm/m).

**3.** ***Determine the total deflection of the shaft***

The total deflection of the shaft is the sum of the deflections caused by the concentrated and uniform loads, or $\Delta_t = 0.198 + 0.0129 = 0.2109$ in (5.36 mm). The total deflection per foot of length is $0.033 + 0.00214 = 0.03514$ in/ft (2.93 mm/m).

Usual design practice limits the transverse deflection of a shaft of any diameter to 0.01 in/ft (0.83 mm/m) of shaft length. The deflection of this shaft is 3½ times this limit. Therefore, the shaft diameter must be increased if this limit is not to be exceeded.

Using a 3-in (7.6-cm) diameter shaft weighing 25 lb/ft (37.2 kg/m) and computing the deflection in the same way, we find the total transverse deflection is 0.0453 in (1.15 mm), and the total deflection per foot of shaft length is 0.00755 in/ft (0.629 mm/m). This is within the desired limits. By reducing the assumed shaft diameter in ⅛-in (0.32-cm) increments and computing the deflection per foot of length, a deflection closer to the limit can be obtained.

**4.** ***Compute the total deflection for the noncentral load***

For a noncentral load, $\Delta = (Wc'/3EIl)[(cl/3 + cc'/3)^3]^{0.5}$, where $c$ = distance of concentrated load from left-hand bearing, in; $c'$ = distance of concentrated load from right-hand bearing, in.

Thus $c + c' = 1$, and for this shaft $c = 24$ in (61.0 cm) and $c' = 48$ in (121.9 cm). Then $\Delta = [600 \times 48/(3 \times 30 \times 10^6 \times 0.7854 \times 72)][(24 \times 72/x + 24 \times 48/3)^3]^{0.5} = 0.169$ in (4.29 mm).

The deflection caused by the weight of the shaft is the same as computed in step 2, or 0.0129 in (0.328 mm). Hence, the total shaft deflection is 0.169 + 0.0129 = 0.1819 in (4.62 mm). The deflection per foot of shaft length is 0.1819/6 = 0.0303 in (2.53 mm/m). Again, this exceeds 0.01 in/ft (0.833 mm/m).

Using a 3-in (7.6-cm) diameter shaft as in step 3 shows that the deflection can be reduced to within the desired limits.

**Related Calculations:** Use this procedure for any metal shaft—aluminum, brass, bronze, etc.—that is uniformly loaded or carries a concentrated load.

## SELECTION OF KEYS FOR MACHINE SHAFTS

Select a key for a 4-in (10.2-cm) diameter shaft transmitting 1000 hp (745.7 kW) at 1000 r/min. The allowable shear stress in the key is 15,000 lb/in$^2$ (103,425.0 kPa), and the allowable compressive stress is 30,000 lb/in$^2$ (206,850.0 kPa). What type of key should be used if the allowable shear stress is 5000 lb/in$^2$ (34,475.0 kPa) and the allowable compressive stress is 20,000 lb/in$^2$ (137,900.0 kPa)?

### Calculation Procedure:

#### 1. *Compute the torque acting on the shaft*

The torque acting on the shaft is $T = 63{,}000hp/R$, or $T = 63{,}000(1000/1000) = 63{,}000$ lb·in (7118.0 N·m).

#### 2. *Determine the shear force acting on the key*

The shear force $F_s$ lb acting on a key is $F_s = T/r$, where $T$ = torque acting on shaft, lb·in; $r$ = radius of shaft, in. Thus, $T = 63{,}000/2 = 31{,}500$ lb (140,118.9 N).

#### 3. *Select the type of key to use*

When a key is designed so that its allowable shear stress is approximately one-half its allowable compressive stress, a square key (i.e., a key having its height equal to its width) is generally chosen. For other values of the stress ratio, a flat key is generally used.

Determine the dimensions of the key from Baumeister and Marks—*Standard Handbook for Mechanical Engineers*. This handbook shows that a 4-in (10.2-cm) diameter shaft should have a square key 1 in wide × 1 in (2.5 cm × 2.5 cm) high.

#### 4. *Determine the required length of the key*

The length of a 1-in (2.5-cm) key based on the allowable shear stress is $l = 2F_s/(w_k s_s)$, where $w_k$ = width of key, in. Thus, $l = 31{,}500/[(1)(15{,}000)] = 2.1$ in (5.3 cm), say 2⅛ in (5.4 cm).

#### 5. *Check key length for the compressive load*

The length of a 1-in (2.5-cm) key based on the allowable compressive stress is $l = 2F_s/(ts_c)$, where $t$ = key thickness, in; $s_c$ = allowable compressive stress, lb/in$^2$. Thus, $l = 2(13{,}500)/[(1)(30{,}000)] = 2.1$ in (5.3 cm). This agrees with the key length based on the allowable shear stress. The key length found in steps 4 and 5 should agree if the key is square in cross section.

#### 6. *Determine the key size for other stress values*

When the allowable shear stress does not equal one-half the allowable compressive stress for a shaft key, a flat key is generally used. A flat key has a width greater than its height.

Find the recommended dimensions for a flat key from Baumeister and Marks—*Standard Handbook for Mechanical Engineers*. This handbook shows that a 4-in (10.2-cm) diameter shaft will use a 1-in (2.5-cm) wide by ¾-in (1.9-cm) thick flat key.

The length of the key based on the allowable shear stress is $l = F_s/(w_k s_s) = 31{,}500/[(l)(5000)] = 6.31$ in (16.0 cm). Use a 6⁵⁄₁₆-in (16.0-cm) long key.

Checking the key length based on the allowable compressive stress yields $l = 2F_s/(ts_c) = 2(31{,}500)/[(0.75)(20{,}000)] = 4.2$ in (10.7 cm). Use the longer length, 6⁵⁄₁₆ in (16.0 cm), because the shorter key would be overloaded in compression.

**Related Calculations:** Use this procedure for shafts and keys made of any metal (steel, bronze, brass, stainless steel, etc.). The dimensions of shaft keys can also be found in ANSI Standard B17f, Woodruff Keys, Keyslots and Cutters. Woodruff keys are used only for light-torque applications.

## SELECTING A LEATHER BELT FOR POWER TRANSMISSION

Choose a leather belt to transmit 50 hp (37.3 kW) from a 1750-r/min squirrel-cage compensator-starting motor through a 12-in (30.5-cm) diameter pulley in an oily atmosphere. What belt width is needed with a 50-hp (37.3-kW) internal-combustion engine fitted with a 1750-r/min 12-in (30.5-cm) diameter pulley operating in an oily atmosphere?

### Calculation Procedure:

**1. *Determine the belt speed***

The speed of a belt $S$ is found from $S = \pi RD$, where $R$ = rpm of driving or driven pulley; $D$ = diameter, ft, of driving or driven pulley. Thus, for this belt, $S = \pi(1750)(12/12) = 5500$ ft/min (27.9 m/s).

**2. *Determine the belt thickness needed***

Use the National Industrial Leather Association recommendations. Enter Table 1 at the bottom at a belt speed of 5500 ft/min (27.9 m/s), i.e., between 4000 and 6000 ft/min (20.3 and 30.5 m/s); and project horizontally to the next smaller pulley diameter than that actually used. Thus, by entering at the line marked 4000–6000 ft/min (20.3–30.5 m/s) and projecting to the 10-in (25.4-cm) minimum diameter pulley, since a 12-in (30.5-cm) pulley is used, we see that a 23/64-in (0.91-cm) thick double-ply heavy belt should be used. Read the belt thickness and type at the top of the column in which the next smaller pulley diameter appears.

**3. *Determine the belt capacity factors***

Enter the body of Table 1 at a belt speed of 5500 ft/min (27.9 m/s), i.e., between 4000 and 6000 ft/min (20.3 and 30.5 m/s); then project to the double-ply heavy column. Interpolating by eye gives a belt capacity factor of $K_c = 14.8$.

**4. *Determine the belt correction factors***

Table 2 lists motor, pulley diameter, and operating correction factors, respectively. Thus, from Table 2, the motor correction factor $M = 1.5$ for a squirrel-cage compensator-starting motor.

**TABLE 1** Leather-Belt Capacity Factors

<table>
<tr><th colspan="2">Belt speed</th><th colspan="4">Double ply</th></tr>
<tr><th rowspan="2">ft/min</th><th rowspan="2">m/s</th><th colspan="2">20/64 in (7.9 mm)</th><th colspan="2">23/64 in (9.1 mm)</th></tr>
<tr><th colspan="2">Medium</th><th colspan="2">Heavy</th></tr>
<tr><td>4000</td><td>20.3</td><td colspan="2">10.9</td><td colspan="2">12.6</td></tr>
<tr><td>5000</td><td>25.4</td><td colspan="2">12.5</td><td colspan="2">14.3</td></tr>
<tr><td>6000</td><td>30.5</td><td colspan="2">13.2</td><td colspan="2">15.2</td></tr>
<tr><td colspan="6">Minimum pulley diameters</td></tr>
<tr><td>Up to 2500</td><td>Up to 12.7</td><td>5 in*</td><td>12.7 cm*</td><td>8 in*</td><td>20.3 cm*</td></tr>
<tr><td>2500–4000</td><td>12.7–20.3</td><td>6*</td><td>15.2*</td><td>9*</td><td>22.9*</td></tr>
<tr><td>4000–6000</td><td>20.3–30.5</td><td>7*</td><td>17.8*</td><td>10*</td><td>25.4*</td></tr>
</table>

*For belts 8 in (20.3 cm) and over, add 2 in (5.1 cm) to pulley diameter.

**TABLE 2** Leather-Belt Correction Factors

| | Correction factor |
|---|---|
| Characteristics or condition of motor and starter: | |
| Squirrel-cage, compensator-starting motor | $M = 1.5$ |
| Squirrel-cage, line-starting | $M = 2.0$ |
| Slip-ring, high starting torque | $M = 2.5$ |
| Diameter of small pulley, in (cm): | |
| 4 and under (10.2 and under) | $P = 0.5$ |
| 4.5 to 8 (11.4 to 20.3) | $P = 0.6$ |
| 9 to 12 (22.9 to 30.5) | $P = 0.7$ |
| 13 to 16 (33.0 to 40.6) | $P = 0.8$ |
| 17 to 30 (43.2 to 76.2) | $P = 0.9$ |
| Over 30 (over 76.2) | $P = 1.0$ |
| Operating conditions: | |
| Oily, wet, or dusty atmosphere | $F = 1.35$ |
| Vertical drives | $F = 1.2$ |
| Jerky loads | $F = 1.2$ |
| Shock and reversing loads | $F = 1.4$ |

Also from Table 2, the smaller pulley diameter correction factor $P = 0.7$; and $F = 1.35$ for an oily atmosphere.

**5. *Compute the required belt width***

The required belt width, in, is $W = hpMF/(K_cP)$, where $hp$ = horsepower transmitted by the belt; the other factors are as given above. For this belt, then, $W = (50)(1.5)(1.35)/[14.8(0.7)] = 9.7$ in (24.6 cm). Thus, a 10-in (25.4-cm) wide belt would be used because belts are commercially available in 1-in (2.5-cm) increments.

**6. *Determine the belt width for the engine drive***

For a double-ply belt driven by a driver other than an electric motor, $W = 2750hp/dR$, where $d$ = driving pulley diameter, in; $R$ = driving pulley, r/min. Thus, $W = 2750(50)/[(12)(1750)] = 6.54$ in (16.6 cm). Hence, a 7-in (17.8-cm) wide belt would be used.

For a single-ply belt the above equation becomes $W = 1925\ hp/dR$.

**Related Calculations:** Note that the relations in steps 1, 5, and 6 can be solved for any unknown variable when the other factors in the equations are known. Where the hp rating of a belt material is available from the manufacturer's catalog or other published data, find the required width from $W = hp_bF/K_cP$, where $hp_b$ = hp rating of the belt material, as stated by the manufacturer; other symbols as before. To find the tension $T_b$ lb in a belt, solve $T_b = 33{,}000hp/S$ where $S$ = belt speed, ft/min. The tension per inch of belt width is $T_{bi} = T_b/W$. Where the belt speed exceeds 6000 ft/min (30.5 m/s), consult the manufacturer.

## SELECTING A RUBBER BELT FOR POWER TRANSMISSION

Choose a rubber belt to transmit 15 hp (11.2 kW) from a 7-in (17.8-cm) diameter pulley driven by a shunt-wound dc motor. The pulley speed is 1300 r/min, and the belt drives an electric generator. The arrangement of the drive is such that the arc of contact of the belt on the pulley is 220°.

### Calculation Procedure:

**1. *Determine the belt service factor***

The belt *service factor* allows for the typical conditions met in the use of a belt with a given driver and driven machine or device. Table 3 lists typical service factors $S_f$ used by the B. F.

**TABLE 3** Service Factor $S$

| Application | Squirrel-cage ac motor | | Wound rotor ac motor (slip ring) | Single-phase capacitor motor | Dc shunt-wound motor | Diesel engine, four or more cylinders, above 700 r/min |
|---|---|---|---|---|---|---|
| | Normal torque, line start | High torque | | | | |
| Agitators | 1.0–1.2 | 1.2–1.4 | 1.2 | | | |
| Compressors | 1.2–1.4 | ...... | 1.4 | 1.2 | 1.2 | 1.2 |
| Belt conveyors (ore, coal, sand) | ...... | 1.4 | ...... | .. | 1.2 | |
| Screw conveyors | ...... | 1.8 | ...... | .. | 1.6 | |
| Crushing machinery | ...... | 1.6 | 1.6 | .. | ...... | 1.4–1.6 |
| Fans, centrifugal | 1.2 | ...... | 1.4 | .. | 1.4 | 1.4 |
| Fans, propeller | 1.4 | 2.0 | 1.6 | .. | 1.6 | 1.6 |
| Generators and exciters | 1.2 | ...... | ...... | .. | 1.2 | 2.0 |
| Line shafts | 1.4 | ...... | 1.4 | 1.4 | 1.4 | 1.6 |
| Machine tools | 1.0–1.2 | ...... | 1.2–1.4 | 1.0 | 1.0–1.2 | |
| Pumps, centrifugal | 1.2 | 1.4 | 1.4 | 1.2 | 1.2 | |
| Pumps, reciprocating | 1.2–1.4 | ...... | 1.4–1.6 | .. | ...... | 1.8–2.0 |

Goodrich Company. Entering Table 3 at the type of driver, a shunt-wound dc motor, and projecting downward to the driven machine, an electric generator, shows that $S_f = 1.2$.

### 2. *Determine the arc-of-contact factor*

A rubber belt can contact a pulley in a range from about 140 to 220°. Since the hp capacity ratings for belts are based on an arc of contact of 180°, a correction factor must be applied for other arcs of contact.

Table 4 lists the arc-of-contact correction factor $C_c$. Thus, for an arc of contact of 220°, $C_c = 1.12$.

### 3. *Compute the belt speed*

The belt speed is $S = \pi RD$, where $S$ = belt speed, ft/min; $R$ = pulley rpm; $D$ = pulley diameter, ft. For this pulley, $S = \pi(1300)(7/12) = 2380$ ft/min (12.1 m/s).

### 4. *Choose the minimum pulley diameter and belt ply*

Table 5 lists minimum recommended pulley diameters, belt material, and number of plies for various belt speeds. Choose the pulley diameter and number of plies for the next higher belt speed when the computed belt speed falls between two tabulated values. Thus, for a belt speed of 2380 ft/min (12.1 m/s), use a 7-in (17.8-cm) diameter pulley as listed under 2500 ft/min (12.7 m/s). The corresponding material specifications are found in the left-hand column and are four plies, 32-oz (0.9-kg) fabric.

### 5. *Determine the belt power rating*

Enter Table 6 at 32 oz (0.9 kg) four-ply material specifications, and project horizontally to the belt speed. This occurs between the tabulated speeds of 2000 and 2500 ft/min (10.2 and 12.7 m/s). Interpolating, we find $[(2500 - 2380)/(2500 - 2000)](4.4 - 3.6) = 0.192$. Hence, the power rating of the belt $hp_{bi}$ is $4.400 - 0.192 = 4.208$ hp/in (1.2 kW/cm) of width.

**TABLE 4** Arc of Contact Factor $K$—Rubber Belts

| Arc of contact, ° | 140 | 160 | 180 | 200 | 220 |
|---|---|---|---|---|---|
| Factor $K$ | 0.82 | 0.93 | 1.00 | 1.06 | 1.12 |

**TABLE 5** Minimum Pulley Diameters, in (cm)—Rubber Belts of 32-oz (0.9-kg) Hard Fabric

| Ply | Belt speed, ft/min (m/s) | | | |
|---|---|---|---|---|
| | 2000 (149.4) | 2500 (186.4) | 3000 (223.7) | 4000 (298.3) |
| 3 | 4 (10.2) | 4 (10.2) | 4 (10.2) | 4 (10.2) |
| 4 | 5 (12.7) | 6 (15.2) | 6 (15.2) | 7 (17.8) |
| 5 | 8 (20.3) | 8 (20.3) | 9 (22.9) | 10 (25.4) |
| 6 | 11 (27.9) | 11 (27.9) | 12 (30.5) | 13 (33.0) |
| 7 | 15 (38.1) | 15 (38.1) | 16 (40.6) | 17 (43.2) |
| 8 | 18 (45.7) | 19 (48.3) | 20 (50.8) | 21 (53.3) |
| 9 | 22 (55.9) | 23 (58.4) | 24 (61.0) | 25 (63.5) |
| 10 | 26 (66.0) | 27 (68.6) | 28 (71.1) | 29 (73.7) |

**6. *Determine the required belt width***

The required belt width $W = hpS_f/(hp_{bi}C_c)$, or $W = (15)(1.2)/[(4.208)(1.12)] = 3.82$ in (9.7 cm). Use a 4-in (10.2-cm) wide belt.

**Related Calculations:** Use this procedure for rubber-belt drives of all types. For additional service factors, consult the engineering data published by B. F. Goodrich Company, The Goodyear Tire and Rubber Company, United States Rubber Company, etc.

## SELECTING A V BELT FOR POWER TRANSMISSION

Choose a V belt to drive a 0.75-hp (559.3-W) stoker at about 900 r/min from a 1750-r/min motor. The stoker is fitted with a 3-in (7.6-cm) diameter sheave and the motor with a 6-in (15.2-cm) diameter sheave. The distance between the sheave shaft centerlines is 18 in (45.7 cm). The stoker handles soft coal free of hard lumps.

### Calculation Procedure:

**1. *Determine the design horsepower for the belt***

V-belt manufacturers publish service factors for belts used in various applications. Table 7 shows that a stoker is classed as heavy service and has a service factor of 1.4 to 1.6. By using the lower value, because the stoker handles soft coal free of hard lumps, the design horsepower for the belt

**TABLE 6** Power Ratings of Rubber Belts [32-oz (0.9-kg) Hard Fabric]

*(Hp = hp/in of belt width for 180° wrap)*
*(Power = kW/cm of belt width for 180° wrap)*

| Ply | Belt speed, ft/min (m/s) | | | |
|---|---|---|---|---|
| | 2000 (10.2) | 2500 (12.7) | 3000 (15.2) | 4000 (20.3) |
| 3 | 2.9 (0.85) | 3.5 (1.03) | 4.1 (1.20) | 5.1 (1.50) |
| 4 | 3.9 (1.14) | 4.7 (1.38) | 5.5 (1.61) | 6.8 (2.00) |
| 5 | 4.9 (1.44) | 5.9 (1.73) | 6.9 (2.03) | 8.5 (2.50) |
| 6 | 5.9 (1.73) | 7.1 (2.08) | 8.3 (2.44) | 10.2 (2.99) |
| 7 | 6.9 (2.03) | 8.3 (2.44) | 9.7 (2.85) | 11.9 (3.49) |
| 8 | 7.9 (2.32) | 9.5 (2.79) | 11.1 (3.26) | 13.6 (3.99) |
| 9 | 8.9 (2.61) | 10.6 (3.11) | 12.4 (3.64) | 15.3 (4.49) |
| 10 | 9.8 (2.88) | 11.7 (3.43) | 13.7 (4.02) | 17.0 (4.99) |

**TABLE 7** Service Factors for V-Belt Drives

| Typical machines | Type of service | Service factors |
|---|---|---|
| Domestic washing machines, domestic ironers, advertising display fixtures, small fans and blowers | Light | 1.0–1.2 |
| Fans and blowers (heavy rotors), centrifugal pumps, oil burners, home workshop machines | Medium | 1.2–1.4 |
| Stokers, reciprocating pumps and compressors, refrigerators, drill presses, grinders, lathes, meat slicers, machines for industrial use | Heavy | 1.4–1.6 |

is found by taking the product of the rated horsepower of the device driven by the belt and the service factor, or (0.75 hp)(1.4 service factor) = 1.05 hp (783.0 W). The belt must be capable of transmitting this, or a greater, horsepower.

### 2. *Determine the belt speed and arc of contact*

The belt speed $S = \pi RD$, where $R$ = sheave rpm; $D$ = sheave pitch diameter, ft = (sheave outside diameter, in − $2X$)/12, where $2X$ = sheave dimension from Table 8. Before solving this equation, an assumption about the cross-sectional width of the belt must be made because $2X$ varies from 0.10 to 0.30 in (2.5 to 7.6 mm), and the exact cross section of the belt that will be used is not yet known. A value of $X$ = 0.15 in (3.8 mm) is usually a safe assumption. It corresponds to a $3L$ belt cross section. Using $X$ = 0.15 and the diameter and speed of the larger sheave, we see $S = \pi(1750)(6.0 - 0.15)/12 = 2675$ ft/min (13.6 m/s).

Compute the belt arc of contact from arc of contact, degrees $= 180 - [60(d_l - d_s)/l]$, where $d_l$ = large sheave nominal diameter, in; $d_s$ = small sheave nominal diameter, in; $l$ = distance between shaft centers, in. For this drive, arc $= 180 - [60(6 - 3)/18] = 170°$. An arc-of-contact correction factor must be applied in computing the belt power capacity. Read this correction

**TABLE 8** Sheave Dimensions—Light-Duty V Belt

| Belt cross section | Sheave effective OD | | Groove angle, deg | W | | D | | 2X | |
|---|---|---|---|---|---|---|---|---|---|
| | in | cm | | in | mm | in | mm | in | mm |
| $2L$ | Under 1.5 | Under 3.8 | 32 | 0.240 | 6.10 | 0.250 | 6.4 | 0.10 | 2.5 |
| | 1.5–1.99 | 3.8–5.05 | 34 | 0.243 | 6.17 | | | | |
| | 2.0–2.5 | 5.08–6.4 | 36 | 0.246 | 6.25 | | | | |
| | Over 2.5 | Over 6.4 | 38 | 0.250 | 6.35 | | | | |
| $3L$ | Under 2.2 | Under 5.6 | 32 | 0.360 | 9.14 | 0.406 | 10.3 | 0.15 | 3.8 |
| | 2.2–3.19 | 5.6–8.10 | 34 | 0.364 | 9.25 | | | | |
| | 3.2–4.2 | 8.13–10.7 | 36 | 0.368 | 9.35 | | | | |
| | Over 4.20 | Over 10.7 | 38 | 0.372 | 9.45 | | | | |
| $4L$ | Under 2.65 | Under 6.7 | 30 | 0.485 | 12.32 | 0.490 | 12.4 | 0.20 | 5.1 |
| | 2.65–3.24 | 6.7–8.23 | 32 | 0.490 | 12.45 | | | | |
| | 3.25–5.65 | 8.26–14.4 | 34 | 0.494 | 12.55 | | | | |
| | Over 5.65 | Over 14.4 | 38 | 0.504 | 12.80 | | | | |

**TABLE 9** Correction Factors for Arc of Contact—V-Belt Drives

| Arc of contact, deg | Correction factor | | Arc of contact, deg | Correction factor | |
|---|---|---|---|---|---|
| | V to V | V to flat* | | V to V | V to flat* |
| 180 | 1.00 | 0.75 | 130 | 0.86 | 0.86 |
| 170 | 0.98 | 0.77 | 120 | 0.82 | 0.82 |
| 160 | 0.95 | 0.80 | 110 | 0.78 | 0.78 |
| 150 | 0.92 | 0.82 | 100 | 0.74 | 0.74 |
| 140 | 0.89 | 0.84 | 90 | 0.69 | 0.69 |

*A V-to-flat drive has a small sheave and a larger-diameter flat pulley.

factor from Table 9 as $C_c$ = 0.98 for a V-sheave to V-sheave drive and a 170° arc of contact. *Note:* If desired, the pitch diameters can be used in the above relation in place of the nominal diameters.

### 3. *Select the belt to be used*

The $2X$ value used in step 3 corresponds to a $3L$ cross section belt. Check the power capacity of this belt by entering Table 10 at a belt speed of 2800 ft/min (14.2 m/s), the next larger tabulated speed, and projecting across to the appropriate small-sheave diameter—3 in (7.6 cm) or larger. Read the belt horsepower rating as 0.87 hp (648.8 W). This is considerably less than the required capacity of 1.05 hp (783.0 W) computed in step 1. Therefore, the $3L$ belt is unsatisfactory.

Try a $4L$ belt, Table 11, following the same procedure. A $4L$ belt with a 3-in (7.6-cm) diameter small sheave has a rating of 1.16 hp (865.0 W). Correct this for the actual arc of contact by multiplying by $C_c$, or (1.16)(0.98) = 1.137 (847.9 W). Thus, the belt is suitable for the design hp value of 1.05 hp (783.0 W).

As a final check, compute the actual belt speed using the actual $2X$ value from Table 8. Thus, for a $4L$ belt on a 6-in (15.2-cm) sheave, $2X = 0.20$, and $S = \pi(1750)(6 - 0.20)/12 = 2660$ ft/min (13.5 m/s). Hence, use of 2800 ft/min (14.2 m/s) in selecting the belt was a safe assumption. Note that the difference between the belt speed based on the assumed value of $2X$, 2675 ft/min (13.6 m/s), and the actual belt speed, 2660 ft/min (13.5 m/s), is about 0.5 percent. This is negligible.

**Related Calculations:** Use this procedure when choosing a single V belt for a drive. Where multiple belts are used, follow the steps given in the next calculation procedure. The data presented for single V belts is abstracted from *Standards for Light-duty or Fractional-horsepower V-Belts*, published by the Rubber Manufacturers Association.

**TABLE 10** Power Ratings of $3L$ Cross Section V Belts
*(Based on 180° arc of contact on small sheave)*

| Belt speed | | Effective OD of small sheave | | | | | | | |
|---|---|---|---|---|---|---|---|---|---|
| | | 1½ in (3.8 cm) | | 2 in (5.1 cm) | | 2½ in (6.4 cm) | | 3 in (7.6 cm) or more | |
| ft/min | m/s | hp | W | hp | W | hp | W | hp | W |
| 2200 | 11.2 | 0.12 | 89.5 | 0.44 | 328.1 | 0.64 | 477.2 | 0.77 | 574.2 |
| 2400 | 12.2 | 0.10 | 74.6 | 0.45 | 335.6 | 0.66 | 492.2 | 0.81 | 604.0 |
| 2600 | 13.2 | 0.07 | 52.2 | 0.46 | 343.0 | 0.69 | 514.5 | 0.84 | 626.4 |
| 2800 | 14.2 | 0.04 | 29.8 | 0.46 | 343.0 | 0.70 | 522.0 | 0.87 | 648.8 |
| 3000 | 15.2 | 0.01 | 7.5 | 0.45 | 335.6 | 0.72 | 536.9 | 0.89 | 663.7 |

**TABLE 11** Power Ratings of 4*L* Cross Section V Belts

*(Based on 180° arc of contact on small sheave)*

| Belt speed | | Effective OD of small sheave | | | | | | | |
|---|---|---|---|---|---|---|---|---|---|
| | | 2½ in (6.4 cm) | | 3 in (7.6 cm) | | 3½ in (8.9 cm) | | 4 in (10.2 cm) or more | |
| ft/min | m/s | hp | W | hp | W | hp | W | hp | W |
| 2200 | 11.2 | 0.67 | 499.6 | 1.08 | 805.4 | 1.37 | 1021.6 | 1.58 | 1178.2 |
| 2400 | 12.2 | 0.68 | 507.1 | 1.12 | 835.2 | 1.43 | 1066.4 | 1.66 | 1237.9 |
| 2600 | 13.2 | 0.66 | 492.2 | 1.16 | 865.0 | 1.50 | 1118.5 | 1.75 | 1305.0 |
| 2800 | 14.2 | 0.65 | 484.7 | 1.18 | 879.9 | 1.54 | 1148.4 | 1.81 | 1349.7 |
| 3000 | 15.2 | 0.63 | 469.8 | 1.19 | 887.4 | 1.58 | 1178.2 | 1.87 | 1394.5 |

## SELECTING MULTIPLE V BELTS FOR POWER TRANSMISSION

Choose the type and number of V belts needed to drive an air compressor from a 5-hp (3.7-kW) wound-rotor ac motor when the motor speed is 1800 r/min and the compressor speed is 600 r/min. The pitch diameter of the large sheave is 20 in (50.8 cm); and the distance between shaft centers is 36.0 in (91.4 cm).

### Calculation Procedure:

#### 1. *Choose the V-belt section*

Determine the design horsepower of the drive by finding the product of the service factor and the rated horsepower. Use Table 3 to find the service factor. The value of this factor is 1.4 for a compressor driven by a wound-rotor ac motor. Thus, the design horsepower = (5.0 hp)(1.4 service factor) = 7.0 hp (5.2 kW).

Enter Fig. 3 at 7.0 hp (5.2 kW), and project up to the small sheave speed, 600 r/min. Read the belt cross section as type B.

#### 2. *Determine the small-sheave pitch diameter*

Use the speed ratio of the shafts to determine the diameter of the small sheave. The speed ratio of the shafts is the ratio of the speed of the high-speed shaft to that of the low-speed shaft, or 1800/600 = 3.0. The sheave pitch diameters have the same ratio, or $20/PD_s = 3$; $PD_s = 20/3 = 6.67$ in (16.9 cm).

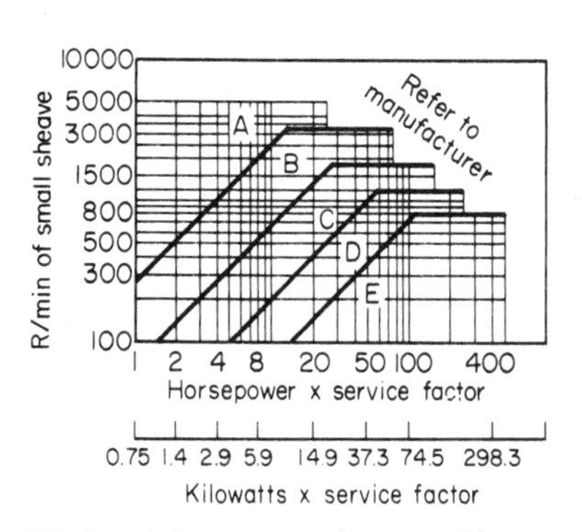


**FIG. 3** V-belt cross section for required hp rating.

#### 3. *Compute the belt speed*

The belt speed is $S = \pi RD$, where $R$ = small-sheave rpm; $D$ = small-sheave pitch diameter, ft. Thus, $S = \pi(1800)(6.67/12) = 3140$ ft/min (16.0 m/s).

#### 4. *Determine the belt horsepower rating*

A tabulation of allowable belt horsepower ratings is used to determine the rating of a specific belt. To enter this table, the belt speed and the small-sheave equivalent diameter must be known.

Find the equivalent diameter $d_e$ of the small sheave by taking the product of the small-sheave pitch diameter and the diameter factor, Table 12. Thus, for a speed range of 3.0 the small-diameter factor = 1.14, from Table 12. Hence, $d_e = (6.67)(1.14) = 7.6$ in (19.3 cm).

Enter Table 12A at a belt speed of 3200 ft/min (16.3 m/s) and $d_e$ = 7.6 in (19.3 cm). In the last column read the belt horsepower rating as 6.5 hp (4.85 kW). This rating must be corrected for the arc of contact and the belt length.

The arc of contact = $180 - [60(d_l - d_s)/l]$, where $d_l$ and $d_s$ = large- and small-sheave pitch diameters, respectively, in; $l$ = distance between sheave shaft centers, in. Thus, arc of contact = $180 - [60(20 - 6.67)/36]$ = 157.8°. Using Table 4 and interpolating, we find the arc-of-contact correction factor $C_c$ = 0.94.

Compute the belt pitch length from $L = 2l + 1.57(d_l + d_s) + (d_l - d_s)^2/(4l)$, where all the symbols are as given earlier. Thus, $L = 2(36) + 1.57(20 + 6.67) + (20 - 6.67)^2/[4(36)]$ = 115.1 in (292.4 cm).

Enter Table 13 by interpolating between the standard belt lengths of 105 and 120 in (266.7 and 304.8 cm), and find the length correction factor for a *B* cross section belt as 1.06.

**TABLE 12** Small-Diameter Factors—Multiple V Belts

| Speed ratio range | Small-diameter factor |
|---|---|
| 1.000–1.019 | 1.00 |
| 1.020–1.032 | 1.01 |
| 1.033–1.055 | 1.02 |
| 1.056–1.081 | 1.03 |
| 1.082–1.109 | 1.04 |
| 1.110–1.142 | 1.05 |
| 1.143–1.178 | 1.06 |
| 1.179–1.222 | 1.07 |
| 1.223–1.274 | 1.08 |
| 1.275–1.340 | 1.09 |
| 1.341–1.429 | 1.10 |
| 1.430–1.562 | 1.11 |
| 1.563–1.814 | 1.12 |
| 1.815–2.948 | 1.13 |
| 2.949 and over | 1.14 |

Find the product of the rated horsepower of the belt and the two correction factors—arc of contact and belt length, or (6.5)(0.94)(1.06) = 6.47 hp (4.82 kW).

### 5. *Choose the number of belts required*

The design horsepower, step 1, is 7.0 hp (5.2 kW). Thus, 7.0 design hp/6.47 rated belt hp = 1.08 belts; use two belts. Choose the next *larger* number of belts whenever a fractional number is indicated.

**Related Calculations:** The tables and data used here are based on engineering information which is available from and updated by the Mechanical Power Transmission Association and the Rubber Manufacturers Association. Similar engineering data are published by the various V-belt manufacturers. Data presented here may be used when manufacturer's engineering data are not available.

## SELECTION OF A WIRE-ROPE DRIVE

Choose a wire-rope drive for a 3000-lb (1360.8-kg) traction-type freight elevator designed to lift freight or passengers totaling 4000 lb (1814.4 kg). The vertical lift of the elevator is 500 ft (152.4 m), and the rope velocity is 750 ft/min (3.8 m/s). The traction-type elevator sheaves are designed

**TABLE 12A** Power Ratings for Premium-Quality *B*-Section V Belts

| Belt speed | | Equivalent diameter $d_e$ | | | |
|---|---|---|---|---|---|
| | | 6.6 in (16.8 cm) | | 7.0+ in (17.8+ cm) | |
| ft/min | m/s | hp | kW | hp | kW |
| 3000 | 15.2 | 5.90 | 4.40 | 6.26 | 4.67 |
| 3200 | 16.3 | 6.12 | 4.56 | 6.50 | 4.85 |
| 3400 | 17.3 | 6.31 | 4.71 | 6.73 | 5.02 |

**TABLE 13** Length Correction Factors—Multiple V Belts

| Standard length designation | | Belt cross section | | | |
|---|---|---|---|---|---|
| in | cm | *A* | *B* | *C* | *D* |
| 96 | 243.8 | 1.08 | . . . | 0.92 | |
| 105 | 266.7 | 1.10 | 1.04 | 0.94 | |
| 120 | 304.8 | 1.13 | 1.07 | 0.97 | 0.86 |
| 136 | 345.4 | . . . | 1.09 | 0.99 | |

to accelerate the car to full speed in 60 ft (18.3 m) when it starts from a stopped position. A 48-in (1.2-m) diameter sheave is used for the elevator.

## Calculation Procedure:

### 1. *Select the number of hoisting ropes to use*

The number of ropes required for an elevator is usually fixed by state or city laws. Check the local ordinances before choosing the number of ropes. Usual laws require at least four ropes for a freight elevator. Assume four ropes are used for this elevator.

### 2. *Select the rope size and strength*

Standard "blue-center" steel hoisting rope is a popular choice, as is "plow-steel" and "mild plow-steel" rope. Assume that four 9/16-in (14.3-mm) six-strand 19-wires-per-strand blue-center steel ropes will be suitable for this car. The 6 × 19 rope is commonly used for freight and passenger elevators. Once the rope size is chosen, its strength can be checked against the actual load. The breaking strength of 9/16 in (14.3 mm), 6 × 19 blue-center steel rope is 13.5 tons (12.2 t), and its weight is 0.51 lb/ft (0.76 kg/m). These values are tabulated in Baumeister and Marks—*Standard Handbook for Mechanical Engineers* and in rope manufacturers' engineering data.

### 3. *Compute the total load on each rope*

The weight of the car and its contents is 3000 + 4000 = 7000 lb (3175.1 kg). With four ropes, the load per rope is 7000/[4(2000 lb·ton)] = 0.875 ton (0.794 t).

With a 500-ft (152.4-m) lift, the length of each rope would be equal to the lift height. Hence, with a rope weight of 0.51 lb/ft (0.76 kg/m), the total weight of the rope = (0.51)(500)/2000 = 0.127 ton (0.115 t).

Acceleration of the car from the stopped condition places an extra load on the rope. The rate of acceleration of the car is found from $a = v^2/(2d)$, where $a$ = car acceleration, ft/s$^2$; $v$ = final velocity of the car, ft/s; $d$ = distance through which the acceleration occurs, ft. For this car, $a = (750/60)^2/[2(60)] = 1.3$ ft/s$^2$ (39.6 cm/s$^2$). The value 60 in the numerator of the above relation converts from feet per minute to feet per second.

The rope load caused by acceleration of the car is $L_r = Wa/$(number of ropes)(2000 lb/ton) [$g$ = 32.2 ft/s$^2$ (9.8 m/s$^2$)], where $L_r$ = rope load, tons; $W$ = weight of car and load, lb. Thus, $L_r$ = (7000)(1.3)/[(4)(2000)(32.2)] = 0.03351 ton (0.03040 t) per rope.

The rope load caused by acceleration of the rope is $L_r = Wa/32.2$, where $W$ = weight of rope, tons. Or, $L_r$ = (0.127)(1.3)/32.2 = 0.0512 ton (0.0464 t).

When the rope bends around the sheave, another load is produced. This bending load is, in pounds, $F_b = AE_r d_w/d_s$, where $A$ = rope area, in$^2$; $E_r$ = modulus of elasticity of the whole rope = 12 × 10$^6$ lb/in$^2$ (82.7 × 10$^6$ kPa) for steel rope; $d_w$ = rope diameter, in; $d_s$ = sheave diameter, in. Thus, for this rope, $F_b$ = (0.0338)(12 × 10$^6$)(0.120/48) = 1014 lb, or 0.507 ton (0.460 t).

The total load on the rope is the sum of the individual loads, or 0.875 + 0.127 + 0.0351 + 0.507 + 0.051 = 1.545 tons (1.4 t). Since the rope has a breaking strength of 13.5 tons (12.2 t), the factor of safety FS = breaking strength, tons/rope load, tons = 13.5/1.545 = 8.74. The usual minimum acceptable FS for elevator ropes is 8.0. Hence, this rope is satisfactory.

**Related Calculations:** Use this general procedure when choosing wire-rope drivers for mine hoists, inclined-shaft hoists, cranes, derricks, car pullers, dredges, well drilling, etc. When standard hoisting rope is chosen, which is the type most commonly used, the sheave diameter should not be less than $30d_w$; the recommended diameter is $45d_w$. For *haulage rope* use $42d_w$ and $72d_w$, respectively; for special flexible *hoisting rope*, use $18d_w$ and $27d_w$ sheaves.

## SPEEDS OF GEARS AND GEAR TRAINS

A gear having 60 teeth is driven by a 12-tooth gear turning at 800 r/min. What is the speed of the driven gear? What would be the speed of the driven gear if a 24-tooth idler gear were placed between the driving and driven gear? What would be the speed of the driven gear if two 24-tooth idlers were used? What is the direction of rotation of the driven gear when one and two idlers are used? A 24-tooth driving gear turning at 600 r/min meshes with a 48-tooth compound gear. The second gear of the compound gear has 72 teeth and drives a 96-tooth gear. What are the speed and direction of rotation of the 96-tooth gear?

### Calculation Procedure:

#### 1. *Compute the speed of the driven gear*

For any two meshing gears, the speed ratio $R_D/R_d = N_d/N_D$, where $R_D$ = rpm of driving gear; $R_d$ = rpm of driven gear; $N_d$ = number of teeth in driven gear; $N_D$ = number of teeth in driving gear. By substituting the given values, $R_D/R_d = N_d/N_D$, or $800/R_d = 60/12$; $R_d = 160$ r/min.

#### 2. *Determine the effect of one idler gear*

An idler gear has *no* effect on the speed of the driving or driven gear. Thus, the speed of each gear would remain the same, regardless of the number of teeth in the idler gear. An idler gear is generally used to reduce the required diameter of the driving and driven gears on two widely separated shafts.

#### 3. *Determine the effect of two idler gears*

The effect of more than one idler is the same as that of a single idler—i.e., the speed of the driving and driven gears remains the same, regardless of the number of idlers used.

#### 4. *Determine the direction of rotation of the gears*

Where an odd number of gears are used in a gear train, the first and last gears turn in the *same* direction. Thus, with one idler, one driver, and one driven gear, the driver and driven gear turn in the *same* direction because there are three gears (i.e., an odd number) in the gear train.

Where an even number of gears is used in a gear train, the first and last gears turn in the *opposite* direction. Thus, with two idlers, one driver, and one driven gear, the driver and driven gear turn in the *opposite* direction because there are four gears (i.e., an even number) in the gear train.

#### 5. *Determine the compound-gear output speed*

A compound gear has two gears keyed to the same shaft. One of the gears is driven by another gear; the second gear of the compound set drives another gear. In a compound gear train, the product of the number of teeth of the driving gears and the rpm of the first driver equals the product of the number of teeth of the driven gears and the rpm of the last driven gear.

In this gearset, the first driver has 24 teeth and the second driver has 72 teeth. The rpm of the first driver is 600. The driven gears have 48 and 96 teeth, respectively. Speed of the final gear is unknown. Applying the above rule gives (24)(72)(600)2(48)(96)($R_d$); $R_d = 215$ r/min.

Apply the rule in step 4 to determine the direction of rotation of the final gear. Since the gearset has an even number of gears, four, the final gear revolves in the opposite direction from the first driving gear.

**Related Calculations:** Use the general procedure given here for gears and gear trains having spur, bevel, helical, spiral, worm, or hypoid gears. Be certain to determine the correct number of teeth and the gear rpm before substituting values in the given equations.

## SELECTION OF GEAR SIZE AND TYPE

Select the type and size of gears to use for a 100-ft$^3$/min (0.047-m$^3$/s) reciprocating air compressor driven by a 50-hp (37.3-kW) electric motor. The compressor and motor shafts are on parallel axes 21 in (53.3 cm) apart. The motor shaft turns at 1800 r/min while the compressor shaft turns at 300 r/min. Is the distance between the shafts sufficient for the gears chosen?

### Calculation Procedure:

#### 1. *Choose the type of gears to use*

Table 14 lists the kinds of gears in common use for shafts having parallel, intersecting, and nonintersecting axes. Thus, Table 14 shows that for shafts having parallel axes, spur or helical, external or internal, gears are commonly chosen. Since external gears are simpler to apply than internal gears, the external type is chosen wherever possible. Internal gears are the planetary type and are popular for applications where limited space is available. Space is not a consideration in this application; hence, an external spur gearset will be used.

Table 15 lists factors to consider in selecting gears by the characteristics of the application. As

**TABLE 14** Types of Gears in Common Use*

| Parallel axes | Intersecting axes | Nonintersecting parallel axes |
|---|---|---|
| Spur, external | Straight bevel | Crossed helical |
| Spur, internal | Zerol† bevel | Single-enveloping worm |
| Helical, external | Spiral bevel | Double-enveloping worm |
| Helical, internal | Face gear | Hypoid |

*From Darle W. Dudley—*Practical Gear Design*, McGraw-Hill, 1954.
†Registered trademark of the Gleason Works.

**TABLE 15** Gear Drive Selection by Application Characteristics*

| Characteristic | Type of gearbox | Kind of teeth | Range of use |
|---|---|---|---|
| | Simple, branched, or epicyclic | Helical | Up to 40,000 hp (29,828 kW) per single mesh; over 60,000 hp (44,742 kW) in MDT designs; up to 40,000 hp (29,282 kW) in epicyclic units |
| High power | Simple, branched, or epicyclic | Spur | Up to 4000 hp (2983 kW) per single mesh; up to 10,000 hp (7457 kW) in an epicyclic |
| | Simple | Spiral bevel | Up to 15,000 hp (11,186 kW) per single mesh |
| | | Zerol bevel | Up to 1000 hp (745.7 kW) per single mesh |
| High efficiency | Simple | Spur, helical or bevel | Over 99 percent efficiency in the most favorable cases—98 percent efficiency is typical |
| Light weight | Epicyclic | Spur or helical | Outstanding in airplane and helicopter drives |
| | Branched-MDT | Helical | Very good in marine main reductions |
| | Differential | Spur or helical | Outstanding in high-torque-actuating devices |
| | | Bevel | Automobiles, trucks, and instruments |
| Compact | Epicyclic | Spur or helical | Good in aircraft nacelles |
| | Simple | Worm-gear | Good in high-ratio industrial speed reducers |
| | Simple | Spiroid | Good in tools and other applications |
| | Simple | Hypoid | Good in auto and truck rear ends plus other applications |
| | Simple | Worm-gear | Widely used in machine-tool index drives |
| | Simple | Hypoid | Used in certain index drives for machine tools |
| Precision | Simple or branched | Helical | A favorite for high-speed, high-accuracy power gears |
| | Simple | Spur | Widely used in radar pedestal gearing, gun control drives, navigation instruments, and many other applications |
| | Simple | Spiroid | Used where precision and adjustable backlash are needed |

**Mechanical Engineering*, November 1965.

**TABLE 16** Gear Drive Selection for the Convenience of the User°

| Consideration | Kind of teeth | Typical applications | Comments |
|---|---|---|---|
| Cost | Spur | Toys, clocks, instruments, industrial drives, machine tools, transmissions, military equipment, household applications, rocket boosters | Very widely used in all manner of applications where power and speed requirements are not too great—parts are often mass-produced at very low cost per part |
| Ease of use | Spur or helical | Change gears in machine tools, vehicle transmissions where gear shifting occurs | Ease of changing gears to change ratio is important |
| | Worm-gear | Speed reducers | High ratio drive obtained with only two gear parts |
| Simplicity | Crossed helical | Light power drives | No critical positioning required in a right-angle drive |
| | Face gear | Small power drives | Simple and easy to position for a right-angle drive |
| | Helical | Marine main drive units for ships, generator drives in power plants | Helical teeth with good accuracy and a design that provides good axial overlap mesh very smoothly |
| | Spiral bevel | Main drive units for aircraft, ships, and many other applications | Helical type of tooth action in a right-angle power drive |
| Noise | Hypoid | Automotive rear axle | Helical type of tooth provides high overlap |
| | Worm-gear | Small power drives in marine, industrial, and household appliance applications | Overlapping, multiple tooth contacts |
| | Spiroid-Gear | Portable tools, home appliances | Overlapping, multiple tooth contacts |

°*Mechanical Engineering*, November 1965.

with Table 14, the data in Table 15 indicate that spur gears are suitable for this drive. Table 16, based on the convenience of the user, also indicates that spur gears are suitable.

### 2. *Compute the pitch diameter of each gear*

The distance between the driving and driven shafts is 21 in (53.3 cm). This distance is approximately equal to the sum of the driving gear pitch radius $r_D$ in and the driven gear pitch radius $r_d$ in. Or $d_D + r_d = 21$ in (53.3 cm).

In this installation the driving gear is mounted on the motor shaft and turns at 1800 r/min. The driven gear is mounted on the compressor shaft and turns at 300 r/min. Thus, the speed ratio of the gears ($R_D$, driver rpm/$R_d$, driven rpm) = 1800/300 = 6. For a spur gear, $R_D/R_d = r_d/r_D$, or $6 = r_d/r_D$, and $r_d = 6r_D$. Hence, substituting in $r_D + r_d = 21$, $r_D + 6r_D = 21$; $r_D = 3$ in (7.6 cm). Then $3 + r_d = 21$, $r_d = 18$ in (45.7 cm). The respective pitch diameters of the gears are $d_D = 2 \times 3 = 6.0$ in (15.2 cm); $d_d = 2 \times 18 = 36.0$ in (91.4 cm).

### 3. *Determine the number of teeth in each gear*

The number of teeth in a spur gearset, $N_D$ and $N_d$, can be approximated from the ratio $R_D/R_d = N_d/N_D$, or $1800/300 = N_d/N_D$; $N_d = 6N_D$. Hence, the driven gear will have approximately six times as many teeth as the driving gear.

As a trial, assume that $N_d = 72$ teeth; then $N_D = N_d/6 = 72/6 = 12$ teeth. This assumption must now be checked to determine whether the gears will give the desired output speed. Since

**TABLE 17** Gear Drive Selection by Arrangement of Driving and Driven Equipment*

| Kind of teeth | Axes | Gearbox type | Type of tooth contact | Generic family |
|---|---|---|---|---|
| Spur† | Parallel | Simple (pinion and gear), epicyclic (planetary, star, solar), branched systems, idler for reverse | Line | Coplanar |
| Helical† (single or double helical, herringbone) | Parallel | Simple, epicyclic, branched | Overlapping line | Coplanar |
| Bevel | Right-angle or angular, but intersecting | Simple, epicyclic, branched | (Straight) line, (Zerol)‡ line, (spiral) overlapping | Coplanar |
| Worm | Right-angle, nonintersecting | Simple | (Cylindrical) overlapping line, (double-enveloping)§ overlapping line | Nonplanar |
| Crossed helical | Right-angle or skew, nonintersecting | Simple | Point | Nonplanar |
| Face gear | Right-angle, intersecting | Simple | Line (overlapping if helical) | Coplanar |
| Hypoid | Right-angle or angular, nonintersecting | Simple | Overlapping line | Nonplanar |
| Spiroid, helicon, planoid | Right-angle, nonintersecting | Simple | Overlapping line | Nonplanar |

* *Mechanical Engineering*, November 1965.
†These kinds of teeth are often used to change rotary motion to linear motion by use of a pinion and rack.
‡Zerol is a registered trademark of the Gleason Works, Rochester, New York.
§The most widely used double-enveloping worm gear is the cone-drive type.

$R_D/R_d = N_d/N_D$, or $1800/300 = 72/12$; $6 = 6$. Thus, the gears will provide the desired speed change.

The distance between the shafts is 21 in (53.3 cm) $= r_D + r_d$. This means that there is no clearance when the gears are meshed. Since all gears require some clearance, the shafts will have to be moved apart slightly to provide this clearance. If the shafts cannot be moved apart, the gear diameter must be reduced. In this installation, however, the electric-motor driver can probably be moved a fraction of an inch to provide the desired clearance.

### 4. *Choose the final gear size*

Refer to a catalog of stock gears. From this catalog choose a driving and a driven gear having the required number of teeth and the required pitch diameter. If gears of the exact size required are not available, pick the nearest suitable stock sizes.

Check the speed ratio, using the procedure in step 3. As a general rule, stock gears having a slightly different number of teeth or a somewhat smaller or larger pitch diameter will provide nearly the desired speed ratio. When suitable stock gears are not available in one catalog, refer to one or more other catalogs. If suitable stock gears are still not available, and if the speed ratio is a critical factor in the selection of the gear, custom-sized gears may have to be manufactured.

**Related Calculations:** Use this general procedure to choose gear drives employing any of the 12 types of gears listed in Table 14. Table 17 lists typical gear selections based on the arrangement of the driving and driven equipment. These tables are the work of Darle W. Dudley.

## GEAR SELECTION FOR LIGHT LOADS

Detail a generalized gear-selection procedure useful for spur, rack, spiral miter, miter, bevel, helical, and worm gears. Assume that the drive horsepower and speed ratio are known.

## Calculation Procedure:

### 1. *Choose the type of gear to use*

Use Table 14 of the previous calculation procedure as a general guide to the type of gear to use. Make a tentative choice of the gear type.

### 2. *Select the pitch diameter of the pinion and gear*

Compute the pitch diameter of the pinion from $d_p = 2c/(R + 2)$, where $d_p$ = pitch diameter, in, of the pinion, which is the *smaller* of the two gears in mesh; $c$ = center distance between the gear shafts, in; $R$ = gear ratio = larger rpm, number of teeth, or pitch diameter + smaller rpm, smaller number of teeth, or smaller pitch diameter.

Compute the pitch diameter of the gear, which is the *larger* of the two gears in mesh, from $d_g = d_pR$.

### 3. *Determine the diametral pitch of the drive*

Tables 18 to 21 show typical diametral pitches for various horsepower ratings and gear materials. Enter the appropriate table at the horsepower that will be transmitted, and select the diametral pitch of the pinion.

### 4. *Choose the gears to use*

Enter a manufacturer's engineering tabulation of gear properties, and select the pinion and gear for the horsepower and rpm of the drive. Note that the rated horsepower of the pinion and the gear must equal, or exceed, the rated horsepower of the drive at this specified input and output rpm.

### 5. *Compute the actual center distance*

Find half the sum of the pitch diameter of the pinion and the pitch diameter of the gear. This is the actual center-to-center distance of the drive. Compare this value with the available space. If the actual center distance exceeds the allowable distance, try to rearrange the drive or select another type of gear and pinion.

**TABLE 18** Spur-Gear Pitch Selection Guide*

*(20° pressure angle)*

| Gear diametral pitch | | Pinion | | Gear | |
|---|---|---|---|---|---|
| in | cm | hp | W | hp | W |
| 20 | 50.8 | 0.04–1.69 | 29.8–1,260 | 0.13–0.96 | 96.9–715.9 |
| 16 | 40.6 | 0.09–2.46 | 67.1–1,834 | 0.22–1.61 | 164.1–1,200 |
| 12 | 30.5 | 0.24–5.04 | 179.0–3,758 | 0.43–3.16 | 320.8–2,356 |
| 10 | 25.4 | 0.46–6.92 | 343.0–5,160 | 0.70–5.12 | 522.0–3,818 |
| 8 | 20.3 | 0.88–10.69 | 656.2–7,972 | 1.11–7.87 | 827.7–5,869 |
| 6 | 15.2 | 1.84–16.63 | 1,372–12,401 | 2.28–12.39 | 1,700–9,239 |
| 5 | 12.7 | 3.04–24.15 | 2,267–18,009 | 3.75–17.19 | 2,796–12,819 |
| 4 | 10.2 | 5.29–34.83 | 3,945–25,973 | 6.36–25.17 | 4,743–18,769 |
| 3 | 7.6 | 13.57–70.46 | 10,119–52,542 | 15.86–51.91 | 11,831–38,709 |

*Morse Chain Company.

**TABLE 19** Miter and Bevel-Gear Pitch Selection Guide*

*(20° pressure angle)*

| Gear diametral pitch | | Hardened gear | | Unhardened gear | |
|---|---|---|---|---|---|
| in | cm | hp | kW | hp | kW |
| | | Steel spiral miter | | | |
| 18 | 45.7 | 0.07–0.70 | 0.053–0.522 | 0.04–0.42 | 0.030–0.313 |
| 12 | 30.5 | 0.15–1.96 | 0.112–1.462 | 0.09–1.17 | 0.067–0.873 |
| 10 | 25.4 | 0.50–4.53 | 0.373–3.378 | 0.30–2.70 | 0.224–2.013 |
| 8 | 20.3 | 1.56–7.15 | 1.163–5.331 | 0.93–4.26 | 0.694–3.177 |
| 7 | 17.8 | 1.93–9.30 | 1.439–6.935 | 1.15–5.54 | 0.858–4.131 |
| | | Steel miter | | | |
| 20 | 50.8 | ........ | .......... | 0.01–0.12 | 0.008–0.090 |
| 16 | 40.6 | 0.07–0.73 | 0.053–0.544 | 0.02–0.72 | 0.015–0.537 |
| 14 | 35.6 | ........ | .......... | 0.04–0.37 | 0.030–0.276 |
| 12 | 30.5 | 0.14–2.96 | 0.104–2.207 | 0.07–1.77 | 0.052–1.320 |
| 10 | 25.4 | 0.39–3.47 | 0.291–2.588 | 0.23–2.07 | 0.172–1.544 |

Steel and cast-iron bevel gears

| Ratio | hp | W |
|---|---|---|
| 1.5:1 | 0.04–2.34 | 29.8–1744.9 |
| 2:1 | 0.01–12.09 | 7.5–9015.5 |
| 3:1 | 0.04–8.32 | 29.8–6204.2 |
| 4:1 | 0.05–10.60 | 37.3–7904.4 |
| 6:1 | 0.07–2.16 | 52.2–1610.7 |

*Morse Chain Company.

### 6. *Check the drive speed ratio*

Find the actual speed ratio by dividing the number of teeth in the gear by the number of teeth in the pinion. Compare the actual ratio with the desired ratio. If there is a major difference, change the number of teeth in the pinion or gear or both.

**Related Calculations:** Use this general procedure to select gear drives for loads up to the ratings shown in the accompanying tables. For larger loads, use the procedures given elsewhere in this section.

## SELECTION OF GEAR DIMENSIONS

A mild-steel 20-tooth 20° full-depth-type spur-gear pinion turning at 900 r/min must transmit 50 hp (37.3 kW) to a 300-r/min mild-steel gear. Select the number of gear teeth, diametral pitch of the gear, width of the gear face, the distance between the shaft centers, and the dimensions of the gear teeth. The allowable stress in the gear teeth is 800 lb/in² (55,160 kPa).

### Calculation Procedure:

### 1. *Compute the number of teeth on the gear*

For any gearset, $R_D/R_d, = N_d/N_D$, where $R_D$ = rpm of driver; $R_d$ = rpm of driven gear; $N_d$ = number of teeth on the driven gear; $N_D$ = number of teeth on driving gear. Thus, $900/300 = N_d/20$; $N_d$ = 60 teeth.

**TABLE 20** Helical-Gear Pitch Selection Guide*

| Gear diametral pitch | | Hardened-steel gear | |
|---|---|---|---|
| in | cm | hp | W |
| 20 | 50.8 | 0.04–1.80 | 29.8–1,342.3 |
| 16 | 40.6 | 0.08–2.97 | 59.7–2,214.7 |
| 12 | 30.5 | 0.22–5.87 | 164.1–4,377.3 |
| 10 | 25.4 | 0.37–8.29 | 275.9–6,181.9 |
| 8 | 20.3 | †0.66–11.71 | 492.2–8,732.1 |
| | | ‡0.49–9.07 | 365.4–6,763.5 |
| 6 | 15.2 | §1.44–19.15 | 1,073.8–14,280.2 |
| | | †1.15–15.91 | 857.6–11,864.1 |

*Morse Chain Company.
†1-in (2.5-cm) face.
‡¾-in (1.9-cm) face.
§1½-in (3.8-cm) face.

### 2. *Compute the diametral pitch of the gear*

The diametral pitch of the gear must be the same as the diametral pitch of the pinion if the gears are to run together. If the diametral pitch of the pinion is known, assume that the diametral pitch of the gear equals that of the pinion.

When the diametral pitch of the pinion is not known, use a modification of the Lewis formula, shown in the next calculation procedure, to compute the diametral pitch. Thus, $P = (\pi\ SaYv/33{,}000\ \text{hp})^{0.5}$, where all the symbols are as in the next calculation procedure, except that $a = 4$ for machined gears. Obtain $Y = 0.421$ for 60 teeth in a 20° full-depth gear from Baumeister and

**TABLE 21** Worm-Gear Pitch Selection Guide*

| Gear diametral pitch | | Bronze gears | | | | | |
|---|---|---|---|---|---|---|---|
| | | Single | | Double | | Quadruple | |
| in | cm | hp | W | hp | W | hp | W |
| 12 | 30.5 | 0.04–0.64 | 29.8–477.2 | 0.05–1.21 | 37.3–902.3 | 0.05–3.11 | 37.3–2319 |
| 10 | 25.4 | 0.06–0.97 | 44.7–723.3 | 0.08–2.49 | 59.7–1856 | 0.13–4.73 | 96.9–3527 |
| 8 | 20.3 | 0.11–1.51 | 82.0–1126 | 0.15–3.95 | 111.9–2946 | 0.08–7.69 | 59.7–5734 |
| | | | | Triple | | | |
| 5 | 12.7 | 0.51–4.61 | 380.3–3437 | 1.10–10.53 | 820.3–7852 | | |
| 4 | 10.2 | 0.66–6.74 | 492.2–5026 | | | | |

*Morse Chain Company.

Marks—*Standard Handbook for Mechanical Engineers*. Assume that $v$ = pitch-line velocity = 1200 ft/min (6.1 m/s). This is a typical reasonable value for $v$. Then $P = [\pi \times 8000 \times 4 \times 0.421 \times 1200/(33{,}000 \times 50)]^{0.5} = 5.56$, say 6, because diametral pitch is expressed as a whole number whenever possible.

**3. *Compute the gear face width***

Spur gears often have a face width equal to about four times the circular pitch of the gear. Circular pitch $p_c = \pi/P = \pi/6 = 0.524$. Hence, the face width of the gear = 4 × 0.524 = 2.095 in, say 2⅛ in (5.4 cm).

**4. *Determine the distance between the shaft centers***

Find the exact shaft centerline distance from $d_c = (N_p + N_g)/2P)$, where $N_p$ = number of teeth on pinion gear; $N_g$ = number of teeth on gear. Thus, $d_c = (20 + 60)/[2(6)] = 6.66$ in (16.9 cm).

**5. *Compute the dimensions of the gear teeth***

Use AGMA *Standards*, Dudley—*Gear Handbook*, or the engineering tables published by gear manufacturers. Each of these sources provides a list of factors by which either the circular or diametral pitch can be multiplied to obtain the various dimensions of the teeth in a gear or pinion. Thus, for a 20° full-depth spur gear, using the circular pitch of 0.524 computed in step 3, we have the following:

| | | Factor | | Circular pitch | | Dimension, in (mm) |
|---|---|---|---|---|---|---|
| Addendum | = | 0.3183 | × | 0.524 | = | 0.1668 (4.2) |
| Dedendum | = | 0.3683 | × | 0.524 | = | 0.1930 (4.9) |
| Working depth | = | 0.6366 | × | 0.524 | = | 0.3336 (8.5) |
| Whole depth | = | 0.6866 | × | 0.524 | = | 0.3598 (9.1) |
| Clearance | = | 0.05 | × | 0.524 | = | 0.0262 (0.67) |
| Tooth thickness | = | 0.50 | × | 0.524 | = | 0.262 (6.7) |
| Width of space | = | 0.52 | × | 0.524 | = | 0.2725 (6.9) |
| Backlash = width of space − tooth thickness | = | | | 0.2725 − 0.262 | = | 0.0105 (0.27) |

The dimensions of the pinion teeth are the same as those of the gear teeth.

**Related Calculations:** Use this general procedure to select the dimensions of helical, herringbone, spiral, and worm gears. Refer to the AGMA *Standards* for suitable factors and typical allowable working stresses for each type of gear and gear material.

## HORSEPOWER RATING OF GEARS

What are the strength horsepower rating, durability horsepower rating, and service horsepower rating of a 600-r/min 36-tooth 1.75-in (4.4-cm) face-width 14.5° full-depth 6-in (15.2-cm) pitch-diameter pinion driving a 150-tooth 1.75-in (4.4-cm) face width 14.5° full-depth 25-in (63.5-cm) pitch-diameter gear if the pinion is made of SAE 1040 steel 245 BHN and the gear is made of cast steel 0.35/0.45 carbon 210 BHN when the gearset operates under intermittent heavy shock loads for 3 h/day under fair lubrication conditions? The pinion is driven by an electric motor.

### Calculation Procedure:

**1. *Compute the strength horsepower, using the Lewis formula***

The widely used Lewis formula gives the strength horsepower, $hp_s = SYFK_v v/(33{,}000P)$, where $S$ = allowable working stress of gear material, lb/in²; $Y$ = tooth form factor (also called the Lewis factor); $F$ = face width, in; $K_v$ = dynamic load factor = $600/(600 + v)$ for metal gears, $0.25 + 150/(200 + v)$ for nonmetallic gears; $v$ = pitchline velocity, ft/min = (pinion pitch diameter, in)(pinion rpm)(0.262); $P$ = diametral pitch, in = number of teeth/pitch diameter, in. Obtain values of $S$ and $Y$ from tables in Baumeister and Marks—*Standard Handbook for Mechanical*

*Engineers*, or AGMA *Standards Books*, or gear manufacturers' engineering data. Compute the strength horsepower for the pinion and gear separately.

Using one of the above references for the pinion, we find $S = 25{,}000$ lb/in$^2$ (172,368.9 kPa) and $Y = 0.298$. The pitchline velocity for the metal pinion is $v = (6.0)(600)(0.262) = 944$ ft/min (4.8 m/s). Then $K_v = 600/(600 + 944) = 0.388$. The diametral pitch of the pinion is $P = N_p/d_p$, where $N_p$ = number of teeth on pinion; $d_p$ = diametral pitch of pinion, in. Or $P = 36/6 = 6$.

Substituting the above values in the Lewis formula gives $hp_s = (25{,}000)(0.298)(1.75)(0.388)(944)/[(33{,}000)(6)] = 24.117$ hp (17.98 kW) for the pinion.

Using the Lewis formula and the same procedure for the 150-tooth gear, $hp_s = (20{,}000)(0.374)(1.75)(0.388)(944)/[(33{,}000)(6)] = 24.2$ hp (18.05 kW). Thus, the strength horsepower of the gear is greater than that of the pinion.

### 2. *Compute the durability horsepower*

The durability horsepower of spur gears is found from $hp_d = F_iK_rD_oC_r$ for 20° pressure-angle full-depth or stub teeth. For 14.5° full-depth teeth, multiply $hp_d$ by 0.75. In this relation, $F_i$ = face-width and built-in factor from AGMA *Standards;* $K_r$ = factor for tooth form, materials, and ratio of gear to pinion from AGMA *Standards;* $D_o = (d_p^2R_p/158{,}000)(1 - v^{0.5}/84)$, where $d_p$ = pinion pitch diameter, in; $R_p$ = pinion rpm; $v$ = pinion pitchline velocity, ft/min, as computed in step 1; $C_r$ = factor to correct for increased stress at the start of single-tooth contact as given by AGMA *Standards*.

Using appropriate values from these standards for low-speed gears of double speed reductions yields $hp_d = (0.75)(1.46)(387)(0.0865)(1.0) = 36.6$ hp (27.3 kW).

### 3. *Compute the gearset service rating*

Determine, by inspection, which is the lowest computed value for the gearset—the strength or durability horsepower. Thus, step 1 shows that the strength horsepower $hp_s = 13.78$ hp (10.3 kW) of the pinion is the lowest computed value. Use this lowest value in computing the gear-train service rating.

Using the AGMA *Standards*, determine the service factors for this installation. The load service factor for heavy shock loads and 3 h/day intermittent operation with an electric-motor drive is 1.5 from the *Standards*. The lubrication factor for a drive operating under fair conditions is, from the *Standards*, 1.25. To find the service rating, divide the lowest computed horsepower by the product of the load and lubrication factors; or, service rating $= 13.78/[(1.5)(1.25)] = 7.35$ hp (5.5 kW).

Were this gearset operated only occasionally (0.5 h or less per day), the service rating could be determined by using the lower of the two computed strength horsepowers, in this case 13.78 hp (10.3 kW). Apply only the load service factor, or 1.25 for occasional heavy shock loads. Thus, the service rating for these conditions $= 13.78/1.25 = 11$ hp (8.2 kW).

**Related Calculations:** Similar AGMA gear construction-material, tooth-form, face-width, tooth-stress, service, and lubrication tables are available for rating helical, double-helical, herringbone, worm, straight-bevel, spiral-bevel, and Zerol gears. Follow the general procedure given here. Be certain, however, to use the applicable values from the appropriate AGMA tables. In general, choose suitable stock gears first; then check the horsepower rating as detailed above.

## MOMENT OF INERTIA OF A GEAR DRIVE

A 12-in (30.5-cm) outside-diameter 36-tooth steel pinion gear having a 3-in (7.6-cm) face width is mounted on a 2-in (5.1-cm) diameter 36-in (91.4-cm) long steel shaft turning at 600 r/min. The pinion drives a 200-r/min 36-in (91.4-cm) outside-diameter 108-tooth steel gear mounted on a 12-in (30.5-cm) long 2-in (5.1-cm) diameter steel shaft that is solidly connected to a 24-in (61.0-cm) long 4-in (10.2-cm) diameter shaft. What is the moment of inertia of the high-speed and low-speed assemblies of this gearset?

### Calculation Procedure:

### 1. *Compute the moment of inertia of each gear*

The moment of inertia of a cylindrical body about its longitudinal axis is $I_i = WR^2$, where $I_i$ = moment of inertia of a cylindrical body, in$^4$/in of length; $W$ = weight of cylindrical material, lb/in$^3$; $R$ = radius of cylinder to its outside surface, in. For a steel shaft or gear, this relation can be simplified to $I_i = D^4/35.997$, where $D$ = shaft or gear diameter, in. When you are computing $I$ for a gear, treat it as a solid blank of material. This is a safe assumption.

Thus, for the 12-in (30.5-cm) diameter pinion, $I = 12^4/35.997 = 576.05$ in$^4$/in (9439.8 cm$^4$/cm) of length. Since the gear has a 3-in (7.6-cm) face width, the moment of inertia for the total length is $I_t$ = (3.0)(576.05) = 1728.15 in$^4$ (71,931.0 cm$^4$).

For the 36-in (91.4-cm) gear, $I_i = 36^4/35.997 = 46{,}659.7$ in$^4$/in (764,615.5 cm$^4$/cm) of length. With a 3-in (7.6-cm) face width, $I_t$ = (3.0)(46,659.7) = 139,979.1 in$^4$ (5,826,370.0 cm$^4$).

## 2. *Compute the moment of inertia of each shaft*

Follow the same procedure as in step 1. Thus for the 36-in (91.4-cm) long 2-in (5.1-cm) diameter pinion shaft, $I_t = (2^4/35.997)(36) = 16.0$ in$^4$ (666.0 cm$^4$).

For the 12-in (30.5-cm) long 2-in (5.1-cm) diameter portion of the gear shaft, $I_t = (2^4/35.997)(12) = 5.33$ in$^4$ (221.9 cm$^4$). For the 24-in (61.0-cm) long 4-in (10.2-cm) diameter portion of the gear shaft, $I_t = (4^4/35.997)(24) = 170.69$ in$^4$ (7104.7 cm$^4$). The total moment of inertia of the gear shaft equals the sum of the individual moments, or $I_t = 5.33 + 170.69 = 176.02$ in$^4$ (7326.5 cm$^4$).

## 3. *Compute the high-speed-assembly moment of inertia*

The effective moment of inertia at the high-speed assembly input = $I_{thi} = I_{th} + I_{tl}/(R_h/R_l)^2$, where $I_{th}$ = moment of inertia of high-speed assembly, in$^4$; $I_{tl}$ = moment of inertia of low-speed assembly, in$^4$; $R_h$ = high speed, r/min; $R_l$ = low speed, r/min. To find $I_{th}$ and $I_{tl}$, take the sum of the shaft and gear moments of inertia for the high- and low-speed assemblies, respectively. Or, $I_{th} = 16.0 + 1728.5 = 1744.15$ in$^4$ (72,597.0 cm$^4$); $I_{tl} = 176.02 + 139{,}979.1 = 140{,}155.1$ in$^4$ (5,833,695.7 cm$^4$).

Then $I_{thi} = 1744.15 + 140{,}155.1/(600/200)^2 = 17{,}324.2$ in$^4$ (721,087.6 cm$^4$).

## 4. *Compute the low-speed-assembly moment of inertia*

The effective moment of inertia at the low-speed assembly output is $I_{tlo} = I_{tl} + I_{th}(R_h/R_l)^2 = 140{,}155.1 + (1744.15)(600/200)^2 = 155{,}852.5$ in$^4$ (6,487,070.8 cm$^4$).

Note that $I_{thi} \neq I_{tlo}$. One value is approximately nine times that of the other. Thus, in stating the moment of inertia of a gear drive, be certain to specify whether the given value applies to the high- or low-speed assembly.

**Related Calculations:** Use this procedure for shafts and gears made of any metal—aluminum, brass, bronze, chromium, copper, cast iron, magnesium, nickel, tungsten, etc. Compute $WR^2$ for steel, and multiply the result by the weight of shaft material, lb/in$^3$/0.283.

# BEARING LOADS IN GEARED DRIVES

A geared drive transmits a torque of 48,000 lb·in (5423.3 N·m). Determine the resulting bearing load in the drive shaft if a 12-in (30.5-cm) pitch-radius spur gear having a 20° pressure angle is used. A helical gear having a 20° pressure angle and a 14.5° spiral angle transmits a torque of 48,000 lb·in (5423.2 N·m). Determine the bearing load it produces if the pitch radius is 12 in (30.5 cm). Determine the bearing load in a straight bevel gear having the same proportions as the helical gear above, except that the pitch cone angle is 14.5°. A worm having an efficiency of 70 percent and a 30° helix angle drives a gear having a 20° normal pressure angle. Determine the bearing load when the torque is 48,000 lb·in (5423.3 N·m) and the worm pitch radius is 12 in (30.5 cm).

## Calculation Procedure:

## 1. *Compute the spur-gear bearing load*

The tangential force acting on a spur-gear tooth is $F_t = T/r$, where $F_t$ = tangential force, lb; $T$ = torque, lb·in; $r$ = pitch radius, in. For this gear, $F_t = T/r = 48{,}000/12 = 4000$ lb (17,792.9 N). This force is tangent to the pitch-diameter circle of the gear.

The separating force acting on a spur-gear tooth perpendicular to the tangential force is $F_s = F_t \tan \alpha$, where $\alpha$ = pressure angle, degrees. For this gear, $F_s = (4000)(0.364) = 1456$ lb (6476.6 N).

Find the resultant force $R_f$ lb from $R_f = (F_t^2 + F_s^2)^{0.5} = (4000^2 + 1456^2)^{0.5} = 4260$ lb (18,949.4 N). This is the bearing load produced by the gear.

### 2. *Compute the helical-gear load*

The tangential force acting on a helical gear is $F_t = T/r = 48{,}000/12 = 4000$ lb (17,792.9 N). The separating force, acting perpendicular to the tangential force, is $F_s = F_t \tan \alpha/\cos \beta$, where $\beta$ = the spiral angle. For this gear, $F_s = (4000)(0.364)/0.986 = 1503$ lb (6685.7 N). The resultant bearing load, which is a side thrust, is $R_f = (4000^2 + 1503^2)^{0.5} = 4380$ lb (19,483.2 N).

Helical gears produce an end thrust as well as the side thrust just computed. This end thrust is given by $F_e = F_t \tan \beta$, or $F_e = (4000)(0.259) = 1036$ lb (4608.4 N). The end thrust of the driving helical gear is equal and opposite to the end thrust of the driven helical gear when the teeth are of the opposite hand in each gear.

### 3. *Compute the bevel-gear load*

The tangential force acting on a bevel gear is $F_t = T/r = 48{,}000/12 = 4000$ lb (17,792.9 N). The separating force is $F_s = F_t \tan \alpha \cos \theta$, where $\theta$ = pitch cone angle. For this gear, $F_s = (4000)(0.364)(0.968) = 1410$ lb (6272.0 N).

Bevel gears produce an end thrust similar to helical gears. This end thrust is $F_e = F_t \tan \alpha \sin \theta$, or $F_e = (4000)(0.364)(0.25) = 364$ lb (1619.2 N). The side thrust in a bevel gear is $F_t = 4000$ lb (17,792.9 N) and acts tangent to the pitch-diameter circle. The resultant is an end thrust produced by $F_s$ and $F_e$, or $R_f = (F_s^2 + F_e^2)^{0.5} = (1410^2 + 364^2)^{0.5} = 1458$ lb (6485.5 N). In a bevel-gear drive, $F_t$ is common to both gears, $F_s$ becomes $F_e$ on the mating gear, and $F_e$ becomes $F_s$ on the mating gear.

### 4. *Compute the worm-gear bearing load*

The worm tangential force $F_t = T/r = 48{,}000/12 = 4000$ lb (17,792.9 N). The separating force is $F_s = F_t E \tan \alpha/\sin \phi$, where $E$ = worm efficiency expressed as a decimal; $\phi$ = worm helix or lead angle. Thus, $F_s = (4999)(0.70)(0.364)/0.50 = 2040$ lb (9074.4 N).

The worm end thrust force is $F_e = F_t E \cot \phi = (4000)(0.70)(1.732) = 4850$ lb (21,573.9 N). This end thrust acts perpendicular to the separating force. Thus the resultant bearing load $R_f = (F_s^2 + F_e^2)^{0.5} = (2040^2 + 4850^2)^{0.5} = 5260$ lb (23,397.6 N).

Forces developed by the gear are equal and opposite to those developed by the worm tangential force if cancelled by the gear tangential force.

**Related Calculations:** Use these procedures to compute the bearing loads in any type of geared drive—open, closed, or semiclosed—serving any type of load. Computation of the bearing load is a necessary step in bearing selection.

## FORCE RATIO OF GEARED DRIVES

A geared hoist will lift a maximum load of 1000 lb (4448.2 N). The hoist is estimated to have friction and mechanical losses of 5 percent of the maximum load. How much force is required to lift the maximum load if the drum on which the lifting cable reels is 10 in (25.4 cm) in diameter and the driving gear is 50 in (127.0 cm) in diameter? If the load is raised at a velocity of 100 ft/min (0.5 m/s), what is the hp output? What is the driving-gear tooth load if the gear turns at 191 r/min? A 15-in (38.1-cm) triple-reduction hoist has three driving gears with 48-, 42-, and 36-in (121.9-, 106.7-, and 91.4-cm) diameters, respectively, and two pinions of 12- and 10-in (30.5- and 25.4-cm) diameter. What force is required to lift a 1000-lb (4448.2-N) load if friction and mechanical losses are 10 percent?

### Calculation Procedure:

### 1. *Compute the total load on the hoist*

The friction and mechanical losses *increase* the maximum load on the drum. Thus, the total load on the drum = maximum lifting load, lb + friction and mechanical losses, lb = 1000 + 1000(0.05) = 1050 lb (4670.6 N).

### 2. *Compute the required lifting force*

Find the lifting force from $L/D_g = F/d_d$, where $L$ = total load on hoist, lb; $D_g$ = diameter of driving gear, in; $F$ = lifting force required, lb; $d_d$ = diameter of lifting drum, in. For this hoist, $1050/50 = F/10$; $F = 210$ lb (934.1 N).

### 3. *Compute the horsepower input*

Find the horsepower input from $hp = Lv/33{,}000$, where $v$ = load velocity, ft/min. Thus, $hp = (1050)(100)/33{,}000 = 3.19$ hp (2.4 kW).

Where the mechanical losses are not added to the load before the horsepower is computed, use the equation $hp = Lv/(1.00 - \text{losses})(33{,}000)$. Thus, $hp = (1000)(100)/(1 - 0.05)(33{,}000) = 3.19$ hp (2.4 kW), as before.

### 4. *Compute the driving-gear tooth load*

Assume that the entire load is carried by one tooth. Then the tooth load $L_t$ lb $= 33{,}000\ hp/v_g$, where $v_g$ = peripheral velocity of the driving gear, ft/min. With a diameter of 50 in (127.0 cm) and a speed of 191 r/min, $v_g = \pi D_g R/12$, where $R$ = gear rpm. Or, $v_g = \pi(50)(191)/12 = 2500$ ft/min (12.7 m/s). Then $L_t = (33{,}000)(3.19)/2500 = 42.1$ lb (187.3 N). This is a nominal tooth-load value.

### 5. *Compute the triple-reduction hoisting force*

Use the equation from step 2, but substitute the product of the three driving-gear diameters for $D_g$ and the three driven-gear diameters for $d_d$. The total load $= 1000 + 0.10(1000) = 1100$ lb (4893.0 N). Then $L/D_g = F/d_d$, or $1100/(48 \times 42 \times 36) = F/(15 \times 12 \times 10)$; $F = 27.2$ lb (121.0 N). Thus, the triple-reduction hoist reduces the required lifting force to about one-tenth that required by a double-reduction hoist (step 2).

**Related Calculations:** Use this procedure for geared hoists of all types. Where desired, the number of gear teeth can be substituted for the driving- and driven-gear diameters in the force equation in step 2.

## DETERMINATION OF GEAR BORE DIAMETER

Two helical gears transmit 500 hp (372.9 kW) at 3600 r/min. What should the bore diameter of each gear be if the allowable stress in the gear shafts is 12,500 lb/in$^2$ (86,187.5 kPa)? How should the gears be fastened to the shafts? The shafts are solid in cross section.

### Calculation Procedure:

### 1. *Compute the required hub bore diameter*

The hub bore diameter must at least equal the outside diameter of the shaft, unless the gear is press- or shrink-fitted on the shaft. Regardless of how the gear is attached to the shaft, the shaft must be large enough to transmit the rated torque at the allowable stress.

Use the method of step 2 of "Solid and Hollow Shafts in Torsion" in this section to compute the required shaft diameter, after finding the torque by using the method of step 1 in the same procedure. Thus, $T = 63{,}000hp/R = (63{,}000)(500)/3600 = 8750$ lb·in (988.6 N·m). Then $d = 1.72(T/s)^{1/3} = 1.72(8750/12{,}500)^{1/3} = 1.526$ in (3.9 cm).

### 2. *Determine how the gear should be fastened to the shaft*

First decide whether the gears are to be permanently fastened or removable. This decision is usually based on the need for gear removal for maintenance or replacement. Removable gears can be fastened by a key, setscrew, spline, pin, clamp, or a taper and screw. Large gears transmitting 100 hp (74.6 kW) or more are usually fitted with a key for easy removal. See "Selection of Keys for Machine Shafts" in this section for the steps in choosing a key.

Permanently fastened gears can be shrunk, pressed, cemented, or riveted to the shaft. Shrink-fit gears generally transmit more torque before slippage occurs than do press-fit gears. With either type of fastening, interference is necessary; i.e., the gear bore is made smaller than the shaft outside diameter.

Baumeister and Marks—*Standard Handbook for Mechanical Engineers* shows that press- or shrink-fit gears on shafts of 1.19- to 1.58-in (3.0- to 4.0-cm) diameter should have an interference ranging from 0.3 to 4.0 thousandths of an inch (0.8 to 10.2 thousandths of a centimeter) on the diameter, depending on the class of fit desired.

**Related Calculations:** Use this general procedure for any type of gear—spur, helical, herringbone, worm, etc. Never reduce the shaft diameter below that required by the stress equation, step 1. Thus, if interference is provided by the shaft diameter, *increase* the diameter; do not reduce it.

## TRANSMISSION GEAR RATIO FOR A GEARED DRIVE

A four-wheel vehicle must develop a drawbar pull of 17,500 lb (77,843.9 N). The engine, which develops 500 hp (372.8 kW) and drives through a gear transmission a 34-tooth spiral bevel pinion gear which meshes with a spiral bevel gear having 51 teeth. This gear is keyed to the drive shaft of the 48-in (121.9-cm) diameter rear wheels of the vehicle. What transmission gear ratio should be used if the engine develops maximum torque at 1500 r/min? Select the axle diameter for an allowable torsional stress of 12,500 lb/in$^2$ (86,187.5 kPa). The efficiency of the bevel-gear differential is 80 percent.

### Calculation Procedure:

**1.** ***Compute the torque developed at the wheel***

The wheel torque = (drawbar pull, lb)(moment arm, ft), where the moment arm = wheel radius, ft. For this vehicle having a wheel radius of 24 in (61.0 cm), or 24/12 = 2 ft (0.6 m), the wheel torque = (17,500)(2) = 35,000 lb·ft (47,453.6 N·m).

**2.** ***Compute the torque developed by the engine***

The engine torque $T = 63{,}000\ hp/R$, or $T = (63{,}000)(500)/1500 = 21{,}000$ lb·ft (28,472.2 N·m), where $R$ = rpm.

**3.** ***Compute the differential speed ratio***

The differential speed ratio = $N_g/N_p = 51/34 = 1.5$, where $N_g$ = number of gear teeth; $N_p$ = number of pinion teeth.

**4.** ***Compute the transmission gear ratio***

For any transmission gear, its ratio = (output torque, lb·ft)/[(input torque, lb·ft)(differential speed ratio)(differential efficiency)], or transmission gear ratio = 35,000/[(21,000)(1.5)(0.80)] = 1.388. Thus, a transmission with a 1.388 ratio will give the desired output torque at the rated engine speed.

**5.** ***Determine the required shaft diameter***

Use the relation $d = 1.72(T/s)^{1/3}$ from the previous calculation procedure to determine the axle diameter. Since the axle is transmitting a total torque of 35,000 lb·ft (47,453.6 N·m), each of the two rear wheels develops a torque of 35,000/2 = 17,500 lb·ft (23,726.8 N·m), and $d = 1.72(17{,}500/12{,}500)^{1/3} = 1.34$ in (3.4 cm).

**Related Calculations:** Use this general procedure for any type of differential—worm gear, herringbone gear, helical gear, or spiral gear—connected to any type of differential. The output torque can be developed through a wheel, propeller, impeller, or any other device. Note that although this vehicle has two rear wheels, the total drawbar pull is developed by *both* wheels. Either wheel delivers *half* the drawbar pull. If the total output torque were developed by only one wheel, its shaft diameter would be $d = 1.72(35{,}000/12{,}500)^{1/3} = 1.69$ in (4.3 cm).

## EPICYCLIC GEAR TRAIN SPEEDS

Figure 4 shows several typical arrangements of epicyclic gear trains. The number of teeth and the rpm of the driving arm are indicated in each diagram. Determine the driven-member rpm for each set of gears.

### Calculation Procedure:

**1.** ***Compute the spur-gear speed***

For a gear arranged as in Fig. 4*a*, $R_d = R_D(1 + N_s/N_d)$, where $R_d$ = driven-member rpm; $R_D$ = driving-member rpm; $N_s$ = number of teeth on the stationary gear; $N_d$ = number of teeth on the driven gear. Given the values given for this gear and since the arm is the driving member, $R_d = 40(1 + 84/21)$; $R_d = 200$ r/min.

Note how the driven-gear speed is attained. During one planetary rotation around the stationary gear, the driven gear will rotate axially on its shaft. The number of times the driven gear rotates on its shaft $= N_s/N_d = 84/21 = 4$ times per planetary rotation about the stationary gear. While rotating on its shaft, the driven gear makes a planetary rotation around the fixed gear. So while rotating axially on its shaft four times, the driven gear makes one additional planetary rotation about the stationary gear. Its total axial and planetary rotation is $4 + 1 = 5$ r/min per rpm of the arm. Thus, the gear ratio $G_r = R_D/R_d = 40/200 = 1{:}5$.

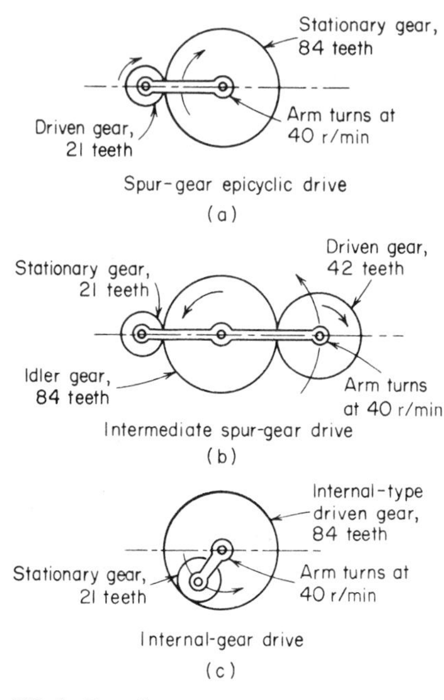


**FIG. 4** Epicyclic gear trains.

### 2. *Compute the idler-gear train speed*

The idler gear, Fig. 4*b*, turns on its shaft while the arm rotates. Movement of the idler gear causes rotation of the driven gear. For an epicyclic gear train of this type, $R_d = R_D(1 - N_s/N_d)$, where the symbols are as defined in step 1. Thus, $R_d = 40(1 - 21/42) = 20$ r/min.

### 3. *Compute the internal gear drive speed*

The arm of the internal gear drive, Fig. 4*c*, turns and carries the stationary gear with it. For a gear train of this type, $R_d = R_D(1 - N_s/N_d)$, or $R_d = 40(1 - 21/84) = 30$ r/min.

Where the internal gear is the driving gear that turns the arm, making the arm the driven member, the velocity equation becomes $r_d = R_D N_D/(N_D + N_s)$, where $R_D$ = driving-member rpm; $N_D$ = number of teeth on the driving member.

**Related Calculations:** The arm was the driving member for each of the gear trains considered here. However, any gear can be made the driving member if desired. Use the same relations as given above, but substitute the gear rpm for $R_D$. Thus, a variety of epicyclic gear problems can be solved by using these relations. Where unusual epicyclic gear configurations are encountered, refer to Dudley—*Gear Handbook* for a tabular procedure for determining the gear ratio.

## PLANETARY-GEAR-SYSTEM SPEED RATIO

Figure 5 shows several arrangements of important planetary-gear systems using internal ring gears, planet gears, sun gears, and one or more carrier arms. Determine the output rpm for each set of gears.

## Calculation Procedure:

### 1. *Determine the planetary-gear output speed*

For the planetary-gear drive, Fig. 5*a*, the gear ratio $G_r = (1 + N_4N_2/N_3N_1)/(1 - N_4N_2/N_5N_1)$, where $N_1, N_2, \ldots, N_5$ = number of teeth, respectively, on each of gears 1, 2, . . . , 5. Also, for any gearset, the gear ratio $G_r$ = input rpm/output rpm, or $G_r$ = driver rpm/driven rpm.

With ring gear 2 fixed and ring gear 5 the output gear, Fig. 5*a*, and the number of teeth shown, $G_r = \{1 + (33)(74)/[(9)(32)]\}/\{1 - (33)(74)/[(175)(32)]\} = -541.667$. The minus sign indicates that the output shaft revolves in a direction *opposite* to the input shaft. Thus, with an input speed of 5000 r/min, $G_r$ = input rpm/output rpm; output rpm = input rpm/$G_r$, or output rpm $= 5000/541.667 = 9.24$ r/min.

**2. *Determine the coupled planetary drive output speed***

The drive, Fig. 5*b*, has the coupled ring gear 2, the sun gear 3, the coupled planet carriers *C* and *C*′, and the fixed ring gear 4. The gear ratio is $G_r = (1 - N_2N_4/N_1N_3)$, where the symbols are the same as before. Find the output speed for any given number of teeth by first solving for $G_r$ and then solving $G_r$ = input rpm/output rpm.

With the number of teeth shown, $G_r = 1 - (75)(75)/[(32)(12)] = -13.65$. Then output rpm = input rpm/$G_r$ = 1200/13.65 = 87.9 r/min.

Two other arrangements of coupled planetary drives are shown in Fig. 5*c* and *d*. Compute the output speed in the same manner as described above.

**3. *Determine the fixed-differential output speed***

Figure 5*e* and *f* shows two typical fixed-differential planetary drives. Compute the output speed in the same manner as step 2.

**4. *Determine the triple planetary output speed***

Figure 5*g* shows three typical triple planetary drives. Compute the output speed in the same manner as step 2.

**5. *Determine the output speed of other drives***

Figure 5*h*, *i*, *j*, *k*, and *l* shows the gear ratio and arrangement for the following drives: compound spur-bevel gear, plancentric, wobble gear, double eccentric, and Humpage's bevel gears. Compute the output speed for each in the same manner as step 2.

**Related Calculations:** Planetary and sun-gear calculations are simple once the gear ratio is determined. The gears illustrated here[1] comprise an important group in the planetary and sun-gear field. For other gear arrangements, consult Dudley—*Gear Handbook*.

## SELECTION OF A RIGID FLANGE-TYPE SHAFT COUPLING

Choose a steel flange-type coupling to transmit a torque of 15,000 lb·in (1694.4 N·m) between two 2½-in (6.4-cm) diameter steel shafts. The load is uniform and free of shocks. Determine how many bolts are needed in the coupling if the allowable bolt shear stress is 3000 lb/in² (20,685.0 kPa). How thick must the coupling flange be, and how long should the coupling hub be if the allowable stress in bearing for the hub is 20,000 lb/in² (137,900.0 kPa) and in shear 6000 lb/in² (41,370.0 kPa)? The allowable shear stress in the key is 12,000 lb/in² (82,740.0 kPa). There is no thrust force acting on the coupling.

### Calculation Procedure:

**1. *Choose the diameter of the coupling bolt circle***

Assume a bolt-circle diameter for the coupling. As a first choice, assume the bolt-circle diameter is three times the shaft diameter, or $3 \times 2.5 = 7.5$ in (19.1 cm). This is a reasonable first assumption for most commercially available couplings.

**2. *Compute the shear force acting at the bolt circle***

The shear force $F_s$ lb acting at the bolt-circle radius $r_b$ in is $F_s = T/r_b$, where $T$ = torque on shaft, lb·in. Or, $F_s = 15{,}000/(7.5/2) = 4000$ lb (17,792.9 N).

**3. *Determine the number of coupling bolts needed***

When the allowable shear stress in the bolts is known, compute the number of bolts $N$ required from $N = 8F_s/(\pi d^2 s_s)$, where $d$ = diameter of each coupling bolt, in; $s_s$ = allowable shear stress in coupling bolts, lb/in².

The usual bolt diameter in flanged, rigid couplings ranges from ¼ to 2 in (0.6 to 5.1 cm), depending on the torque transmitted. Assuming that ½-in (1.3-cm) diameter bolts are used in this coupling, we see that $N = 8(4000)/[\pi(0.5)^2(3000)] = 13.58$, say 14 bolts.

Most flanged, rigid couplings have two to eight bolts, depending on the torque transmitted. A coupling having 14 bolts would be a poor design. To reduce the number of bolts, assume a larger

[1]John H. Glover, "Planetary Gear Systems," *Product Engineering*, Jan. 6, 1964.

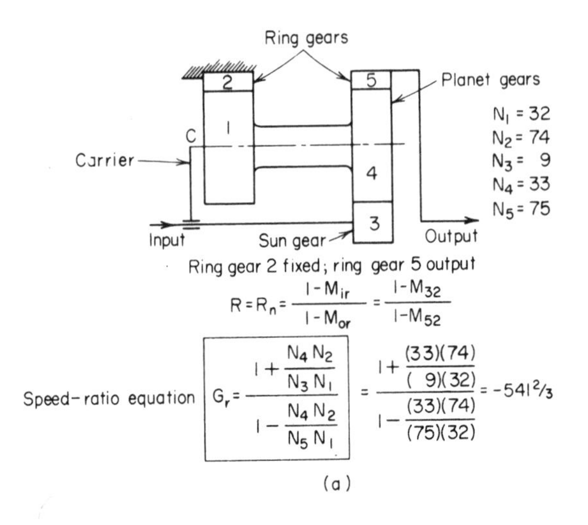


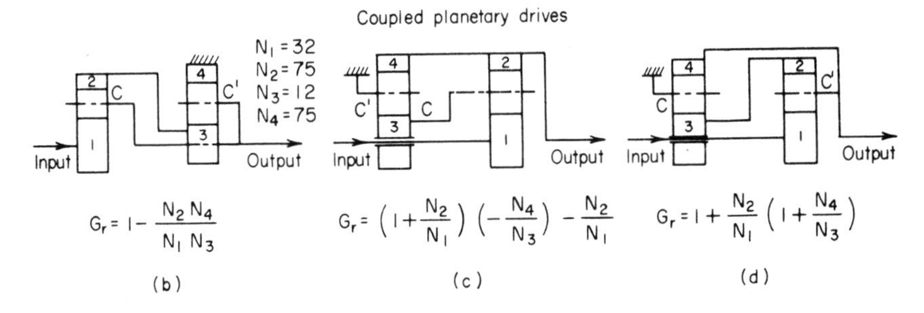


Fixed-differential drives

Output is difference between speeds of two parts leading to high reduction ratios

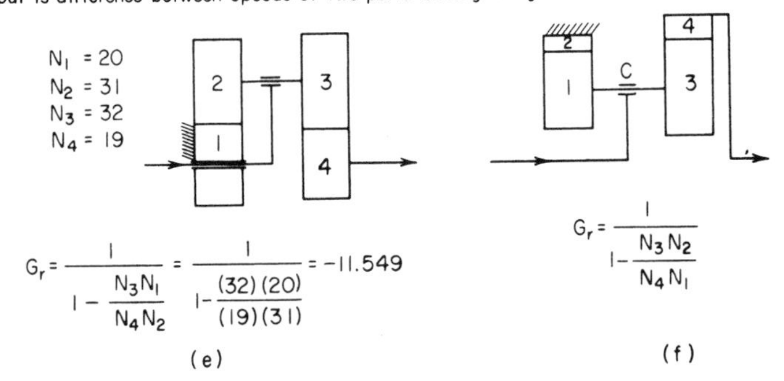


**FIG. 5** Planetary gear systems. *(Product Engineering.)*

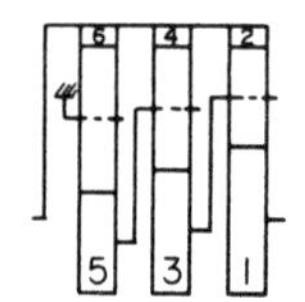


Input to gear 1, ouput from gear 6

$$G_r = \left(1 + \frac{N_2}{N_1}\right)\left[\left(1 + \frac{N_4}{N_3}\right)\left(-\frac{N_6}{N_5}\right) - \frac{N_4}{N_3}\right] - \frac{N_2}{N_1}$$

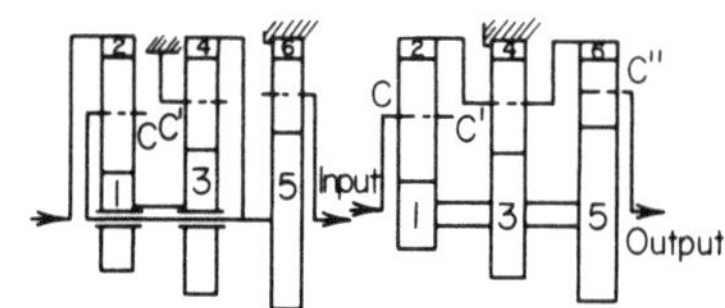


$$G_r = \left[1 + \frac{N_1}{N_2}\left(1 + \frac{N_4}{N_3}\right)\right]\left(1 + \frac{N_6}{N_5}\right)$$

$$G_r = \left[1 + \frac{N_4/N_3}{1 + (N_2/N_1)}\right] \Big/ \left[1 + \frac{N_4/N_3}{1 + (N_6/N_5)}\right]$$

(g)

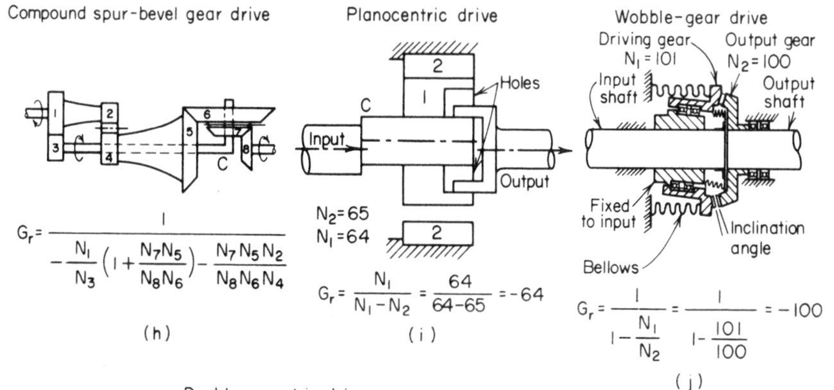


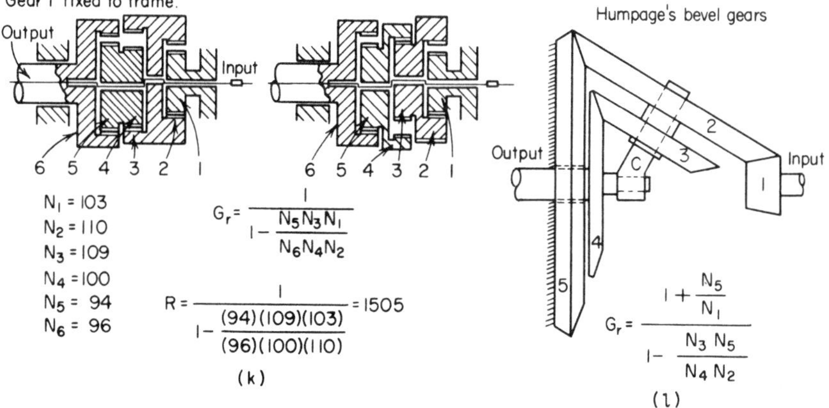

diameter, say 0.75 in (1.9 cm). Then $N = 8(4000)/[\pi(0.75)^2(3000)] = 6.03$, say eight bolts, because an odd number of bolts are seldom used in flanged couplings.

Determine the shear stress in the bolts by solving the above equation for $s_s = 8F_s/(\pi d^2 N) = 8(4000)/[\pi(0.75)^2(8)] = 2265$ lb/in² (15,617.2 kPa). Thus, the bolts are not overstressed, because the allowable stress is 3000 lb/in² (20,685.0 kPa).

### 4. *Compute the coupling flange thickness required*

The flange thickness $t$ in for an allowable bearing stress $s_b$ lb/in² is $t = 2F_s/(Nds_b) = 2(4000)/[(8)(0.75)(20{,}000)] = 0.0666$ in (0.169 cm). This thickness is much less than the usual thickness used for flanged couplings manufactured for off-the-shelf use.

### 5. *Determine the hub length required*

The hub length is a function of the key length required. Assuming a ¾-in (1.9-cm) square key, compute the hub length $l$ in from $l = 2F_{ss}/(t_k s_t)$, where $F_{ss}$ = force acting at shaft outer surface, lb; $t_k$ = key thickness, in. The force $F_{ss} = T/r_h$, where $r_h$ = inside radius of hub, in = shaft radius = 2.5/2 = 1.25 in (3.2 cm) for this shaft. Then $F_{ss} = 15{,}000/1.25 = 12{,}000$ lb·in (1355.8 N·m). Then $l = 2(12{,}000)/[(0.75)(20{,}000)] = 1.6$ in (4.1 cm).

When the allowable design stress for bearing, 20,000 lb/in² (137,895.1 kPa) here, is less than half the allowable design stress for shear, 12,000 lb/in² (82,740.0 kPa) here, the longest key length is obtained when the bearing stress is used. Thus, it is not necessary to compute the thickness needed to resist the shear stress for this coupling. If it is necessary to compute this thickness, find the force acting at the surface of the coupling hub from $F_h = T/r_h$, where $r_h$ = hub radius, in. Then $t_s = F_h/\pi d_h s_s$, where $d_h$ = hub diameter, in; $s_s$ = allowable hub shear stress, lb/in².

**Related Calculations:** Couplings offered as standard parts by manufacturers are usually of sufficient thickness to prevent fatigue failure.

Since each half of the coupling transmits the total torque acting, the length of the key must be the same in each coupling half. The hub diameter of the coupling is usually 2 to 2.5 times the shaft diameter, and the coupling lip is generally made the same thickness as the coupling flange. The procedure given here can be used for couplings made of any metallic material.

## SELECTION OF A FLEXIBLE COUPLING FOR A SHAFT

Choose a stock flexible coupling to transmit 15 hp (11.2 kW) from a 1000-r/min four-cylinder gasoline engine to a dewatering pump turning at the same rpm. The pump runs 8 h/day and is an uneven load because debris may enter the pump. The pump and motor shafts are each 1.0 in (2.5 cm) in diameter. Maximum misalignment of the shafts will not exceed 0.5°. There is no thrust force acting on the coupling, but the end float or play may reach 1/16 in (0.2 cm).

## Calculation Procedure:

### 1. *Choose the type of coupling to use*

Consult Table 22 or the engineering data published by several coupling manufacturers. Make a tentative choice from Table 22 of the type of coupling to use, based on the maximum misalignment expected and the tabulated end-float capacity of the coupling. Thus, a roller-chain-type coupling (one in which the two flanges are connected by a double roller chain) will be chosen

**TABLE 22** Allowable Flexible Coupling Misalignment

| Coupling type | Angular misalignment | Parallel misalignment | | End float | |
|---|---|---|---|---|---|
| | | USCS | SI | USCS | SI |
| Plastic chain | Up to 1.0° | 0.005 in | 0.1 mm | 1/16 in | 2 mm |
| Roller chain | Up to 0.5° | 2% of chain pitch | 2% of chain pitch | 1/16 in | 2 mm |
| Silent chain | Up to 0.5° | 2% of chain pitch | 2% of chain pitch | ¼ to ¾ in | 0.6 to 1.9 cm |
| Neoprene biscuit | Up to 5.0° | 0.01 to 0.05 in | 0.3 to 1.3 mm | Up to ½ in | Up to 1.3 cm |
| Radial | Up to 0.5° | 0.01 to 0.02 in | 0.3 to 0.5 mm | Up to 1/16 in | Up to 2 mm |

**TABLE 23** Flexible Coupling Service Factors*

| Type of drive | | | |
|---|---|---|---|
| Engine,† less than six cylinders | Engine, six cylinders or more | Electric motor; steam turbine | Type of load |
| 2.0 | 1.5 | 1.0 | Even load, 8 h/day; nonreversing, low starting torque |
| 2.5 | 2.0 | 1.5 | Uneven load, 8 h/day; moderate shock or torque, nonreversing |
| 3.0 | 2.5 | 2.0 | Heavy shock load, 8 h/day; reversing under full load, high starting torque |

*Morse Chain Company.
†Gasoline or diesel.

from Table 22 for this drive because it can accommodate 0.5° of misalignment and an end float of up to 1/16 in (0.2 cm).

**2. *Choose a suitable service factor***

Table 23 lists typical service factors for roller-chain-type flexible couplings. Thus, for a four-cylinder gasoline engine driving an uneven load, the service factor SF = 2.5.

**3. *Apply the service factor chosen***

Multiply the horsepower or torque to be transmitted by the service factor to obtain the coupling design horsepower or torque. Or, coupling design *hp* = (15)(2.5) = 37.5 hp (28.0 kW).

**4. *Select the coupling to use***

Refer to the coupling design horsepower rating table in the manufacturer's engineering data. Enter the table at the shaft rpm, and project to a design horsepower slightly greater than the value computed in step 3. Thus, in Table 24 a typical rating tabulation shows that a coupling design horsepower rating of 38.3 hp (28.6 kW) is the next higher value above 37.5 hp (28.0 kW).

**5. *Determine whether the coupling bore is suitable***

Table 24 shows that a coupling suitable for 38.3 hp (28.6 kW) will have a maximum bore diameter up to 1.75 in (4.4 cm) and a minimum bore diameter of 0.625 in (1.6 cm). Since the engine and pump shafts are each 1.0 in (2.5 cm) in diameter, the coupling is suitable.

The usual engineering data available from manufacturers include the stock keyway sizes, coupling weight, and principal dimensions of the coupling. Check the overall dimensions of the coupling to determine whether the coupling will fit the available space. Where the coupling bore diameter is too small to fit the shaft, choose the next larger coupling. If the dimensions of the coupling make it unsuitable for the available space, choose a different type or make a coupling.

**TABLE 24** Flexible Coupling hp Ratings*

| r/min | | | Bore diameter, in (cm) | |
|---|---|---|---|---|
| 800 | 1000 | 1200 | Maximum | Minimum |
| 16.7 | 19.9 | 23.2 | 1.25 (3.18) | 0.5 (1.27) |
| 32.0 | 38.3 | 44.5 | 1.75 (4.44) | 0.625 (1.59) |
| 75.9 | 90.7 | 105.0 | 2.25 (5.72) | 0.75 (1.91) |

*Morse Chain Company.

**Related Calculations:** Use the general procedure given here to select any type of flexible coupling using flanges, springs, roller chain, preloaded biscuits, etc., to transmit torque. Be certain to apply the service factor recommended by the manufacturer. Note that biscuit-type couplings are rated in hp/100 r/min. Thus, a biscuit-type coupling rated at 1.60 hp/100 r/min (1.2 kW/100 r/min) and a maximum allowable speed of 4800 r/min could transmit a maximum of (1.80 hp)(4800/100) = 76.8 hp (57.3 kW).

## SELECTION OF A SHAFT COUPLING FOR TORQUE AND THRUST LOADS

Select a shaft coupling to transmit 500 hp (372.9 kW) and a thrust of 12,500 lb (55,602.8 N) at 100 r/min from a six-cylinder diesel engine. The load is an even one, free of shock.

### Calculation Procedure:

#### 1. *Compute the torque acting on the coupling*

Use the relation $T = 5252hp/R$ to determine the torque, where $T$ = torque acting on coupling, lb·ft; $hp$ = horsepower transmitted by the coupling; $R$ = shaft rotative speed, r/min. For this coupling, $T = (5252)(500)/100 = 26{,}260$ lb·ft (35,603.8 N·m).

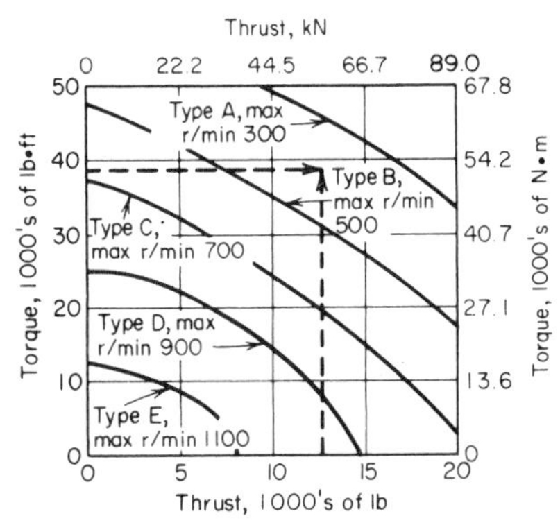


**FIG. 6** Shaft-coupling characteristics.

#### 2. *Find the service torque*

Multiply the torque $T$ by the appropriate service factor from Table 23. This table shows that a service factor of 1.5 is suitable for an even load, free of shock. Thus, the service torque = (26,260 lb·ft)(1.5) = 39,390 lb·ft, say 39,500 lb·ft (53,554.8 N·m).

#### 3. *Choose a suitable coupling*

Enter Fig. 6 at the torque on the left, and project horizontally to the right. Using the known thrust, 12,500 lb (55,602.8 N), enter Fig. 6 at the bottom and project vertically upward until the torque line is intersected. Choose the coupling model represented by the next higher curve. This shows that a type A coupling having a maximum allowable speed of 300 r/min will be suitable. If the plotted maximum rpm is lower than the actual rpm of the coupling, use the next plotted coupling type rated for the actual, or a higher, rpm.

In choosing a specific coupling, use the manufacturer's engineering data. This will resemble Fig. 6 or will be a tabulation of the ranges plotted.

**Related Calculations:** Use this procedure to select couplings for industrial and marine drives where both torque and thrust must be accommodated. See the Marine Engineering section of this handbook for an accurate way to compute the thrust produced on a coupling by a marine propeller. Always check to see that the coupling bore is large enough to accommodate the connected shafts. Where the bore is too small, use the next larger coupling.

## HIGH-SPEED POWER-COUPLING CHARACTERISTICS

Select the type of power coupling to transmit 50 hp (37.3 kW) at 200 r/min if the angular misalignment varies from a minimum of 0 to a maximum of 45°. Determine the effect of angular misalignment on the shaft position, speed, and acceleration at angular misalignments of 30 and 45°

### Calculation Procedure:

#### 1. *Determine the type of coupling to use*

Table 25, developed by N. B. Rothfuss, lists the operating characteristics of eight types of high-speed couplings. Study of this table shows that a universal joint is the only type of coupling among

**TABLE 25** Operating Characteristics of Couplings*

| | Contoured diaphragm | Axial spring | Laminated disk | Universal joint | Ball-race | Gear | Chain | Elastomeric |
|---|---|---|---|---|---|---|---|---|
| Speed range, r/min | 0–60,000 | 0–8,000 | 0–20,000 | 0–8,000 | 0–8,000 | 0–25,000 | 0–6,300 | 0–6,000 |
| Power range, kW/100 r/min | 1–500 | 1–9,000 | 1–100 | 1–100 | 1–100 | 1–2,000 | 1–200 | 0–400 |
| Angular misalignment, degrees | 0–8.0 | 0–2.0 | 0–1.5 | 0–45 | 0–40 | 0–3 | 0–2 | 0–4 |
| Parallel misalignment, mm | 0–2.5 | 0–2.5 | 0–2.5 | None | None | 0–2.5 | 0–2.5 | 0–2.5 |
| Axial movement, cm | 0–0.5 | 0–2.5 | 0–0.5 | None | None | 0–5.1 | 0–2.5 | 0–0.8 |
| Ambient temperature, °C | 900 | Varies | 900 | Varies | Varies | Varies | Varies | Varies |
| Ambient pressure, kPa | Sea level to zero | Varies | Sea level to zero | Varies | Varies | Varies | Varies | Varies |

* *Product Engineering.*

**TABLE 26** Functional Characteristics of Couplings°

| | Contoured diaphragm | Axial spring | Laminated disk | Universal joint | Ball-race | Gear | Chain | Elastomeric |
|---|---|---|---|---|---|---|---|---|
| No lubrication | √ | . . . | √ | . . . | . . . | . . . | . . . | √ |
| No backlash | √ | √ | √ | † | † | . . . | . . . | √ |
| Constant velocity ratio | √ | √ | . . . | ‡ | √ | ‡ | ‡ | √ |
| Containment | √ | . . . | † | . . . | † | | | |
| Angular only | √ | . . . | √ | √ | √ | √ | | |
| Axial and angular | √ | . . . | √ | . . . | √ | √ | √ | √ |
| Axial and parallel | √ | √ | √ | . . . | . . . | √ | √ | √ |
| Axial, angular, and parallel | √ | √ | √ | . . . | . . . | √ | √ | √ |
| High temperature | √ | . . . | √ | | | | | |
| High altitude | √ | . . . | √ | | | | | |
| High torsional spring rate | √ | . . . | √ | √ | √ | √ | √ | |
| Low bending moment | √ | √ | √ | √ | √ | √ | √ | √ |
| No relative movement | √ | . . . | . . . | . . . | . . . | . . . | . . . | √ |

° *Product Engineering.*
†Zero backlash and containment can be obtained by special design.
‡Constant velocity ratio at small angles can be closely approximated.

**TABLE 27** Universal Joint Output Variations*

| Misalignment angle, deg | Maximum position error | Maximum speed error, percent | Ratio $A/\omega^2$ |
|---|---|---|---|
| 5 | 0°06′34″ | 0.382 | 0.011747 |
| 10 | 0°26′18″ | 1.543 | 0.030626 |
| 15 | 0°59′36″ | 3.526 | 0.069409 |
| 20 | 1°46′54″ | 6.418 | 0.124966 |
| 25 | 2°48′42″ | 10.338 | 0.198965 |
| 30 | 4°06′42″ | 15.470 | 0.294571 |
| 35 | 5°42′20″ | 22.077 | 0.417232 |
| 40 | 7°36′43″ | 30.541 | 0.576215 |
| 45 | 9°52′26″ | 41.421 | 0.787200 |

*Caused by misalignment of the shaft. Table from *Machine Design*.

those listed that can handle an angular misalignment of 45°. Further study shows that a universal coupling has a suitable speed and hp range for the load being considered. The other items tabulated are not factors in this application. Therefore, a universal coupling will be suitable. Table 26 compares the functional characteristics of the couplings. Data shown support the choice of the universal joint.

### 2. *Determine the shaft position error*

Table 27, developed by David A. Lee, shows the output variations caused by misalignment between the shafts. Thus, at 30° angular misalignment, the position error is 4°06′42″. This means that the output shaft position shifts from −4°06′42″ to +4°06′42″ twice each revolution. At a 45° misalignment the position error, Table 27, is 9°52′26″. The shift in position is similar to that occurring at 30° angular misalignment.

### 3. *Compute the output-shaft speed variation*

Table 27 shows that at 30° angular misalignment the output-shaft speed variation is ±15.47 percent. Thus, the output-shaft speed varies between 200(1.00 ± 0.1547) = 169.06 and 230.94 r/min. This speed variation occurs *twice* per revolution.

For a 45° angular misalignment the speed variation, determined in the same way, is 117.16 to 282.84 r/min. This speed variation also occurs twice per revolution.

### 4. *Determine output-shaft acceleration*

Table 27 lists the ratio of maximum output-shaft acceleration $A$ to the square of the input speed, $\omega^2$, expressed in radians. To convert r/min to rad/s, use $rps$ = 0.1047 r/min = 0.1047(200) = 20.94 rad/s.

For 30° angular misalignment, from Table 27 $A/\omega^2 = 0.294571$. Thus, $A = \omega^2(0.294571) = (20.94)^2(0.294571) = 129.6\ \text{rad/s}^2$. This means that a constant input speed of 200 r/min produces an output acceleration ranging from −129.6 to +129.6 rad/s$^2$, and back, at a frequency of 2(200 r/min) = 400 cycles/min.

At a 45° angular misalignment, the acceleration range of the output shaft, determined in the same way, is −346 to +346 rad/s$^2$. Thus, the acceleration range at the larger shaft angle misalignment is 2.67 times that at the smaller, 30°, misalignment.

**Related Calculations:** Table 25 is useful for choosing any of seven other types of high-speed couplings. The eight couplings listed in this table are popular for high-horsepower applications. All are classed as rigid types, as distinguished from entirely flexible connectors such as flexible cables.

Values listed in Table 25 are nominal ones that may be exceeded by special designs. These values are guideposts rather than fixed; in borderline cases, consult the manufacturer's engineering data. Table 26 compares the functional characteristics of the couplings and is useful to the designer who is seeking a unit with specific operating characteristics. Note that the values in Table 25 are maximum and not additive. In other words, a coupling *cannot* be operated at the maximum angular and parallel misalignment and at the maximum horsepower and speed simultaneously—although in some cases the combination of maximum angular misalignment, maximum horsepower, and maximum speed would be acceptable. Where shock loads are anticipated, apply a suitable correction factor, as given in earlier calculation procedures, to the horsepower to be transmitted before entering Table 25.

## SELECTION OF ROLLER AND INVERTED-TOOTH (SILENT) CHAIN DRIVES

Choose a roller chain and the sprockets to transmit 6 hp (4.5 kW) from an electric motor to a propeller fan. The speed of the motor shaft is 1800 r/min and of the driven shaft 900 r/min. How long will the chain be if the centerline distance between the shafts is 30 in (76.2 cm)?

### Calculation Procedure:

**1. *Determine, and apply, the load service factor***

Consult the manufacturer's engineering data for the appropriate load service factor. Table 28 shows several typical load ratings (smooth, moderate shock, heavy shock) for various types of driven devices. Use the load rating and the type of drive to determine the service factor. Thus, a propeller fan is rated as a heavy shock load. For this type of load and an electric-motor drive, the load service factor is 1.5, from Table 28.

Apply the load service factor by taking the product of it and the horsepower transmitted, or (1.5)(6 hp) = 9.0 hp (6.7 kW). The roller chain and sprockets must have enough strength to transmit this horsepower.

**2. *Choose the chain and number of teeth in the small sprocket***

Using the manufacturer's engineering data, enter the horsepower rating table at the small-sprocket rpm and project to a horsepower value equal to, or slightly greater than, the required

**TABLE 28** Roller Chain Loads and Service Factors*

Load rating

| Driven device | Type of load |
|---|---|
| Agitators (paddle or propeller) | Smooth |
| Brick and clay machinery | Heavy shock |
| Compressors (centrifugal and rotary) | Moderate shock |
| Conveyors (belt) | Smooth |
| Crushing machinery | Heavy shock |
| Fans (centrifugal) | Moderate shock |
| Fans (propeller) | Heavy shock |
| Generators and exciters | Moderate shock |
| Laundry machinery | Moderate shock |
| Mills | Heavy shock |
| Pumps (centrifugal, rotary) | Moderate shock |
| Textile machinery | Smooth |

Service factor

| | Internal-combustion engine | | |
|---|---|---|---|
| Type of load | Hydraulic drive | Mechanical drive | Electric motor or turbine |
| Smooth | 1.0 | 1.2 | 1.0 |
| Moderate shock | 1.2 | 1.4 | 1.3 |
| Heavy shock | 1.4 | 1.7 | 1.5 |

*Excerpted from Morse Chain Company data.

**TABLE 29** Roller Chain Power Rating°

[*Single-strand, ⅝-in (1.6-cm) pitch roller chain*]

| No. of teeth in small sprocket | Small sprocket rpm | | | | | |
|---|---|---|---|---|---|---|
| | 1500 | | 1800 | | 2100 | |
| | hp | kW | hp | kW | hp | kW |
| 14 | 10.7 | 7.98 | 8.01 | 5.97 | 6.34 | 4.73 |
| 15 | 11.9 | 8.87 | 8.89 | 6.63 | 7.03 | 5.24 |
| 16 | 13.1 | 9.77 | 9.79 | 7.30 | 7.74 | 5.77 |
| 17 | 14.3 | 10.7 | 10.7 | 7.98 | | |

°Excerpted from Morse Chain Company data.

rating. At this horsepower rating, read the number of teeth in the small sprocket, which is also listed in the table. Thus, in Table 29, which is an excerpt from a typical horsepower rating tabulation, 9.0 hp (6.7 kW) is not listed at a speed of 1800 r/min. However, the next higher horsepower rating, 9.79 hp (7.3 kW), will be satisfactory. The table shows that at this power rating, 16 teeth are used in the small sprocket.

This sprocket is a good choice because most manufacturers recommend that at least 16 teeth be used in the smaller sprocket, except at low speeds (100 to 500 r/min).

### 3. *Determine the chain pitch and number of strands*

Each horsepower rating table is prepared for a given chain pitch, number of chain strands, and various types of lubrication. Thus, Table 29 is for standard single-strand ⅝-in (1.6-cm) pitch roller chain. The 9.79-hp (7.3-kW) rating at 1800 r/min for this chain is with type III lubrication—oil bath or oil slinger—with the oil level maintained in the chain casing at a predetermined height. See the manufacturer's engineering data for the other types of lubrication (manual, drip, and oil stream) requirements.

### 4. *Compute the drive speed ratio*

For a roller chain drive, the speed ratio $S_r = R_h/R_l$, where $R_h$ = rpm of high-speed shaft; $R_l$ = rpm of low-speed shaft. For this drive, $S_r = 1800/900 = 2$.

### 5. *Determine the number of teeth in the large sprocket*

To find the number of teeth in the large sprocket, multiply the number of teeth in the small sprocket, found in step 2, by the speed ratio, found in step 4. Thus, the number of teeth in the large sprocket = (16)(2) = 32.

### 6. *Select the sprockets*

Refer to the manufacturer's engineering data for the dimensions of the available sprockets. Thus, one manufacturer supplies the following sprockets for ⅝-in (1.6-cm) pitch single-strand roller chain: 16 teeth, OD = 3.517 in (8.9 cm), bore = ⅝ in (1.6 cm); 32 teeth, OD = 6.721 in (17.1 cm), bore = ⅝ or ¾ in (1.6 or 1.9 cm). When choosing a sprocket, be certain to refer to data for the size and type of chain selected in step 3, because each sprocket is made for a specific type of chain. Choose the type of hub—setscrew, keyed, or taper-lock bushing—based on the torque that must be transmitted by the drive. See earlier calculation procedures in this section for data on key selection.

### 7. *Determine the length of the chain*

Compute the chain length in pitches $L_p$ from $L_p = 2C + (S/2) + K/C$, where $C$ = shaft center distance, in/chain pitch, in; $S$ = sum of the number of teeth in the small and large sprocket; $K$ = a constant from Table 30, obtained by entering this table with the value $D$ = number of teeth in large sprocket − number of teeth in small sprocket. For this drive, $C = 30/0.625 = 48$; $S = 16 + 32 = 48$; $D = 32 - 16 = 16$; $K = 6.48$ from Table 30. Then, $L_p = 2(48) + 48/2 + 6.48/48 = 120.135$ pitches. However, a chain cannot contain a fractional pitch; therefore, use the next higher number of pitches, or $L_p = 121$ pitches.

Convert the length in pitches to length in inches, $L_i$, by taking the product of the chain pitch $p$ in and $L_p$. Or $L_i = L_p p = (121)(0.625) = 75.625$ in (192.1 cm).

**Related Calculations:** At low-speed ratios, large-diameter sprockets can be used to reduce the roller-chain pull and bearing loads. At high-speed ratios, the number of teeth in the high-speed sprocket may have to be kept as small as possible to reduce the chain pull and bearing loads. The Morse Chain Company states: Ratios over 7:1 are generally not recommended for single-width roller chain drives. Very slow-speed drives (10 to 100 r/min) are often practical with as few as 9 or 10 teeth in the small sprocket, allowing ratios up to 12:1. In all cases where ratios exceed 5:1, the designer should consider the possibility of using compound drives to obtain maximum service life.

**TABLE 30** Roller Chain Length Factors*

| $D$ | $K$ | $D$ | $K$ |
|---|---|---|---|
| 1 | 0.03 | 11 | 3.06 |
| 2 | 0.10 | 12 | 3.65 |
| 3 | 0.23 | 13 | 4.28 |
| 4 | 0.41 | 14 | 4.96 |
| 5 | 0.63 | 15 | 5.70 |
| 6 | 0.91 | 16 | 6.48 |
| 7 | 1.24 | 17 | 7.32 |
| 8 | 1.62 | 18 | 8.21 |
| 9 | 2.05 | 19 | 9.14 |
| 10 | 2.53 | 20 | 10.13 |

*Excerpted from Morse Chain Company data.

When you select standard inverted-tooth (silent) chain and high-velocity inverted-tooth silent-chain drives, follow the same general procedures as given above, except for the following changes.

*Standard inverted-tooth silent chain:* (*a*) Use a minimum of 17 teeth, and an odd number of teeth on one sprocket, where possible. This increases the chain life. (*b*) To achieve minimum noise, select sprockets having 23 or more teeth. (*c*) Use the proper service factor for the load, as given in the manufacturer's engineering data. (*d*) Where a long or fixed-center drive is necessary, use a sprocket or shoe idler where the largest amount of slack occurs. (*e*) Do not use an idler to reduce the chain wrap on small-diameter sprockets. (*f*) Check to see that the small-diameter sprocket bore will fit the high-speed shaft. Where the high-speed shaft diameter exceeds the maximum bore available for the chosen smaller sprocket, increase the number of teeth in the sprocket or choose the next larger chain pitch. (This general procedure also applies to roller chain sprockets.) (*g*) Compute the chain design horsepower from (drive hp)(chain service factor). (*h*) Select the chain pitch, number of teeth in the small sprocket, and chain *width* from the manufacturer's rating table. Thus, if the chain design horsepower = 36 hp (26.8 kW) and the chain is rated at 4 hp/in (1.2 kW/cm) of width, the required chain width = 36 hp/(4 hp/in) = 9 in (22.9 cm).

*High-velocity inverted-tooth silent chain:* (*a*) Use a minimum of 25 teeth and an odd number of teeth on one sprocket, where possible. This increases the chain life. (*b*) To achieve minimum noise, select sprockets with 27 or more teeth. (*c*) Use a larger service factor than the manufacturer's engineering data recommends, if trouble-free drives are desired. (*d*) Use a wider chain than needed, if an increased chain life is wanted. Note that the chain width is computed in the same way as described in item *h* above. (*e*) If a longer center distance between the drive shafts is desired, select a larger chain pitch [usual pitches are ¾, 1, 1½, or 2 in (1.9, 2.5, 3.8, or 5.1 cm)]. (*f*) Provide a means to adjust the centerline distance between the shafts. Such an adjustment *must* be provided in vertical drives. (*g*) Try to use an even number of pitches in the chain to avoid an offset link.

## CAM CLUTCH SELECTION AND ANALYSIS

Choose a cam-type clutch to drive a centrifugal pump. The clutch must transmit 125 hp (93.2 kW) at 1800 r/min to the pump, which starts and stops 40 times per hour throughout its 12-h/day, 360-day/year operating period. The life of the pump will be 10 years.

### Calculation Procedure:

#### 1. *Compute the maximum torque acting on the clutch*

Compute the torque acting on the clutch from $T = 5252hp/R$, where the symbols are the same as in the previous calculation procedure. Thus, for this clutch, $T = 5252 \times 125/1800 = 365$ lb·ft (494.9 N·m).

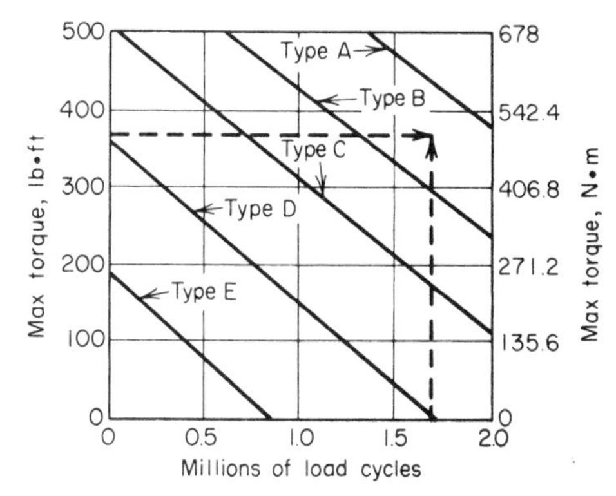


**FIG. 7** Cam-type-clutch selection chart.

#### 2. *Analyze the torque acting on the clutch*

For installations free of shock loads during starting and stopping, the running torque is the maximum torque that acts on the clutch. But if there is a shock load during starting or stopping, or at other times, the shock torque must be added to the running torque to determine the total torque acting. Compute the shock torque using the relation in step 1 and the actual hp and speed developed by the shock load.

#### 3. *Compute the total number of load applications*

With 40 starts and stops (cycles) per hour, a 12-h day, and 360 operating days per year, the number of cycles per year is (40 cycles/h)(12 h/day)(360 days/year) = 172,800. In 10 years, the clutch will undergo (172,800 cycles/year)(10 years) = 1,728,000 cycles.

#### 4. *Choose the clutch size*

Enter Fig. 7 at the maximum torque, 365 lb·ft (494.9 N·m), on the left, and the number of load cycles, 1,728,000, on the bottom. Project horizontally and vertically until the point of intersection is reached. Select the clutch represented by the next higher curve. Thus a type A clutch would be used for this load. (Note that the clutch capacity could be tabulated instead of plotted, but the results would be the same.)

#### 5. *Check the clutch dimensions*

Determine whether the clutch bore will accommodate the shafts. If the clutch bore is too small, choose the next larger clutch size. Also check to see whether the clutch will fit into the available space.

**Related Calculations:** Use this general procedure to select cam-type clutches for business machines, compressors, conveyors, cranes, food processing, helicopters, fans, aircraft, printing machinery, pumps, punch presses, speed reducers, looms, grinders, etc. When choosing a specific clutch, use the manufacturer's engineering data to select the clutch size.

## TIMING-BELT DRIVE SELECTION AND ANALYSIS

Choose a toothed timing belt to transmit 20 hp (14.9 kW) from an electric motor to a rotary mixer for liquids. The motor shaft turns at 1750 r/min and the mixer shaft is to turn at 600 ± 20 r/min. This drive will operate 12 h/day, 7 days/week. Determine the type of timing belt to use and the driving and driven pulley diameters if the shaft centerline distance is about 27 in (68.6 cm).

### Calculation Procedure:

#### 1. *Choose the service factor for the drive*

Timing-belt manufacturers publish service factors in their engineering data based on the type of prime mover, the type of driven machine (compressor, mixer, pump, etc.), type of drive (speedup), and drive conditions (continuous operation, use of an idler, etc.).

**TABLE 31** Typical Timing-Belt Service Factors*

| Type of drive | Type of load | Service factor |
|---|---|---|
| Electric motors, hydraulic motors, internal-combustion engines, line shafts | Shock-free | 2.0 |
| | Shocks | 2.5 |
| | Continuous operation or idler use | 2.7 |
| | Speed-up | 3.0 |

*Use only for preliminary selection of belt. From Morse Chain Company data.

Usual service factors for any type of driver range from 1.3 to 2.5 for various types of driven machines. Correction factors for speed-up drives range from 0 to 0.40; the specific value chosen is *added* to the machine-drive correction factor. Drive conditions, such as 24-h continuous operation or the use of an idler pulley on the drive, cause an additional 0.2 to be added to the correction factor. Seasonal or intermittent operation *reduces* the machine-drive factor by 0.2.

Look up the service factor in Table 31, if the manufacturer's engineering data are not readily available. Table 31 gives safe data for usual timing-belt applications and is suitable for preliminary selection of belts. Where a final choice is being made, use the manufacturer's engineering data.

For a liquid mixer shock-free load, use a service factor of 2.0 from Table 31, since there are no other features which would require a larger value.

### 2. Compute the design horsepower for the belt

The design horsepower $hp_d = hp_l \times \text{SF}$, where $hp_l$ = load horsepower; SF = service factor. Thus, for this drive, $hp_d = (20)(2) = 40$ hp (29.8 kW).

### 3. Compute the drive speed ratio

The drive speed ratio $S_r = R_h/R_l$, where $R_h$ = rpm of high-speed shaft; $R_l$ = rpm of low-speed shaft. For this drive $S_r = 1750/600 = 2.92{:}1$, the rated rpm. If the driven-pulley speed falls 10 r/min, $S_r = 1750/580 = 3.02{:}1$. Thus, the speed ratio may vary between 2.92 and 3.02.

### 4. Choose the timing-belt pitch

Enter Table 32, or the manufacturer's engineering data, at the design horsepower and project to the driver rpm. Where the exact value of the design horsepower is not tabulated, use the next higher tabulated value. Thus, for this 1750-r/min drive having a design horsepower of 40 (29.8 kW), Table 32 shows that a ⅞-in (2.2-cm) pitch belt is required. This value is found by entering Table 32 at the next higher design horsepower, 50 (37.3 kW), and projecting to the 1750-r/min column. If 40 hp (29.8 kW) were tabulated, the table would be entered at this value.

### 5. Choose the number of teeth for the high-speed sprocket

Enter Table 33, or the manufacturer's engineering data, at the timing-belt pitch and project across to the rpm of the high-speed shaft. Opposite this value read the minimum number of sprocket

**TABLE 32** Typical Timing-Belt Pitch*

| Design power | | Speed of high-speed shaft, r/min | | | | | |
|---|---|---|---|---|---|---|---|
| | | 3500 | | 1750 | | 1160 | |
| hp | kW | in | cm | in | cm | in | cm |
| 25 | 18.6 | ½, ⅞ | 1.3, 2.2 | ½, ⅞ | 1.3, 2.2 | ⅞ | 2.2 |
| 50 | 37.3 | ½, ⅞ | 1.3, 2.2 | ⅞ | 2.2 | ⅞, 1¼ | 2.2, 3.2 |
| 60 and up | 44.7 and up | ⅞ | 2.2 | ⅞ | 2.2 | ⅞, 1¼ | 2.2, 3.2 |

*Morse Chain Company.

**TABLE 33** Minimum Number of Sprocket Teeth*

| Belt pitch | | High-speed shaft, r/min | Minimum sprocket pitch distance | | No. of teeth |
|---|---|---|---|---|---|
| in | cm | | in | cm | |
| ½ | 1.3 | 3500 | 3.501 | 8.9 | 20 |
| | | 1750 | 3.183 | 8.1 | 18 |
| | | 1160 | 2.865 | 7.3 | 16 |
| ⅞ | 2.2 | 3500 | 7.241 | 18.4 | 26 |
| | | 1750 | 6.685 | 17.0 | 24 |
| | | 1160 | 6.127 | 15.6 | 22 |
| 1¼ | 3.2 | 3500 | 10.345 | 26.3 | 26 |
| | | 1750 | 9.549 | 24.3 | 24 |
| | | 1160 | 8.753 | 22.2 | 22 |

*Morse Chain Company.

teeth. Thus, for a 1750-r/min ⅞-in (2.2-cm) pitch timing belt, Table 32 shows that the high-speed sprocket should have no less than 24 teeth nor a pitch diameter less than 6.685 in (17.0 cm). (If a smaller diameter sprocket were used, the belt service life would be reduced.)

**6. *Select a suitable timing belt***

Enter Table 34, or the manufacturer's engineering data, at either the exact speed ratio, if tabulated, or the nearest value to the speed-ratio range. For this drive, having a ratio of 2.92:3.02, the nearest value in Table 34 is 3.00. This table shows that with a 24-tooth driver and a 72-tooth driven sprocket, a center distance of 27.17 in (69.0 cm) is obtainable. Since a center distance of about 27 in (68.6 cm) is desired, this belt is acceptable.

Where an exact center distance is specified, several different sprocket combinations may have to be tried before a belt having a suitable center distance is obtained.

**7. *Determine the required belt width***

Each center distance listed in Table 34 corresponds to a specific pitch and type of belt construction. The belt construction is often termed XL, L, H, XH, and XXH. Thus, the belt chosen in step 6 is an XH construction.

Refer now to Table 35 or the manufacturer's engineering data. Table 35 shows that a 2-in (5.1-cm) wide belt will transmit 38 hp (28.3 kW) at 1750 r/min. This is too low, because the design horsepower rating of the belt is 40 hp (29.8 kW). A 3-in (7.6-cm) wide belt will transmit 60 hp (44.7 kW). Therefore, a 3-in (7.6-cm) belt should be used because it can safely transmit the required horsepower.

If five, or less, teeth are in mesh when a timing belt is installed, the width of the belt must be

**TABLE 34** Timing-Belt Center Distances*

| Speed ratio | No. of sprocket teeth | | Center distance | | | | | |
|---|---|---|---|---|---|---|---|---|
| | | | XH | | XH | | XH | |
| | Driver | Driven | in | cm | in | cm | in | cm |
| 2.80 | 30 | 84 | 22.81 | 57.94 | 30.11 | 76.48 | 37.30 | 94.74 |
| 3.00 | 24 | 72 | 27.17 | 69.01 | 34.34 | 87.22 | 41.46 | 105.3 |
| 3.20 | 30 | 96 | 19.19 | 48.74 | 26.84 | 68.17 | 34.19 | 86.84 |

*Morse Chain Company.

**TABLE 35** Belt Power Rating°
[⅞ in (2.2 cm) pitch XH]

| No. of teeth in high-speed sprocket | Belt width | | Sprocket rpm | | | | | |
|---|---|---|---|---|---|---|---|---|
| | | | 1700 | | 1750 | | 2000 | |
| | in | cm | hp | kW | hp | kW | hp | kW |
| 24 | 2 | 5.1 | 37 | 27.6 | 38 | 28.3 | 43 | 32.1 |
| | 3 | 7.6 | 59 | 44.0 | 60 | 44.7 | 67 | 50.0 |
| | 4 | 10.2 | 83 | 61.9 | 85 | 63.4 | 95 | 70.8 |

°Morse Chain Company.

increased to ensure sufficient load-carrying ability. To determine the required belt width to carry the load, divide the belt width by the appropriate factor given below.

| Teeth in mesh | 5 | 4 | 3 | 2 |
|---|---|---|---|---|
| Factor | 0.80 | 0.60 | 0.40 | 0.20 |

Thus, a 3-in (7.6-cm) belt with four teeth in mesh would have to be widened to 3/0.60 = 5.0 in (12.7 cm) to carry the desired load.

**Related Calculations:** Use this procedure to select timing belts for any of these drives: agitators, mixers, centrifuges, compressors, conveyors, fans, blowers, generators (electric), exciters, hammer mills, hoists, elevators, laundry machinery, line shafts, machine tools, paper-manufacturing machinery, printing machinery, pumps, sawmills, textile machinery, woodworking tools, etc. For exact selection of a specific make of belt, consult the manufacturer's tabulated or plotted engineering data.

## GEARED SPEED REDUCER SELECTION AND APPLICATION

Select a speed reducer to lift a sluice gate weighing 200 lb (889.6 N) through a distance of 6 ft (1.8 m) in 5 s or less. The door must be opened and closed 12 times per hour. The drive for the door lifter is a 1150-r/min electric motor that operates 10 h/day.

### Calculation Procedure:

#### 1. *Choose the type of speed reducer to use*

There are many types of speed reducers available for industrial drives. Thus, a roller chain with different size sprockets, a V-belt drive, or a timing-belt drive might be considered for a speed-reduction application because all will reduce the speed of a driven shaft. Where a load is to be raised, often geared speed reducers are selected because they provide a positive drive without slippage. Also, modern geared drives are compact, efficient units that are easily connected to an electric motor. For these reasons, a right-angle worm-gear speed reducer will be tentatively chosen for this drive. If upon investigation this type of drive proves unsuitable, another type will be chosen.

#### 2. *Determine the torque that the speed reducer must develop*

A convenient way to lift a sluice door is by means of a roller chain attached to a bracket on the door and driven by a sprocket keyed to the speed reducer output shaft. As a trial, assume that a 12-in (30.5-cm) diameter sprocket is used.

The torque $T$ lb·in developed by sprocket = $T = Wr$, where $W$ = weight lifted, lb; $r$ = sprocket radius, in. For this sprocket, by assuming that the starting friction in the sluice-door guides produces an additional load of 50 lb (222.4 N), $T = (200 + 50)(6) = 1500$ lb·in (169.5 N·m).

### 3. *Compute the required rpm of the output shaft*

The door must be lifted 6 ft (1.8 m) in 5 s. This is a speed of (6 ft × 60 s/min)/5 s = 72 ft/min (0.4 m/s). The circumference of the sprocket is $\pi d = \pi(1.0) = 3.142$ ft (1.0 m). To lift the door at a speed of 72 ft/min (0.4 m/s), the output shaft must turn at a speed of (ft/min)/(ft/r) = 72/3.142 = 22.9 r/min. Since a slight increase in the speed of the door is not objectionable, assume that the output shaft turns at 23 r/min.

### 4. *Apply the drive service factor*

The AGMA *Standard Practice for Single and Double Reduction Cylindrical Worm and Helical Worm Speed Reducers* lists service factors for geared speed reducers driven by electric motors and internal-combustion engines. These factors range from a low of 0.80 for an electric motor driving a machine producing a uniform load for occasional 0.5-h service to a high of 2.25 for a single-cylinder internal-combustion engine driving a heavy shock load 24 h/day. The service factor for this drive, assuming a heavy shock load during opening and closing of the sluice gate, would be 1.50 for 10-h/day operation. Thus, the drive must develop a torque of at least (load torque, lb·in)(service factor) = (1500)(1.5) = 2250 lb·in (254.2 N·m).

### 5. *Choose the speed reducer*

Refer to Table 36 or the manufacturer's engineering data. Table 36 shows that a single-reduction worm-gear speed reducer having an input of 1.24 hp (924.7 W) will develop 2300 lb·in (254.2 N·m) of torque at 23 r/min. This is an acceptable speed reducer because the required ouput torque is 2250 lb·in (254.2 N·m) at 23 r/min. Also, the allowable overhung load, 1367 lb (6080.7 N), is adequate for the sluice-gate weight. A 1.5-hp (1118.5-W) motor would be chosen for this drive.

**Related Calculations:** Use this general procedure to select geared speed reducers (single- or double-reduction worm gears, single-reduction helical gears, gear motors, and miter boxes) for machinery drives of all types, including pumps, loaders, stokers, welding positioners, fans, blowers, and machine tools. The starting friction load, applied to the drive considered in this procedure, is typical for applications where a heavy friction load is likely to occur. In rotating machinery of many types, the starting friction load is usually nil, except where the drive is connected to a loaded member, such as a conveyor belt. Where a clutch disconnects the driver from the load, there is negligible starting friction.

Well-designed geared speed reducers generally will not run at temperatures higher than 100°F (55.6°C) *above* the prevailing ambient temperature, measured in the lubricant sump. At higher operating temperatures the lubricant may break down, leading to excessive wear. Fan-cooled speed reducers can carry heavier loads than noncooled reducers without overheating.

**TABLE 36** Speed Reducer Torque Ratings*

*(Single-reduction worm gear)*

| Input power at 1150 r/min | | Drive output | | | | |
|---|---|---|---|---|---|---|
| | | | Torque | | OHL† | |
| hp | kW | r/min | lb·in | N·m | lb | N |
| 1.54 | 1.15 | 28.7 | 2416 | 273.0 | 1367 | 6080.7 |
| 1.24 | 0.92 | 23.0 | 2300 | 259.9 | 1367 | 6080.7 |
| 0.93 | 0.69 | 19.2 | 1970 | 222.6 | 1367 | 6080.7 |

*Extracted from Morse Chain Company data.
†Allowable overhung load on drive.

## POWER TRANSMISSION FOR A VARIABLE-SPEED DRIVE

Choose the power-transmission system for a three-wheeled contractor's vehicle designed to carry a load of 1000 lb (4448.2 N) at a speed of 8 mi/h (3.6 m/s) over rough terrain. The vehicle tires will be 16 in (40.6 cm) in diameter, and the engine driving the vehicle will operate continuously. The empty vehicle weighs 600 lb (2668.9 N), and the engine being considered has a maximum speed of 4200 r/min.

### Calculation Procedure:

**1. *Compute the horsepower required to drive the vehicle***

Compute the required driving horsepower from $hp = 1.25\ Wmph/1750$, where $W$ = total weight of *loaded* vehicle, lb (N); $mph$ = maximum loaded vehicle speed, mi/h (km/h). Thus, for this vehicle, $hp = 1.25(1000 + 600)(8)/1750 = 9.15$ hp (6.8 kW).

**2. *Determine the maximum vehicle wheel speed***

Compute the maximum wheel rpm from $rpm_w$ = (maximum vehicle speed, mi/h) × (5280 ft/mi)/15.72 (tire rolling diameter, in). Or, $rpm_w = (8)(5280)/[(15.72)(16)] = 167.8$ r/min.

**3. *Select the power transmission for the vehicle***

Refer to engineering data published by drive manufacturers. Choose a drive suitable for the anticipated load. The load on a typical contractor's vehicle is one of sudden starts and stops. Also, the drive must be capable of transmitting the required horsepower. A 10-hp (7.5-kW) drive would be chosen for this vehicle.

Small vehicles are often belt-driven by means of an infinitely variable transmission. Such a drive, having an overdrive or speed-increase ratio of 1:1.5 or 1:1, would be suitable for this vehicle. From the manufacturer's engineering data, a drive having an input rating of 10 hp (7.5 kW) will be suitable for momentary overloads of up to 25 percent. The operating temperature of any part of the drive should never exceed 250°F (121.1°C). For best results, the drive should be operated at temperatures well below this limit.

**4. *Compute the required output-shaft speed reduction***

To obtain the maximum power output from the engine, the engine should operate at its maximum rpm when the vehicle is traveling at its highest speed. This prevents lugging of the engine at lower speeds.

The transmission transmits power from the engine to the driving axle. Usually, however, the transmission cannot provide the needed speed reduction between the engine and the axle. Therefore, a speed-reduction gear is needed between the transmission and the axle. The transmission chosen for this drive could provide a 1:1 or a 1:1.5 speed ratio. Assume that the 1:1.5 speed ratio is chosen to provide higher speeds at the maximum vehicle load. Then the speed reduction required = (maximum engine speed, r/min)(transmission ratio)/(maximum wheel rpm) = $(4200)(1.5)/167.8 = 37.6$.

Check the manufacturer's engineering data for the ratios of available geared speed reducers. Thus, a study of one manufacturer's data shows that a speed-reduction ratio of 38 is available by using a single-reduction worm-gear drive. This drive would be suitable if it were rated at 10 hp (7.5 kW) or higher. Check to see that the gear has a suitable horsepower rating before making the final selection.

**Related Calculations:** Use the general procedure given here to choose power transmissions for small-vehicle compressors, hoists, lawn mowers, machine tools, conveyors, pumps, snow sleds, and similar equipment. For nonvehicle drives, substitute the maximum rpm of the driven machine for the maximum wheel velocity in steps 2, 3, and 4.

## BEARING-TYPE SELECTION FOR A KNOWN LOAD

Choose a suitable bearing for a 3-in (7.6-cm) diameter 100-r/min shaft carrying a total radial load of 12,000 lb (53,379 N). A reasonable degree of shaft misalignment must be allowed by the bearing. Quiet operation of the shaft is desired. Lubrication will be intermittent.

## Calculation Procedure:

### 1. *Analyze the desired characteristics of the bearing*

Two major types of bearings are available to the designer, *rolling* and *sliding*. Rolling bearings are of two types, *ball* and *roller*. Sliding bearings are also of two types, *journal* for radial loads and *thrust* for axial loads only or for combined axial and radial loads. Table 37 shows the principal characteristics of rolling and sliding bearings. Based on the data in Table 37, a sliding bearing would be suitable for this application because it has a *fair* misalignment tolerance and a *quiet* noise level. Both factors are key considerations in the bearing choice.

**TABLE 37** Key Characteristics of Rolling and Sliding Bearings*

| | Rolling | Sliding |
|---|---|---|
| Life | Limited by fatigue properties of bearing metal | Unlimited, except for cyclic loading |
| Load: | | |
| Unidirectional | Excellent | Good |
| Cyclic | Good | Good |
| Starting | Excellent | Poor |
| Unbalance | Excellent | Good |
| Shock | Good | Fair |
| Emergency | Fair | Fair |
| Speed limited by: | Centrifugal loading and material surface speeds | Turbulence and temperature rise |
| Starting friction | Good | Poor |
| Cost | Intermediate, but standardized, varying little with quantity | Very low in simple types or in mass production |
| Space requirements (radial bearing): | | |
| Radial dimension | Large | Small |
| Axial dimension | ⅓ to ½ shaft diameter | ¼ to 2 times shaft diameter |
| Misalignment tolerance | Poor in ball bearings except where designed for at sacrifice of load capacity; good in spherical roller bearings; poor in cylindrical roller bearings | Fair |
| Noise | May be noisy, depending on quality and resonance of mounting | Quiet |
| Damping | Poor | Good |
| Low-temperature starting | Good | Poor |
| High-temperature operation | Limited by lubricant | Limited by lubricant |
| Type of lubricant | Oil or grease | Oil, water, other liquids, grease, dry lubricants, air, or gas |
| Lubrication, quantity required | Very small, except where large amounts of heat must be removed | Large, except in low-speed boundary-lubrication types |
| Type of failure | Limited operation may continue after fatigue failure but not after lubricant failure | Often permits limited emergency operation after failure |
| Ease of replacement | Function of type of installation; usually shaft need not be replaced | Function of design and installation; split bearings used in large machines |

* *Product Engineering.*

**TABLE 38** Materials for Sleeve Bearings*

*[Cost figures are for a 1-in (2.54-cm) sleeve bearing ordered in quantity]*

| | Maximum load | | Maximum speed | | PV limit | | Maximum operating temperature | | Cost, $ |
|---|---|---|---|---|---|---|---|---|---|
| | $lb/in^2$ | kPa | ft/min | m/s | $(lb/in^2)(ft/min)$ | kPa·m/s | °F | °C | |
| Porous bronze | 4,000 | 27,579.0 | 1,500 | 7.6 | 50,000 | 1,751.3 | 150 | 65.6 | 0.11 |
| Porous iron | 8,000 | 55,158.1 | 800 | 4.1 | 50,000 | 1,751.3 | 150 | 65.6 | 0.09 |
| Teflon fabric | 60,000 | 413,685.4 | 50 | 0.3 | 25,000 | 875.6 | 500 | 260.0 | 0.04 |
| Phenolic | 6,000 | 41,368.5 | 2,500 | 12.7 | 15,000 | 525.4 | 200 | 93.3 | 0.05 |
| Wood | 6,000 | 41,368.5 | 2,000 | 10.2 | 15,000 | 525.4 | 150 | 65.6 | 0.40 |
| Carbon-graphite | 600 | 4,136.9 | 2,500 | 12.7 | 10,000 | 350.3 | 750 | 398.9 | 0.39 |
| Reinforced Teflon | 2,500 | 17,236.9 | 2,500 | 12.7 | 10,000 | 350.3 | 500 | 260.0 | 0.45 |
| Nylon | 1,000 | 6,894.8 | 1,000 | 5.1 | 3,000 | 105.1 | 200 | 93.3 | 0.04 |
| Delrin | 1,000 | 6,894.8 | 1,000 | 5.1 | 3,000 | 105.1 | 180 | 82.2 | 0.03 |
| Lexan | 1,000 | 6,894.8 | 1,000 | 5.1 | 3,000 | 105.1 | 220 | 104.4 | 0.05 |
| Teflon | 500 | 3,447.4 | 100 | 0.5 | 1,000 | 35.0 | 500 | 260.0 | 1.00 |

* *Product Engineering.*

### 2. *Choose the bearing materials*

Table 38 shows that a porous-bronze bearing, suitable for intermittent lubrication, can carry a maximum pressure load of 4000 lb/in$^2$ (27,580.0 kPa) at a maximum shaft speed of 1500 ft/min (7.62 m/s). By using the relation $l = L/(Pd)$, where $l$ = bearing length, in, $L$ = load, lb, $d$ = shaft diameter, in, the required length of this sleeve bearing is $l = L/(Pd) = 12{,}000/[(4000)(3)] = 1$ in (2.5 cm).

Compute the shaft surface speed $V$ ft/min from $V = \pi dR/12$, where $d$ = shaft diameter, in; $R$ = shaft rpm. Thus, $V = \pi(3)(100)/12 = 78.4$ ft/min (0.4 m/s).

With the shaft speed known, the PV, or pressure-velocity, value of the bearing can be computed. For this bearing, with an operating pressure of 4000 lb/in$^2$ (27,580.0 kPa), PV = 4000 × 78.4 = 313,600 (lb/in$^2$) (ft/min) (10,984.3 kPa·m/s). This is considerably in excess of the PV limit of 50,000 (lb/in$^2$) (ft/min) (1751.3 kPa·m/s) listed in Table 38. To come within the recommended PV limit, the operating pressure of the bearing must be reduced.

Assume an operating pressure of 600 lb/in$^2$ (4137.0 kPa). Then $l = L/(Pd) = 12{,}000/[(600)(3)] = 6.67$ in (16.9 cm), say 7 in (17.8 cm). The PV value of the bearing then is (600)(78.4) = 47,000 (lb/in$^2$) (ft/min) (1646.3 kPa·m/s). This is a satisfactory value for a porous-bronze bearing because the recommended limit is 50,000 (lb/in$^2$) (ft/min) (1751.3 kPa·m/s).

### 3. *Check the selected bearing size*

The sliding bearing chosen will have a diameter somewhat in excess of 3 in (7.6 cm) and a length of 7 in (17.8 cm). If this length is too great to fit in the allowable space, another bearing material will have to be studied, by using the same procedure. Figure 8 shows the space occupied by rolling and sliding bearings of various types.

Table 39 shows the load-carrying capacity and maximum operating temperatures for oil-film journal sliding bearings that are regularly lubricated. These bearings are termed *full film* because they receive a supply of lubricant at regular intervals. Surface speeds of 20,000 to 25,000 ft/min (101.6 to 127.0 m/s) are common for industrial machines fitted with these bearings. This corresponds closely to the surface speed for ball and roller bearings.

### 4. *Evaluate oil-film bearings*

Oil-film sliding bearings are chosen by the method of the next calculation procedure. The bearing size is made large enough that the maximum operating temperature listed in Table 39 is not exceeded. Table 40 lists typical design load limits for oil-film bearings in various services. Figure 9 shows the typical temperature limits for rolling and sliding bearings made of various materials.

### 5. *Evaluate rolling bearings*

Rolling bearings have lower starting friction (coefficient of friction $f = 0.002$ to 0.005) than sliding bearings ($f = 0.15$ to 0.30). Thus, the rolling bearing is preferred for applications requiring low starting torque [integral-horsepower electric motors up to 500 hp (372.9 kW), jet engines, etc.]. By pumping oil into a sliding bearing, its starting coefficient of friction can be reduced to nearly zero. This arrangement is used in large electric generators and certain mill machines.

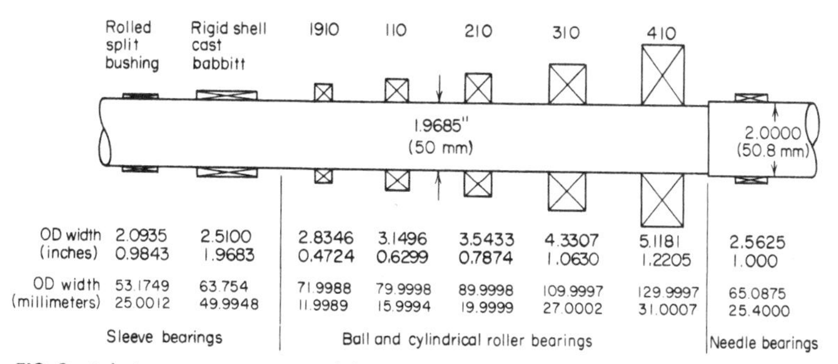


**FIG. 8** Relative space requirements of sleeve and rolling-element bearings to carry the same diameter shaft. *(Product Engineering.)*

**TABLE 39** Oil-Film Journal Bearing Characteristics*

| Bearing material | Load-carrying capacity | | Maximum operating temperature | |
|---|---|---|---|---|
| | lb/in² | kPa | °F | °C |
| Tin-based babbitt | 800–1,500 | 5,516.0–10,342.5 | 300 | 148.9 |
| Lead-based babbitt | 800–1,200 | 5,516.0–8,274.0 | 300 | 148.9 |
| Alkali-hardened steel | 1,200–1,500 | 8,274.0–10,342.5 | 500 | 260.0 |
| Cadmium base | 1,500–2,000 | 10,342.5–13,789.5 | 500 | 260.0 |
| Copper-lead | 1,500–2,500 | 10,342.5–17,236.9 | 350 | 176.7 |
| Tin bronze | 4,000 | 27,580.0 | 500+ | 260.0+ |
| Lead bronze | 3,000–4,000 | 20,685.0–27,580.0 | 450 | 232.2 |
| Aluminum alloy | 4,000 | 27,580.0 | 250 | 121.1 |
| Silver (overplated) | 4,000 | 27,580.0 | 500 | 260.0 |
| Three-component bearings babbitt-surfaced | 2,000–4,000 | 13,790.0–27,580.0 | 225–300 | 107.2–148.9 |

* *Product Engineering.*

**TABLE 40** Typical Design Load Limits for Oil-Film Bearings*

| Bearing | Maximum load on projected area | |
|---|---|---|
| | lb/in² | kPa |
| Electric motors | 200 | 1,379.0 |
| Steam turbines | 300 | 2,068.4 |
| Automotive engines: | | |
| Main bearings | 3,500 | 24,131.6 |
| Connecting rods | 5,000 | 34,473.8 |
| Diesel engines: | | |
| Main bearings | 3,000 | 20,684.7 |
| Connecting rods | 4,500 | 31,026.4 |
| Railroad car axles | 350 | 2,413.2 |
| Steel mill roll necks: | | |
| Steady | 2,000 | 13,789.5 |
| Peak | 5,000 | 24,473.8 |

* *Product Engineering.*

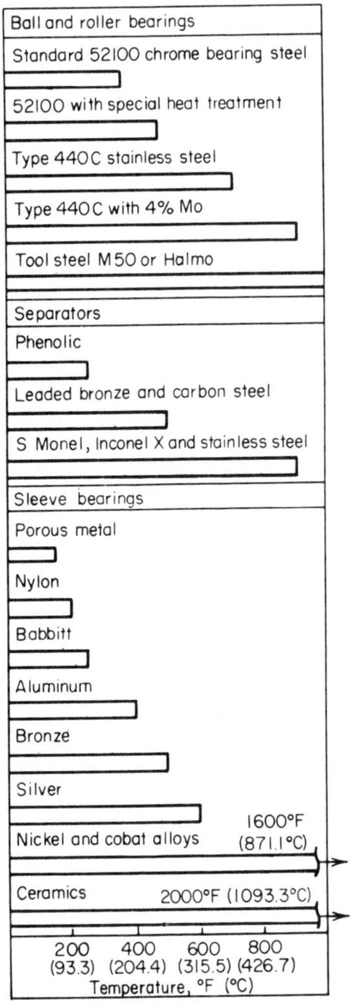

**FIG. 9** Bearing temperature limits. *(Product Engineering.)*

**TABLE 41** Relative Load Capacity, Cost, and Size of Rolling Bearings*

| Bearing type (for 50-mm bore) | Radial capacity | Axial capacity | Cost | Outer diameter | Width |
|---|---|---|---|---|---|
| Ball bearings: | | | | | |
| Deep groove (Conrad) | 1.0 | 1.0 | 1.0 | 1.0 | 1.0 |
| Filling notch | 1.2 | Low | 1.2 | 1.0 | 1.0 |
| Double row | 1.5 | 1.1 | 2.2 | 1.0 | 1.6 |
| Angular contact | 1.1 | 1.9 | 1.6 | 1.0 | 1.0 |
| Duplex | 1.8 | 1.9 | 2.0 | 1.0 | 2.0 |
| Self-aligning | 0.7 | 0.2 | 1.3 | 1.0 | 1.0 |
| Ball thrust | 0 | 0.9 | 0.8 | 0.7 | 0.8 |
| Roller bearings: | | | | | |
| Cylindrical | 1.6 | . . . | 1.9 | 1.0 | 1.0 |
| Tapered | 1.3 | 0.8 | 0.9 | 1.0 | 1.0 |
| Spherical | 3.0 | 1.1 | 5.0 | 1.0 | 1.5 |
| Needle | 1.0 | 0 | 0.3 | 0.5 | 1.6 |
| Flat thrust | 0 | 4.0 | 3.8 | 0.8 | 0.9 |

**Product Engineering.*

The running friction of rolling bearings is in the range of $f = 0.001$ to $0.002$. For oil-film sliding bearings, $f = 0.002$ to $0.005$.

Rolling bearings are more susceptible to dirt than are sliding bearings. Also, rolling bearings are inherently noisy. Oil-film bearings are relatively quiet, but they may allow higher amplitudes of shaft vibration.

Table 41 compares the size, load capacity, and cost of rolling bearings of various types. Briefly, ball bearings and roller bearings may be compared thus: ball bearings (*a*) run at higher speeds without undue heating, (*b*) cost less per pound of load-carrying capacity for light loads, (*c*) have friction torque at light loads, (*d*) are available in a wider variety of sizes, (*e*) can be made in smaller sizes, and (*f*) have seals and shields for easy lubrication. Roller bearings (*a*) can carry heavier loads, (*b*) are less expensive for larger sizes and heavier loads, (*c*) are more satisfactory under shock and impact loading, and (*d*) may have lower friction at heavy loads. Table 42 shows the speed limit, termed the *dR limit* (equals bearing shaft bore *d* in mm multiplied by the shaft rpm *R*), for ball and roller bearings. Speeds higher than those shown in Table 42 may lead to early bearing failures. Since the *dR* limit is proportional to the shaft surface speed, the *dR* value gives an approximate measure of the bearing power loss and temperature rise.

**Related Calculations:** Use this general procedure to select shaft bearings for any type of regular service conditions. For unusual service (i.e., excessively high or low operating temperatures, large loads, etc.) consult the specific selection procedures given elsewhere in this section.

**TABLE 42** Speed Limits for Ball and Roller Bearings*

| Lubrication | DN limit, mm × r/min |
|---|---|
| Oil: | |
| Conventional bearing designs | 300,000–350,000 |
| Special finishes and separators | 1,000,000–1,500,000 |
| Grease: | |
| Conventional bearing designs | 250,000–300,000 |
| Silicone grease | 150,000–200,000 |
| Special finishes and separators high-speed greases | 500,000–600,000 |

**Product Engineering.*

Note that the PV value of a sliding bearing can also be expressed as PV = $L/(dl) \times \pi dR/12 = \pi LR/(12l)$. The bearing load and shaft speed are usually fixed by other requirements of a design. Where the PV equation is solved for the bearing length $l$ and the bearing is too long to fit the available space, select a bearing material having a higher allowable PV value.

## SHAFT BEARING LENGTH AND HEAT GENERATION

How long should a sleeve-type bearing be if the combined weight of the shaft and gear tooth load acting on the bearing is 2000 lb (8896 N)? The shaft is 1 in (2.5 cm) in diameter and is oil-lubricated. What is the rate of heat generation in the bearing when the shaft turns at 600 r/min? How much above an ambient room temperature of 70°F (21.1°C) will the temperature of the bearing rise during operation in still air? In moving air?

### Calculation Procedure:

#### 1. *Compute the required length of the bearing*

The required length $l$ of a sleeve bearing carrying a load of $L$ lb is $l = L/(Pd)$, where $l$ = bearing length, in; $L$ = bearing load, lb = bearing reaction force, lb; $P$ = allowable mean bearing pressure, lb/in$^2$ [ranges from 25 to 2500 lb/in$^2$ (172.4 to 17,237.5 kPa) for normal service and up to 8000 lb/in$^2$ (55,160.0 kPa) for severe service], on the projected bearing area, in$^2$ = $ld$; $d$ = shaft diameter, in. Thus, for this bearing, assuming an allowable mean bearing pressure of 400 lb/in$^2$ (2758.0 kPa), $l = L/(Pd) = 2000/[(400)(1)] = 5$ in (12.7 cm).

#### 2. *Compute the rate of bearing heat generation*

The rate of heat generation in a plain sleeve bearing is given by $h = fLdR/3000$, where $h$ = rate of heat generation in the bearing, Btu/min; $f$ = bearing coefficient of friction for the lubricant used; $R$ = shaft rpm; other symbols as in step 1.

The coefficient of friction for oil-lubricated bearings ranges from 0.005 to 0.030, depending on the lubricant viscosity, shaft rpm, and mean bearing pressure. Given a value of $f = 0.020$, $h = (0.020)(200)(1)(600)/3000 = 0.8$ Btu/min, or $H = 0.8(60 \text{ min/h}) = 48.0$ Btu/h (14.1 kW).

#### 3. *Compute the bearing wall area*

The wall area $A$ of a small sleeve-type bearing, such as a pillow block fitted with a bushing, is $A = (10 \text{ to } 15)dl/144$, where $A$ = bearing wall area, ft$^2$; other symbols as before. For larger bearing pedestals fitted with a cast-iron or steel bearing shell, the factor in this equation varies from 18 to 25.

Since this is a small bearing having a 1-in (2.5-cm) diameter shaft, the first equation with a factor of 15 to give a larger wall area can be used. The value of 15 was chosen to ensure adequate radiating surface. Where space or weight is a factor, the value of 10 might be chosen. Intermediate values might be chosen for other conditions. Substituting yields $A = (15)(1)(5)/144 = 0.521$ ft$^2$ (0.048 m$^2$).

#### 4. *Determine the bearing temperature rise*

In *still air*, a bearing will dissipate $H = 2.2A(t_w - t_a)$ Btu/h, where $t_w$ = bearing wall temperature, °F; $t_a$ ambient air temperature, °F; other symbols as before. Since $H$ and $A$ are known, the temperature rise can be found by solving for $t_w - t_a = H/(2.2A) = 48.0/[(2.2)(0.521)] = 41.9$°F (23.3°C), and $t_w = 41.9 + 70 = 111.9$°F (44.4°C). This is a low enough temperature for safe operation of the bearing. The maximum allowable bearing operating temperature for sleeve bearings using normal lubricants is usually assumed to be 200°F (93.3°C). To reduce the operating temperature of a sleeve bearing, the bearing wall area must be increased, the shaft speed decreased, or the bearing load reduced.

In *moving air*, the heat dissipation from a sleeve-type bearing is $H = 6.5A(t_w - t_a)$. Solving for the temperature rise as before, we get $t_w - t_a = H/(6.5A) = 48.0/[(6.5)(0.521)] = 14.2$°F (7.9°C), and $t_w = 14.2 + 70 = 84.2$°F (29.0°C). This is a moderate operating temperature that could be safely tolerated by any of the popular bearing materials.

**Related Calculations:** Use this procedure to analyze sleeve-type bearings used for industrial line shafts, marine propeller shafts, conveyor shafts, etc. Where the ambient temperature varies during bearing operation, use the highest ambient temperature expected, in computing the bearing operating temperature.

## ROLLER-BEARING OPERATING-LIFE ANALYSIS

A machine must have a shaft of about 5.5 in (14.0 cm) in diameter. Choose a roller bearing for this 5.5-in (14.0-cm) diameter shaft that turns at 1000 r/min while carrying a radial load of 20,000 lb (88,964 N). What is the expected life of this bearing?

### Calculation Procedure:

#### 1. *Determine the bearing life in revolutions*

The operating life of rolling-type bearings is often stated in millions of revolutions. Find this life from $R_L = (C/L)^{10/3}$, where $R_L$ = bearing operating life, millions of revolutions; $C$ = dynamic capacity of the bearing, lb; $L$ = applied radial load on bearing, lb.

Obtain the dynamic capacity of the bearing being considered by consulting the manufacturer's engineering data. Usual values of dynamic capacity range between 2500 lb (11,120.6 N) and 750,000 lb (3,338,166.5 N), depending on the bearing design, type, and bore. For a typical 5.5118-in (14.0-cm) bore roller bearing, $C$ = 92,400 lb (411,015.7 N).

With $C$ known, compute $R_L = (C/L)^{10/3} = (92{,}400/20{,}000)^{10/3} = 162 \times 10^6$.

#### 2. *Determine the bearing life*

The minimum life of a bearing in millions of revolutions, $R_L$, is related to its life in hours, $h$, by the expression $R_L = 60Rh/10^6$, where $R$ = shaft speed, r/min. Solving gives $h = 10^6 R_L/(60R) = (10^6)(162)/[(60)(1000)] = 2700$-h minimum life.

**Related Calculations:** This procedure is useful for those situations where a bearing must fit a previously determined shaft diameter or fit in a restricted space. In these circumstances, the bearing size cannot be varied appreciably, and the machine designer is interested in knowing the minimum probable life that a given size of bearing will have. Use this procedure whenever the bearing size is approximately predetermined by the installation conditions in motors, pumps, engines, portable tools, etc. Obtain the dynamic capacity of any bearing under consideration from the manufacturer's engineering data.

## ROLLER-BEARING CAPACITY REQUIREMENTS

A machine must be fitted with a roller bearing that will operate at least 30,000 h without failure. Select a suitable bearing for this machine in which the shaft operates at 3600 r/min and carries a radial load of 5000 lb (22,241.1 N).

### Calculation Procedure:

#### 1. *Determine the bearing life in revolutions*

Use the relation $R_L = 60Rh/10^6$, where the symbols are the same as in the previous calculation procedure. Thus, $R_L = 60(3600)(30{,}000)/10^6$; $R_L$ = 6480 million revolutions (Mr).

#### 2. *Determine the required dynamic capacity of the bearing*

Use the relation $R_L = (C/L)^{10/3}$, where the symbols are the same as in the previous calculation procedure. So $C = L(R_L)^{3/10} = (5000)(6480)^{3/10} = 69{,}200$ lb (307,187.0 N).

Choose a bearing of suitable bore having a dynamic capacity of 69,200 lb (307,187.0 N) or more. Thus, a typical 5.9055-in (15.0-cm) bore roller bearing has a dynamic capacity of 72,400 lb (322,051.3 N). It is common practice to undercut the shaft to suit the bearing bore, if such a reduction in the shaft does not weaken the shaft. Use the manufacturer's engineering data in choosing the actual bearing to be used.

**Related Calculations:** This procedure shows a situation in which the life of the bearing is of greater importance than its size. Such a situation is common when the reliability of a machine is a key factor in its design. A dynamic rating of a given amount, say 72,400 lb (322,051.3 N), means that if in a large group of bearings of this size each bearing has a 72,400-lb (322,051.3-N) load applied to it, 90 percent of the bearings in the group will complete, or exceed, $10^6$ r before the first evidence of fatigue occurs. This average life of the bearing is the number of revolutions that 50 percent of the bearings will complete, or exceed, before the first evidence of fatigue develops. The average life is about 3.5 times the minimum life.

Use this procedure to choose bearings for motors, engines, turbines, portable tools, etc. Where extreme reliability is required, some designers choose a bearing having a much larger dynamic capacity than calculations show is required.

## RADIAL LOAD RATING FOR ROLLING BEARINGS

A mounted rolling bearing is fitted to a shaft driven by a 4-in (10.2-cm) wide double-ply leather belt. The shaft is subjected to moderate shock loads about one-third of the time while operating at 300 r/min. An operating life of 40,000 h is required of the bearing. What is the required radial capacity of the bearing? The bearing has a normal rated life of 15,000 h at 500 r/min. The weight of the pulley and shaft is 145 lb (644.9 N).

### Calculation Procedure:

**1. *Determine the bearing operating factors***

To determine the required radial capacity of a rolling bearing, a series of operating factors must be applied to the radial load acting on the shaft: life factor $f_L$, operating factor $f_O$, belt tension factor $f_B$, and speed factor $f_S$. Obtain each of the four factors from the manufacturer's engineering data because there may be a slight variation in the factor value between different bearing makers. Where a given factor does not apply to the bearing being considered, omit it from the calculation.

**2. *Determine the bearing life factor***

Rolling bearings are normally rated for a certain life, expressed in hours. If a different life for the bearing is required, a life factor must be applied. The bearing being considered here has a normal rated life of 15,000 h. The manufacturer's engineering data show that for a mounted bearing which must have a life of 40,000 h, a life factor $f_L = 1.340$ should be used. For this particular make of bearing, $f_L$ varies from 0.360 at a 500-h to 1.700 at a 100,000-h life for mounted units. At 15,000 h, $f_L = 1.000$.

**3. *Determine the bearing operating factor***

A rolling-bearing operating factor is used to show the effect of peak and shock loads on the bearing. Usual operating factors vary from 1.00 for steady loads with any amount of overload to 2.00 for bearings with heavy shock loads throughout their operating period. For this bearing with moderate shock loads about one-third of the time, $f_O = 1.32$.

A combined operating factor, obtained by taking the product of two applicable factors, is used in some circumstances. Thus, when the load is an oscillating type, an additional factor of 1.25 must be applied. This type of load occurs in certain linkages and pumps. When the outer race of the bearing revolves, as in sheaves, truck wheels, or gyrating loads, an additional factor of 1.2 is used. To find the combined operating factor, first find the normal operating factor, as described earlier. Then take the product of the normal and the additional operating factors. The result is the combined operating factor.

**4. *Determine the bearing belt-tension factor***

When the bearing is used on a belt-driven shaft, a belt-tension factor must be applied. Usual values of this factor range from 1.0 for a chain drive to 2.30 for a single-ply leather belt. For a double-ply leather belt, $f_B = 2.0$.

**5. *Determine the bearing speed factor***

Rolling bearings are rated at various speeds. When the shaft operates at a speed different from the rated speed, a speed factor $f_S$ must be applied. Since this shaft operates at 300 r/min while the rated speed of the bearing is 500 r/min, $f_S = 0.860$, from the manufacturer's engineering data. For a 500-r/min bearing, $f_S$ varies from 0.245 at 5 r/min to 1.87 at 4000 r/min.

**6. *Determine the radial load on the bearing***

The radial load produced by a leather belt can vary from 130 lb/in (227.7 N/cm) of width for normal-tension belts to 450 lb/in (788.1 N/cm) of width for very tight belts. Assuming normal tension, the radial load for a double-ply leather belt is, from engineering data, 180 lb/in (315.2 N/cm) of width. Since this belt is 4 in (10.2 cm) wide, the radial belt load = 4(180) = 720 lb (3202.7 N). The total radial load $R_T$ is the sum of the belt, shaft, and pulley loads, or $R_T = 720 + 145 = 865$ lb (3847.7 N).

**7. *Compute the required radial capacity of the bearing***

The required radial capacity of a bearing $R_C = R_T f_L f_O f_B f_S = (865)(1.340)(1.32)(2.0)(0.86) = 2630$ lb (11,698.8 N).

**8. *Select a suitable bearing***

Enter the manufacturer's engineering data at the shaft rpm (300 r/min for this shaft) and project to a bearing radial capacity equal to, or slightly greater than, the computed required radial capacity. Thus, one make of bearing, suitable for 2- and 2³⁄₁₆-in (5.1- and 5.6-cm) diameter shafts, has a radial capacity of 2710 at 300 r/min. This is close enough for general selection purposes.

**Related Calculations:** Use this general procedure for any type of rolling bearing. When comparing different makes of rolling bearings, be sure to convert them to the same life expectancies before making the comparison. Use the life-factor table presented in engineering handbooks or a manufacturer's engineering data for each bearing to convert the bearings being considered to equal lives.

## ROLLING-BEARING CAPACITY AND RELIABILITY

What is the required basic load rating of a ball bearing having an equivalent radial load of 3000 lb (13,344.7 N) if the bearing must have a life of $400 \times 10^6$ r at a reliability of 0.92? The ratio of the average life to the rating life of the bearing is 5.0. Show how the required basic load rating is determined for a roller bearing.

### Calculation Procedure:

**1. *Compute the required basic load rating***

Use the Weibull two-parameter equation, which for a life ratio of 5 is $L_e/L_B = (1.898/R_L^{0.333})(\ln 1/R_e)^{0.285}$, where $L_e$ = equivalent radial load on the bearing, lb; $L_B$ = the required basic load rating of the bearing, lb, to give the desired reliability at the stated life; $R_L$ = bearing operating life, Mr; ln = natural or Naperian logarithm to the base $e$; $R_e$ = required reliability, expressed as a decimal.

Substituting gives $3000/L_B = (1.898/400^{0.333})(\ln 1/0.92)^{0.285}$; $L_B = 23{,}425$ lb (104,199.6 N). Thus, a bearing having a basic load rating of at least 23,425 lb (104,199.6 N) would provide the desired reliability. Select a bearing having a load rating equal to, or slightly in excess of, this value. Use the manufacturer's engineering data as the source of load-rating data.

**2. *Compute the roller-bearing basic load rating***

Use the following form of the Weibull equation for roller bearings: $L_e/L_B = (1.780/R_L^{0.30})(\ln 1/R_e)^{0.257}$. Substitute values and solve for $L_B$, as in step 1.

**Related Calculations:** Use the Weibull equation as given here when computing bearing life in the range of 0.9 and higher. The ratio of average life/rating life = 5 is usual for commercially available bearings.[1]

## POROUS-METAL BEARING CAPACITY AND FRICTION

Determine the load capacity $\psi$ and coefficient of friction of a porous-metal bearing for a 1-in (2.5-cm) diameter shaft, 1-in (2.5-cm) bearing length, 0.2-in (0.5-cm) thick bearing, 0.001-in (0.003-cm) radial clearance, metal permeability $\phi = 5 \times 10^{-10}$ in ($1.3 \times 10^{-9}$ cm), shaft speed = 1500 r/min, eccentricity ratio $\epsilon = 0.8$, and an SAE-30 mineral-oil lubricant with a viscosity of $6 \times 10^{-6}$ lb·s/in² ($4.1 \times 10^{-6}$ N·s/cm²).

### Calculation Procedure:

**1. *Sketch the bearing and shaft***

Figure 10 shows the bearing and shaft with the various known dimensions indicated by the identifying symbols given above.

[1]C. Mischke, "Bearing Reliability and Capacity," *Machine Design*, Sept. 30, 1965.

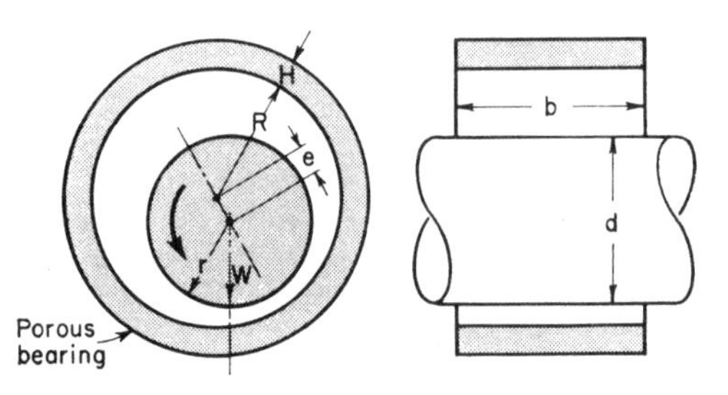


**FIG. 10** Typical porous-metal bearing. *(Product Engineering.)*

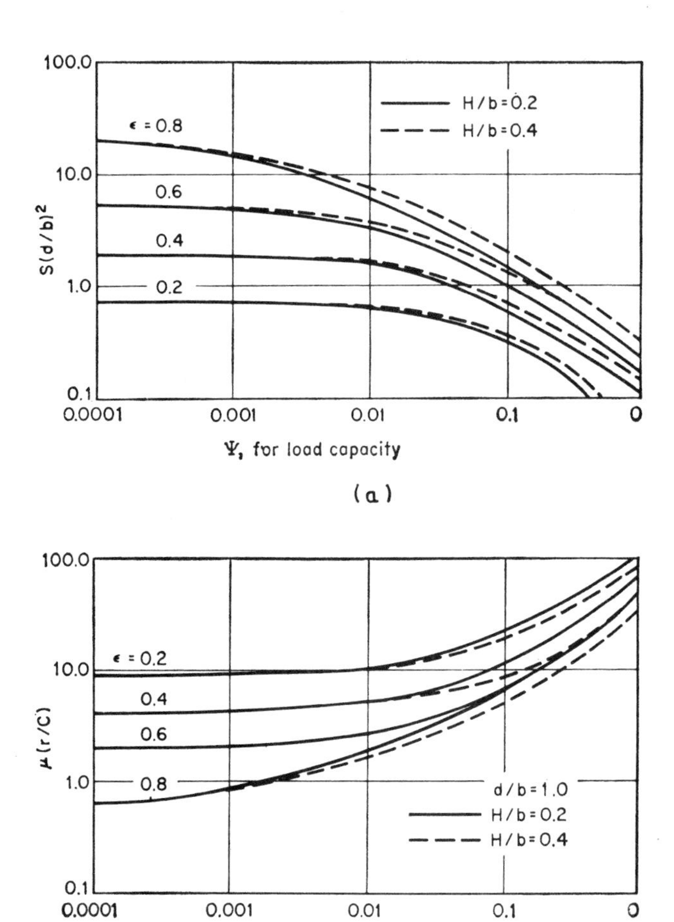


**FIG. 11** (*a*) Bearing load capacity factors; (*b*) bearing friction factors. *(Product Engineering.)*

### 2. *Compute the load capacity factor*

The load capacity factor $\psi = \phi H/C^3$, where $\phi$ = metal permeability, in; $H$ = bearing thickness, in; $C$ = radial clearance of bearing, $R_b - r$ = 0.001 in (0.003 cm) for this bearing. Hence, $\psi = (5 \times 10^{-10})(0.2)/(0.001)^3 = 0.10$.

### 3. *Compute the bearing thickness-length ratio*

The thickness-length ratio = $H/b = 0.2/1.0 = 0.2$.

### 4. *Determine the S(d/b)² value for the bearing*

In the $S(d/b)^2$ value for the bearing, $S$ = the Summerfeld number for the bearing; the other values are as shown in Fig. 10.

Using the $\psi$, $\epsilon$, and $H/b$ values, enter Fig. 11*a*, and read $S(b/d)^2 = 1.4$ for an eccentricity ratio of $\epsilon = e/C = 0.8$. Substitute in this equation the known values for $b$ and $d$ and solve for $S$, or $S(1.0/1.0)^2 = 1.4$; $S = 1.4$.

### 5. *Compute the bearing load capacity*

Find the bearing load capacity from $S = (L/R_i \eta b)(C/r)^2$, where $\eta$ = lubricant viscosity, lb·s/in²; $r$ = shaft radius, in; $R_i$ = shaft velocity, in·s. Solving gives $L = (SR_i \eta b)/(C/r)^2 = (1.4)(78.5)(6 \times 10^{-6})(1.0)/(0.001/0.5)^2 = 164.7$ lb (732.6 N). (The shaft rotative velocity must be expressed in in/s in this equation because the lubricant viscosity is given in lb·s/in².)

### 6. *Determine the bearing coefficient of friction*

Enter Fig. 11*b* with the known values of $\psi$, $\epsilon$, and $H/b$ and read $u(r/C) = 7.4$. Substitute in this equation the known values for $r$ and $C$, and solve for the bearing coefficient of friction $\mu$, or $\mu = 7.4C/r = (7.4)(0.001)/0.5 = 0.0148$.

**Related Calculations:** Porous-metal bearings are similar to conventional sliding-journal bearings except that the pores contain an additional supply of lubricant to replace that which may be lost during operation. The porous-metal bearing is useful in assemblies where there is not enough room for a conventional lubrication system or where there is a need for improved lubrication during the starting and stopping of a machine. The permeability of the finished porous metal greatly influences the ability of a lubricant to work its way through the pores. Porous-metal bearings are used in railroad axle supports, water pumps, generators, machine tools, and other equipment. Use the procedure given here when choosing porous-metal bearings for any of these applications. The method given here was developed by Professor W. T. Rouleau, Carnegie Institute of Technology, and C. A. Rhodes, Senior Research Engineer, Jet Propulsion Laboratory, California Institute of Technology.

## HYDROSTATIC THRUST BEARING ANALYSIS

An oil-lubricated hydrostatic thrust bearing must support a load of 107,700 lb (479,073.5 N). This vertical bearing has an outside diameter of 16 in (40.6 cm) and a recess diameter of 10 in (25.4 cm). What oil pressure and flow rate are required to maintain a 0.006-in (0.15-mm) lubricant film thickness with an SAE-20 oil having an absolute viscosity of $\eta = 42.4 \times 10^{-7}$ lb/(s·in²) [$2.9 \times 10^{-6}$ N/(s·cm²)] if the shaft turns at 750 r/min? What are the pumping loss and the viscous friction loss? What is the optimum lubricant-film thickness?

## Calculation Procedure:

### 1. *Determine the required lubricant-supply pressure*

The design equations and methods developed at Franklin Institute by Dudley Fuller, Professor of Mechanical Engineering, Columbia University, are applicable to vertical hydrostatic bearings, Fig. 12, using oil, grease, or gas lubrication. By substituting the appropriate value for the lubricant viscosity, the same set of design equations can be used for any of the lubricants listed above. These equations are accurate, simple, and reliable; they are therefore used here.

Solve Fuller's applied load equation, $L = (p_i \pi/2)\{r^2 - r_i^2/[\ln(r/r_i)]\}$, for the lubricant-supply inlet pressure. In this equation, $L$ = applied load on the bearing, lb; $p_i$ = lubricant-supply inlet pressure, lb/in², $r$ = shaft radius, in; $r_i$ = recess or step radius, in; ln = natural or Naperian logarithm to the base $e$, Solving gives $p_i = 2L/\pi\{r^2 - r_i^2/[\ln(r/r_i)]\} = 2(107{,}700)/\pi[8^2 - 5^2/(\ln 8/5)]$, or $p_i = 825$ lb/in² (5688.4 kPa).

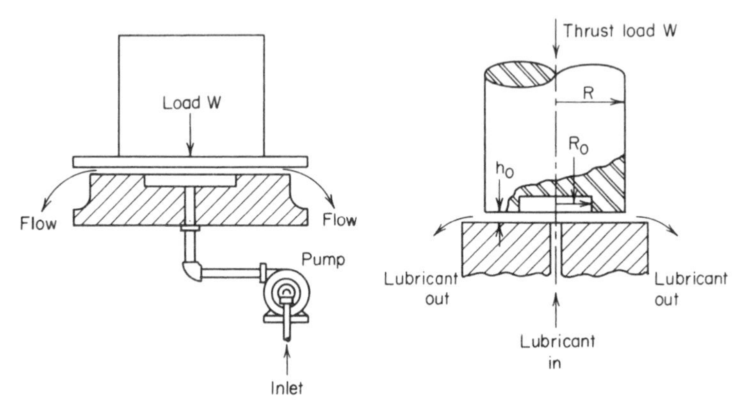


**FIG. 12** Hydrostatic thrust bearings. *(Product Engineering.)*

### 2. *Compute the required lubricant flow rate*

By Fuller's flow-rate equation, $Q = p_i \pi h^3/[6\eta \ln (r/r_i)]$, where $Q$ = lubricant flow rate, in$^3$/s; $h$ = lubricant-film thickness, in; $\eta$ = lubricant absolute viscosity, lb·s/in$^2$; other symbols as in step 1. Thus, with $h = 0.006$ in, $Q = (825\ \pi)(0.006)^3/[6(42.4 \times 10^{-6})(0.470)] = 46.85$ in$^3$/s (767.7 cm$^3$/s).

### 3. *Compute the pumping loss*

The pumping loss results from the work necessary to force the lubricant radially outward through the film space, or $H_p = Q(p_i - p_o)$, where $H_p$ = power required to pump the lubricant = pumping loss, in·lb/s; $p_o$ = lubricant outlet pressure, lb/in$^2$; other symbols as in step 1. For circular thrust bearings it can be assumed that the lubricant outlet pressure $p_0$ is negligible, or $p_0$ = 0. Then $H_p = 46.85(825 - 0) = 38{,}680$ in·lb/s (4370.3 N·m/s) = 38,680/[550 ft·lb/(min·hp)](12 in/ft) = 5.86 hp (4.4 kW).

### 4. *Compute the viscous friction loss*

The viscous-friction-loss equation developed by Fuller is $H_f = [(R^2\eta/(58.05h)](r^4 - r_0^4)$, where $H_f$ = viscous friction loss, in·lb/s; $R$ = shaft rpm; other symbols as in step 1. Thus, $H_f = \{(750)^2(42.4 \times 10^{-7})/[(58.05)(0.006)]\}(8^4 - 5^4)$, or $H_f = 23{,}770$ in·lb/s (2685.6 N·m/s) = 23,770/[(550)(12)] = 3.60 hp (2.7 kW).

### 5. *Compute the optimum lubricant-film thickness*

The film thickness that will produce a minimum combination of pumping loss and friction loss can be evaluated by determining the minimum point of the curve representing the sum of the respective energy losses (pumping and viscous friction) when plotted against film thickness.

With the shaft speed, lubricant viscosity, and bearing dimensions constant at the values given in the problem statement, the viscous-friction loss becomes $H_f = 0.0216/h$ for this bearing. Substitute various values for $h$ ranging between 0.001 and 0.010 in (0.0254 and 0.254 mm) (the usual film thickness range), and solve for $H_f$. Plot the results as shown in Fig. 13.

Combine the lubricant-flow and pumping-loss equations to express $H_p$ in terms of the lubricant-film thickness, or $H_p = (1000h)^3/36.85$, for this bearing with a pump having an efficiency of 100 percent. For a pump with a 50 percent efficiency, this equation becomes $H_p = (1000h)^3/18.42$. Substitute various values of $h$ ranging between 0.001 and 0.010, and plot the results as in Fig. 13 for pumps with 100 and 50 percent efficiencies, respectively. Figure 13 shows that for a 100 percent efficient pump, the minimum total energy loss occurs at a film thickness of 0.004 in (0.102 mm). For 50 percent efficiency, the minimum total energy loss occurs at 0.0035-in (0.09-mm) film thickness, Fig. 13.

**Related Calculations:** Similar equations developed by Fuller can be used to analyze hydrostatic thrust bearings of other configurations. Figures 14 and 15 show the equations for modified square bearings and circular-sector bearings. To apply these equations, use the same general pro-

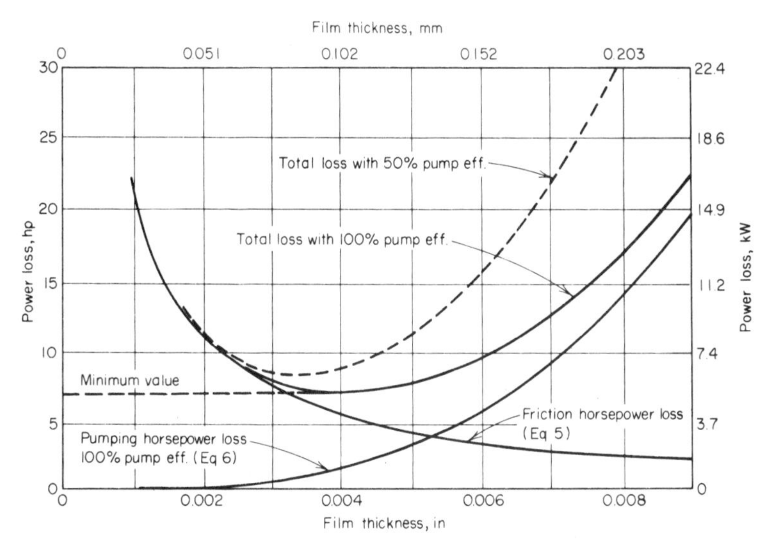


**FIG. 13** Oil-firm thickness for minimum power loss in a hydrostatic thrust bearing. *(Product Engineering.)*

cedures shown above. Note, however, that each equation uses a factor $K$ obtained from the respective design chart.

Also note that a hydrostatic bearing uses an externally fed pressurized fluid to keep two bearing surfaces *completely* separated. Compared with hydrodynamic bearings, in which the pressure is self-induced by the rotation of the shaft, hydrostatic bearings have (1) lower friction, (2) higher load-carrying capacity, (3) a lubricant-film thickness insensitive to shaft speed, (4) a higher spring constant, which leads to a self-centering effect, and (5) a relatively thick lubricant film permitting cooler operation at high shaft speeds. Hydrostatic bearings are used in rolling mills, instruments, machine tools, radar, telescopes, and other applications.

## HYDROSTATIC JOURNAL BEARING ANALYSIS

A 4.000-in (10.160-cm) metal shaft rests in a journal bearing having an internal diameter of 4.012 in (10.190 cm). The lubricant is SAE-30 oil at 100°F (37.8°C) having a viscosity of $152 \times 10^{-7}$ reyn. This lubricant is supplied under pressure through a groove at the lowest point in the bearing. The length of the bearing is 6 in (15.2 cm), the length of the groove is 3 in (7.6 cm), and the load on the bearing is 3600 lb (16,013.6 N). What lubricant-inlet pressure and flow rate are required to raise the shaft 0.002 in (0.051 mm) and 0.004 in (0.102 mm)?

### Calculation Procedure:

#### 1. *Determine the radial clearance and clearance modulus*

The design equations and methods developed at Franklin Institute by Dudley Fuller, Professor of Mechanical Engineering, Columbia University, are applicable to hydrostatic journal bearings using oil, grease, or gas lubrication. These equations are accurate, simple, and reliable; therefore they are used here.

By Fuller's method, the radial clearance $c$, in $= r_b - r_s$, Fig. 16, where $r_b$ = bearing internal radius, in; $r_s$ = shaft radius, in. Or, $c = (4.012/2) - (4.000/2) = 0.006$ in (0.152 mm).

Next, compute the clearance modulus $m$ from $m = c/r_s = 0.006/2 = 0.003$ in/in (0.003 cm/cm). Typical values of $m$ range from 0.005 to 0.003 in/in (0.005 to 0.003 cm/cm) for hydrostatic journal bearings.

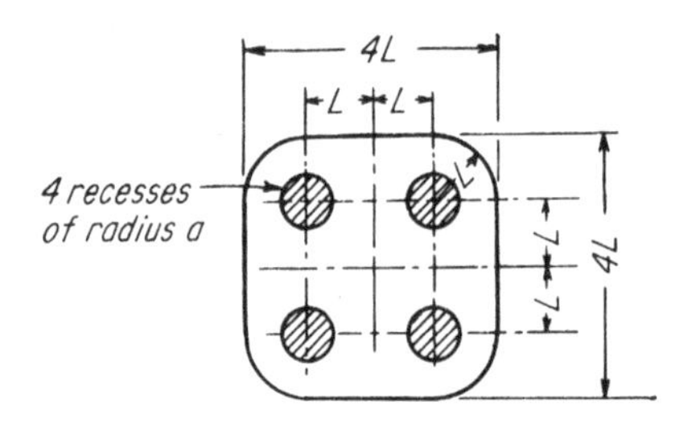


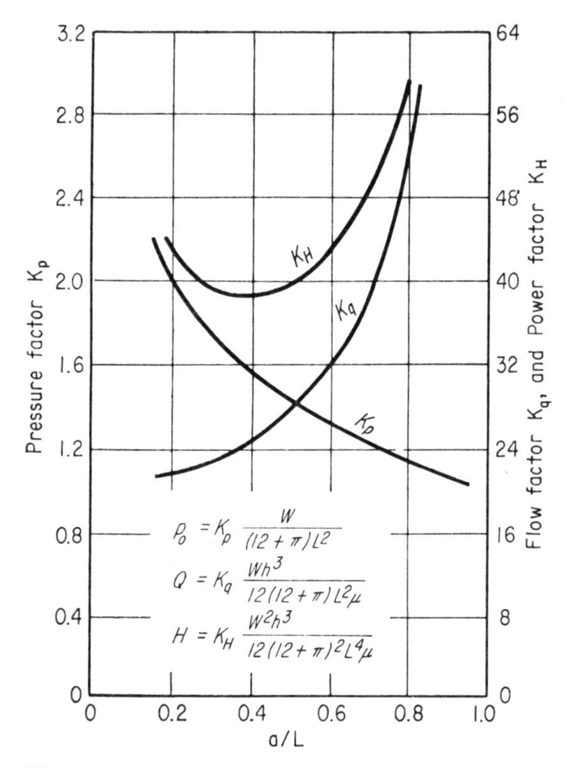


**FIG. 14** Constants and equations for modified square hydrostatic bearings. *(Product Engineering.)*

### 2. *Compute the shaft eccentricity in the clearance space*

The numerical parameter used to describe the eccentricity of the shaft in the bearing clearance space is the ratio $\epsilon = 1 - h/(mr)$, where $h$ = shaft clearance, in, during operation. With a clearance of $h = 0.002$ in (0.051 mm), $\epsilon = 1 - [0.002/(0.003 \times 2)] = 0.667$. With a clearance of $h = 0.004$ in (0.102 mm), $\epsilon = 1 - [0.004/(0.003 \times 2)] = 0.333$.

### 3. *Compute the eccentricity constants*

The eccentricity constant $A_k = 12[2 - \epsilon/(1 - \epsilon)^2] = 12[2 - 0.667/(1 - 0.667)^2] = 144.6$. A second eccentricity constant $B$ is given by $B_k = 12\{\epsilon(4 - \epsilon^2)/[2(1 - \epsilon^2)^2] + 2 + \epsilon^2/(1 - \epsilon^2)^{2.5} \times \arctan[1 + \epsilon/(1 - \epsilon^2)^{0.5}]\}$. Since this relation is awkward to handle, Fig. 17 was developed by Fuller. From Fig. 17, $B_k = 183$.

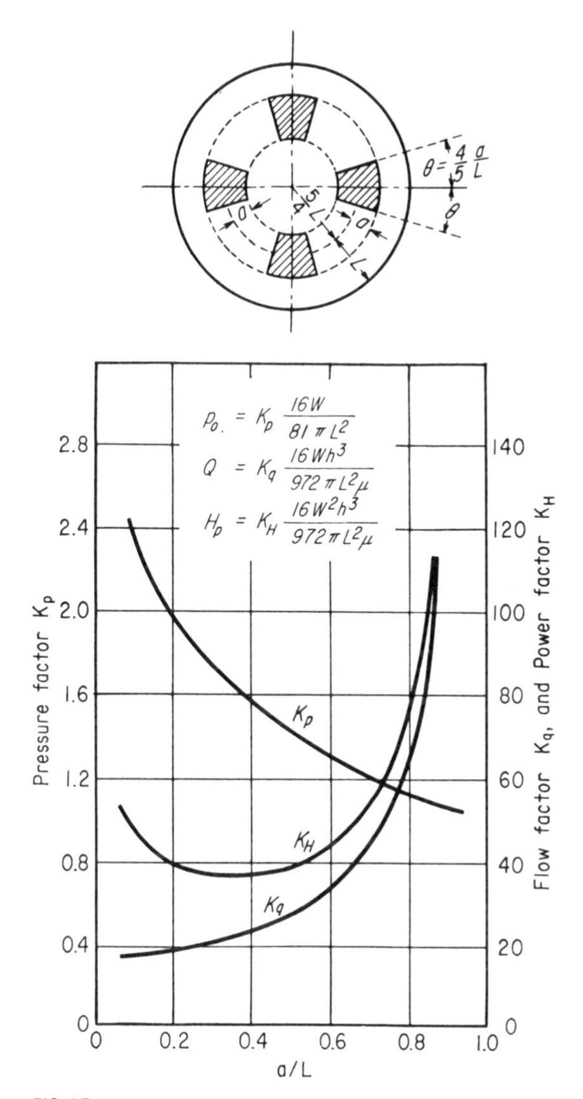


**FIG. 15** Constants and equations for circular-sector hydrostatic bearings. *(Product Engineering.)*

### 4. *Compute the required lubricant flow rate*

The lubricant flow rate is found from $Q$ in$^3$/s $= 2Lm^3r_s/(\eta A_k)$, where $L$ = load acting on shaft, lb; $\eta$ = lubricant viscosity, reyns. For the bearing with $h$ = 0.002 in (0.051 mm), $Q$ = 2(3600)(0.003)$^3$(2)/[(152 × 10$^{-7}$)(144.6)] = 0.177 in$^3$/s (2.9 cm$^3$/s), or 0.0465 gal/min (2.9 mL/s).

### 5. *Compute the required lubricant-inlet pressure*

The lubricant-inlet pressure is found from $p_i = \eta QB/(2bm^3r_s^2)$, where $p_i$ = lubricant inlet pressure, lb/in$^2$; other symbols as before. Thus, $p_i$ = (152 × 10$^{-7}$)(0.177)(183)/[(2)(3)(0.003)$^3$(2)$^2$] = 759 lb/in$^2$ (5233.3 kPa).

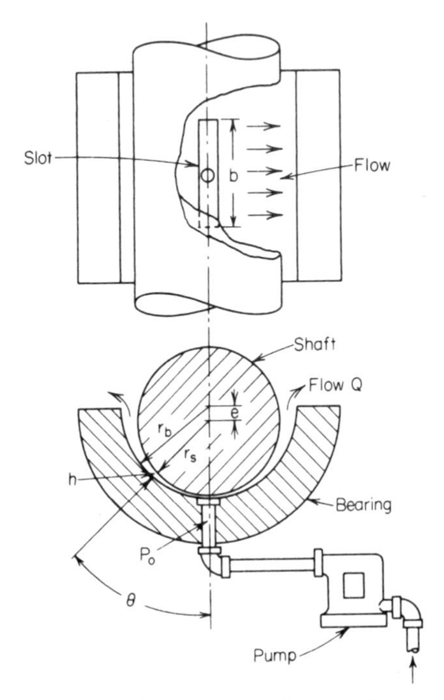


**FIG. 16** Hydrostatic journal bearing. *(Product Engineering.)*

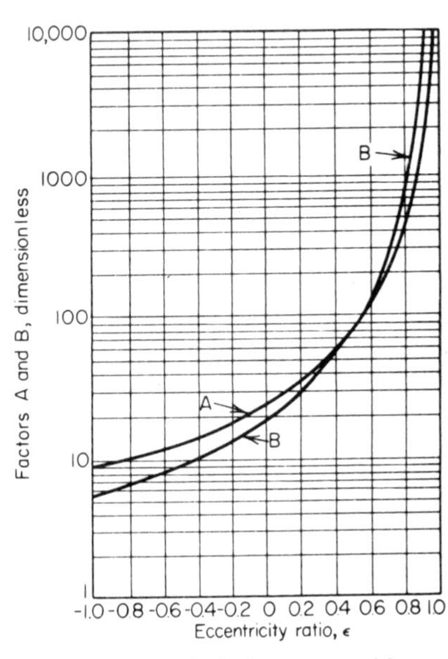


**FIG. 17** Constants for hydrostatic journal bearing oil flow and load capacity. *(Product Engineering.)*

### 6. *Analyze the larger-clearance bearing*

Use the same procedure as in steps 1 through 5. Then $\epsilon = 0.333$; $A = 45.0$; $B = 42.0$; $Q = 0.560$ in$^3$/s (9.2 cm$^3$/s) = 0.1472 gal/min (9.2 mL/s); $p_i = 551$ lb/in$^2$ (3799.1 kPa).

**Related Calculations:** Note that the closer the shaft is to the center of the bearing, the smaller the lubricant pressure required and the larger the oil flow. If the larger flow requirements can be met, the design with the thicker oil film is usually preferred, because it has a greater ability to absorb shock loads and tolerate thermal change.

Use the general design procedure given here for any applications where a hydrostatic journal bearing is applicable and there is no thrust load.

## HYDROSTATIC MULTIDIRECTION BEARING ANALYSIS

Determine the lubricant pressure and flow requirements for the multidirection hydrostatic bearing shown in Fig. 18 if the vertical coplanar forces acting on the plate are 164,000 lb (729,508.4 N) upward and downward, respectively. The lubricant viscosity $\eta = 393 \times 10^{-7}$ reyn film thickness = 0.005 in (0.127 mm), $L = 7$ in (17.8 cm), $a = 3.5$ in (8.9 cm). What would be the effect of decreasing the film thickness $h$ on one side of the plate by 0.001-in (0.025-mm) increments from 0.005 to 0.002 in (0.127 to 0.051 mm)? What is the bearing stiffness?

### Calculation Procedure:

### 1. *Compute the required lubricant-inlet pressure*

By Fuller's method, Fig. 19 shows that the bearing has four pressure pads to support the plate loads. Figure 19 also shows the required pressure, flow, and power equations, and the appropriate constants for these equations. The inlet-pressure equation is $p_i = K_p L_s/(16L^2)$, where $p_i$ =

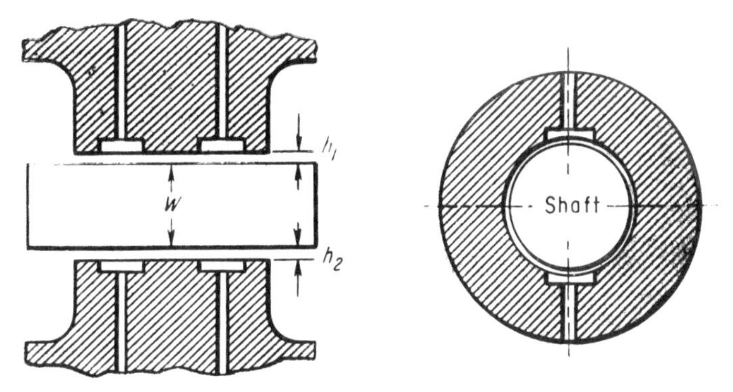


**FIG. 18** Double-acting hydrostatic thrust bearing. *(Product Engineering.)*

required lubricant inlet pressure, lb/in$^2$; $K_p$ = pressure constant from Fig. 19; $L_s$ = plate load, lb; $L$ = bearing length, in.

Find $K_p$ from Fig. 19 after setting up the ratio $a/L = 3.5/7.0 = 0.5$, where $a$ = one-half the pad length, in. Then $K_p = 1.4$. Hence $p_i = (1.4)(164{,}000)/[(16)(7)^2] = 293$ lb/in$^2$ (2020.2 kPa).

## 2. *Compute the required lubricant flow rate*

From Fig. 19, the required lubricant flow rate $Q = K_q L_s h^3/(192L^2\eta)$, where $K_q$ = flow constant; $h$ = lubricant-film thickness, in; other symbols as before. For $a/L = 0.5$, $K_q = 36$. Then $Q = (36)(164{,}000)(0.005)^3/[(192)(7)^2(393 \times 10^{-7})] = 1.99$ in$^3$/s (32.6 cm$^3$/s). This can be rounded off to 2.0 in$^3$/s (32.8 cm$^3$/s) for usual design calculations.

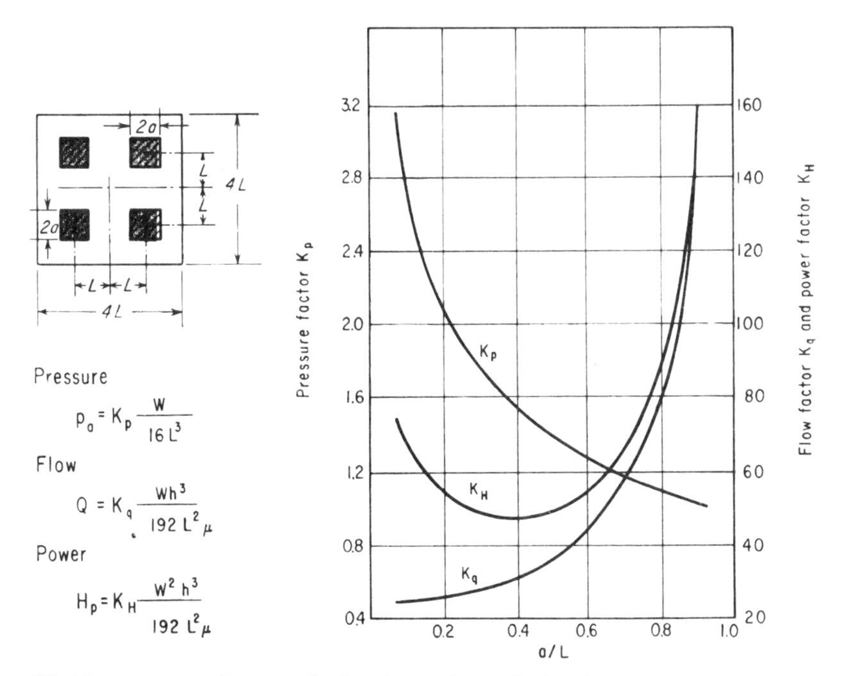


**FIG. 19** Dimensions and equations for thrust-bearing design. *(Product Engineering.)*

**TABLE 43** Load Capacity of Dual-Direction Bearing

| Film thickness, in (cm) | | Inlet pressure, lb/in$^2$ (kPa) | | Load, lb (N) | | Load capacity, lb (N) |
|---|---|---|---|---|---|---|
| $h_2$ | $h_1$ | $p_{i2}$ | $p_{i1}$ | $L_{s2}$ | $L_{s1}$ | $L_{s2} - L_{s1}$ |
| 0.005 | 0.005 | 293 | 293 | 164,000 | 164,000 | 0 |
| (0.127) | (0.127) | (2,020.2) | (2,020.0) | (729,508.4) | (729,508.4) | (0.0) |
| 0.004 | 0.006 | 571 | 170 | 320,000 | 95,000 | 225,000 |
| (0.102) | (0.152) | (3,937.0) | (1,172.2) | (1,423,431.0) | (422,581.1) | (1,000,850.0) |
| 0.003 | 0.007 | 1,360 | 106 | 760,000 | 59,700 | 700,300 |
| (0.076) | (0.178) | (9,377.2) | (730.9) | (3,380,648.7) | (265,558.9) | (3,115,089.9) |
| 0.002 | 0.008 | 4,570 | 71 | 2,560,000 | 40,000 | 2,520,000 |
| (0.051) | (0.203) | (31,510.2) | (489.5) | (11,387,448.3) | (177,928.9) | (11,209,519.4) |

### 3. *Compute the pressure and load for other plate clearances*

The sum of the plate lubricant-film thicknesses, $h_1 + h_2$, Fig. 18, is a constant. For this bearing, $h_1 + h_2 = 0.005 + 0.005 = 0.010$ in (0.254 mm). With no load on the plate, if oil is pumped into both bearing faces at the rate of 2.0 in$^3$/s (32.8 cm$^3$/s), the maximum recess pressure will be 293 lb/in$^2$ (2020.2 kPa). The force developed on each face will be 164,000 lb (729,508.4 N). Since the lower face is pushed up with this force and the top face is pushed down with the same force, the net result is zero.

With a downward external load imposed on the plate such that the lower film thickness $h_2$ is reduced to 0.004 in (0.102 mm), the upper film thickness will become 0.006 in (0.152 mm), since $h_1 + h_2 = 0.010$ in (0.254 mm) = a constant for this bearing. If the lubricant flow rate is held constant at 2.0 in$^3$/s (32.8 cm$^3$/s), then $K_q = 36$ from Fig. 19, since $a/L = 0.5$. With these constants, the load equation becomes $L_s = 0.0205/h^3$, and the inlet-pressure equation becomes $p_i = L_s/560$.

Using these equations, compute the upper and lower loads and inlet pressures for $h_2$ and $h_1$ ranging from 0.004 to 0.002 and 0.006 to 0.008 in (0.102 to 0.051 and 0.152 to 0.203 mm), respectively. Tabulate the results as shown in Table 43. Note that the allowable load is computed by using the respective film thickness for the lower and upper parts of the plate. The same is true of the lubricant-inlet pressure, except that the corresponding load is used instead.

Thus, for $h_2 = 0.004$ in (0.102 mm), $L_{s2} = 0.0205/(0.004)^3 = 320{,}000$ lb (1,423,431.0 N). Then $p_{i2} = 320{,}000/560 = 571$ lb/in$^2$ (3937.0 kPa). For $h_k = 0.006$ in (0.152 mm), $L_{s1} = 0.0205/(0.006)^3 = 95{,}000$ lb (422,581.1 N). Then $p_{i1} = 95{,}000/560 = 169.6$, say 170, lb/in$^2$ (1172.2 kPa).

The load difference $L_{s2} - L_{s1}$ = the bearing load capacity. For the film thicknesses considered above, $L_{s2} - L_{s1} = 320{,}000 - 95{,}000 = 225{,}000$ lb (1,000,850.0 N). Load capacities for various other film thicknesses are also shown in Table 43.

### 4. *Determine the bearing stiffness*

Plot the net load capacity of this bearing vs. the lower film thickness, Fig. 20. A tangent to the curve at any point indicates the stiffness of this bearing. Draw a tangent through the origin where $h_2 = h_1 = 0.005$ in (0.127 mm). The slope of this tangent = vertical value/horizontal value = $(725{,}000 - 0)/(0.005 - 0.002) = 241{,}000{,}000$ lb/in (42,205,571 N/m) = the bearing stiffness. This means that a load of 241,000,000 lb (1,072,021,502 N) would be required to displace the plate 1.0 in (2.5 cm). Since the plate cannot move this far, a load of 241,000 lb (1,072,021.5 N) would move the plate 0.001 in (0.025 mm).

If the lubricant flow rate to each face of the bearing were doubled, to $Q = 4.0$ in$^3$/s (65.5 cm$^3$/s), the stiffness of the bearing would increase to 333,000,000 lb/in (58,317,241 N/m), as shown in Fig. 20. This means that an additional load of 333,000 lb (1,481,257.9 N) would displace the plate 0.001 in (0.025 mm). The stiffness of a hydrostatic bearing can be controlled by suitable design, and a wide range of stiffness values can be designed into the bearing system.

**Related Calculations:** Hydrostatic bearings of the design shown here are useful for a variety

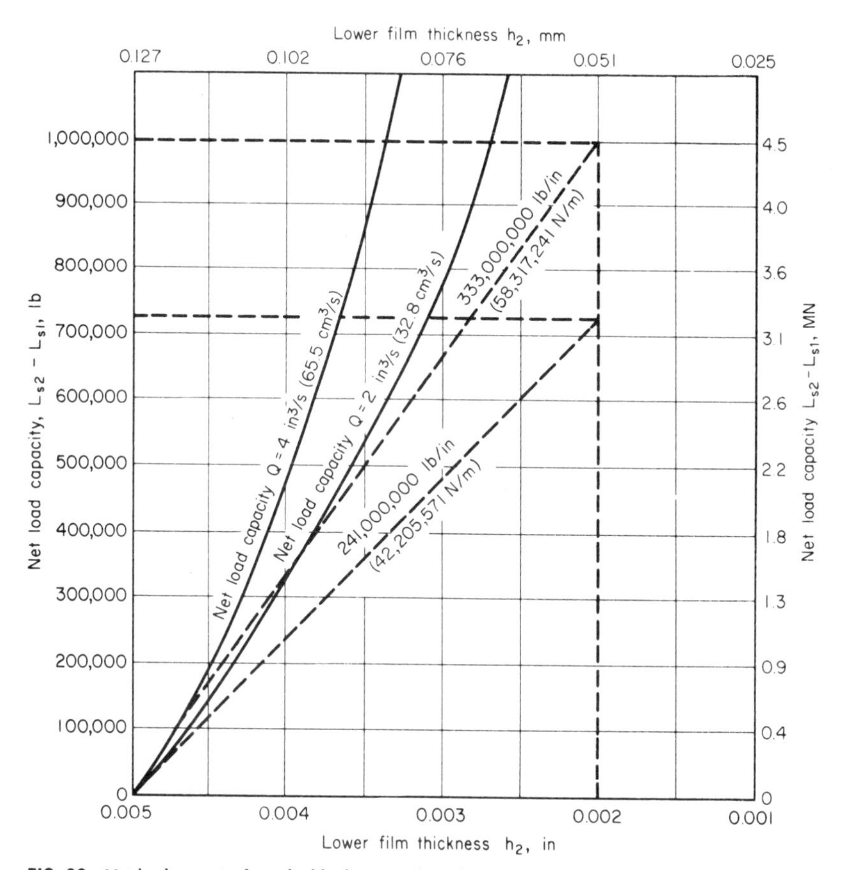


**FIG. 20** Net load capacity for a double-direction thrust bearing. *(Product Engineering.)*

of applications. Journal bearings for multidirectional loads are analyzed in a manner similar to that described here.

## LOAD CAPACITY OF GAS BEARINGS

Determine the load capacity and bearing stiffness of a hydrostatic air bearing, using 70°F (21.1°C) air if the bearing orifice radius = 0.0087 in (0.0221 cm), the radial clearance $h = 0.0015$ in (0.038 mm), the bearing diameter $d = 3.00$ in (7.6 cm), the bearing length $L = 3$ in (7.6 cm), the total number of air orifices $N = 8$, the ambient air pressure $p_a = 14.7$ lb/in$^2$ (abs) (101.4 kPa), the air supply pressure $p_s = 15$ lb/in$^2$ (gage) = 29.7 lb/in$^2$ (abs) (204.8 kPa), the (gas constant)(total temperature) = $RT = 1.322 \times 10^8$ in$^2$/s$^2$ ($8.5 \times 10^8$ cm$^2$/s$^2$), $\epsilon$ = the eccentricity ratio = 0.30, air viscosity $\eta = 2.82 \times 10^{-9}$ lb·s/in$^2$ ($1.94 \times 10^{-9}$ N·s/cm$^2$), the orifice coefficient $\alpha = 0.63$, and the shaft speed $\omega = 2100$ rad/s = 20,000 r/min.

### Calculation Procedure:

#### 1. *Compute the bearing factors* $\Lambda$ *and* $\Lambda_T$

For a hydrostatic gas bearing, $\Lambda = (6\eta\omega/p_a)(d/zh)^2 = [(6)(2.82 \times 10^{-9})(2100)/14.7][3/(2 \times 0.0015)]^2 = 2.41$.

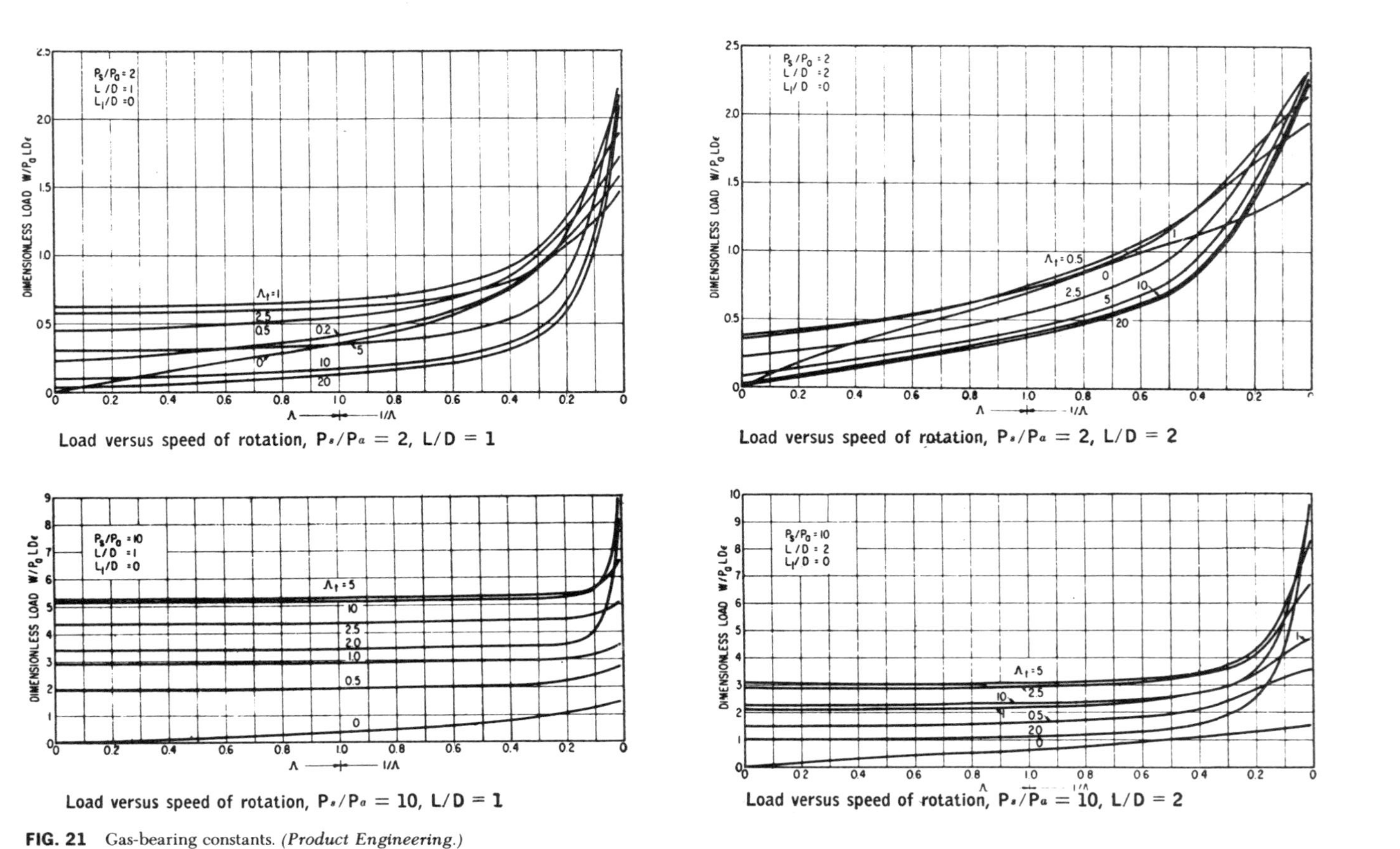


Load versus speed of rotation, $P_s/P_a = 2$, $L/D = 1$

Load versus speed of rotation, $P_s/P_a = 2$, $L/D = 2$

Load versus speed of rotation, $P_s/P_a = 10$, $L/D = 1$

Load versus speed of rotation, $P_s/P_a = 10$, $L/D = 2$

**FIG. 21** Gas-bearing constants. *(Product Engineering.)*

Also, $\Lambda_T = 6\eta Na^2\alpha(RT)^{0.5}/p_a h^3 = (6)(2.82 \times 10^{-9})(8)(0.0087)^2(0.63)(1.322 \times 10^8)^2/[(14.7)(0.0015)^3] = 1.5$.

**2. *Determine the dimensionless load***

Since $L/d = 3/3 = 1$, and $p_s/p_a = 29.7/14.7 \approx 2$, use the first chart[1] in Fig. 21. Before entering the chart, compute $1/\Lambda = 1/2.41 = 0.415$. Then, from the chart, the dimensionless load $= 0.92 = L_d$.

**3. *Compute the bearing load capacity***

The bearing load capacity $L_s = L_d p_a LD\epsilon = (0.92)(14.7)(3.00)(3.00)(0.3) = 36.5$ lb (162.4 N). If there were no shaft rotation, $\Lambda = 0$, $\Lambda_T = 1.5$, and $L_d = 0.65$, from the same chart. Then $L_s = L_d p_a LD\epsilon = (0.65)(14.7)(3.00)(3.00)(0.3) = 25.8$ lb (114.8 N). Thus, rotation of the shaft increases the load-carrying ability of the bearing by $[(36.5 - 25.8)/25.8](100) = 41.5$ percent.

**4. *Compute the bearing stiffness***

For a hydrostatic gas bearing, the bearing stiffness $B_s = L_s/(h\epsilon) = 36.5/[(0.0015)(0.3)] = 81{,}200$ lb/in (142,203.0 N/cm).

**Related Calculations:** Use this procedure for the selection of gas bearings where the four charts presented here are applicable. The data summarized in these charts result from computer solutions of the complex equations for "hybrid" gas bearings. The work was done by Mechanical Technology Inc., headed by Beno Sternlicht.

## SPRING SELECTION FOR A KNOWN LOAD AND DEFLECTION

Give the steps in choosing a spring for a known load and an allowable deflection. Show how the type and size of spring are determined.

### Calculation Procedure:

**1. *Determine the load that must be handled***

A spring may be required to absorb the force produced by a falling load or the recoil of a mass, to mitigate a mechanical shock load, to apply a force or torque, to isolate vibration, to support moving masses, or to indicate or control a load or torque. Analyze the load to determine the magnitude of the force that is acting and the distance through which it acts.

Once the magnitude of the force is known, determine how it might be absorbed—by compression or extension (tension) of a spring. In some applications, either compression or extension of the spring is acceptable.

**2. *Determine the distance through which the load acts***

The load member usually moves when it applies a force to the spring. This movement can be in a vertical, horizontal, or angular direction, or it may be a rotation. With the first type of movement, a *compression*, or *tension*, spring is generally chosen. With a torsional movement, a *torsion-type* spring is usually selected. Note that the movement in either case may be negligible (i.e., the spring applies a large restraining force), or the movement may be large, with the spring exerting only a nominal force compared with the load.

**3. *Make a tentative choice of spring type***

Refer to Table 44, entering at the type of load. Based on the information known about the load, make a tentative choice of the type of spring to use.

**4. *Compute the spring size and stress***

Use the methods given in the following calculation procedures to determine the spring dimensions, stress, and deflection.

**5. *Check the suitability of the spring***

Determine (*a*) whether the spring will fit in the allowable space, (*b*) the probable spring life, (*c*) the spring cost, and (*d*) the spring reliability. Based on these findings, use the spring chosen, if it

[1]"Gas Bearings," *Product Engineering*, July 8, 1963.

**TABLE 44** Metal Spring Selection Guide

| Type of load | Suitable spring type | Relative magnitude of load on spring | Deflection absorbed |
|---|---|---|---|
| Compression | Helical | Small to large | Small to large |
| | Leaf | Large | Moderate |
| | Flat | Small to large | Small to large |
| | Belleville | Small to large | Moderate |
| | Ring | Large | Small |
| Tension | Helical | Small to large | Small to large |
| | Leaf | Large | Moderate |
| | Flat | Small to large | Small to large |
| Torsion | Helical torsion | Small to large | Small to large |
| | Spiral | Moderate | Moderate |
| | Torsion bar | Large | Small |

is satisfactory. If the spring is unsatisfactory, choose another type of spring from Table 44 and repeat the study.

## SPRING WIRE LENGTH AND WEIGHT

How long a wire is needed to make a helical spring having a mean coil diameter of 0.820 in (20.8 mm) if there are five coils in the spring? What will this spring weigh if it is made of oil-tempered spring steel 0.055 in (1.40 mm) in diameter?

### Calculation Procedure:

#### 1. *Compute the spring wire length*

Find the spring length from $l = \pi n d_m$, where $l$ = wire length, in; $n$ = number of coils in the spring, in. Thus, for this spring, $l = \pi(5)(0.820) = 12.9$ in (32.8 cm).

#### 2. *Compute the weight of the spring*

Find the spring weight from $w = 0.224ld^2$, where $w$ = spring weight, lb; $d$ = spring wire diameter, in. For this spring, $w = 0.224(12.9)(0.055)^2 = 0.0087$ lb (0.0387 kg).

**Related Calculations:** The weight equation in step 2 is valid for springs made of oil-tempered steel, chrome vanadium steel, silica-manganese steel, and silicon-chromium steel. For stainless steels, use a constant of 0.228, in place of 0.224, in the equation. The relation given in this procedure is valid for any spring having a continuous coil—helical, spiral, etc. Where a number of springs are to be made, simply multiply the length and weight of each by the number to be made to determine the total wire length required and the weight of the wire.

## HELICAL COMPRESSION AND TENSION SPRING ANALYSIS

Determine the dimensions of a helical compression spring to carry a 5000-lb (22,241.1-N) load if it is made of hard-drawn steel wire having an allowable shear stress of 65,000 lb/in$^2$ (448,175.0 kPa). The spring must fit in a 2-in (5.1-cm) diameter hole. What is the deflection of the spring? The spring operates at atmospheric temperature, and the shear modulus of elasticity is $5 \times 10^6$ lb/in$^2$ ($34.5 \times 10^9$ Pa).

### Calculation Procedure:

#### 1. *Choose the tentative dimensions of the spring*

Since the spring must fit inside a 2-in (5.1-cm) diameter hole, the mean diameter of the coil should not exceed about 1.75 in (4.5 cm). Use this as a trial mean diameter, and compute the wire diameter from $d = [8Ld_m/(\pi s_s)]^{1/3}$, where $d$ = spring wire diameter, in; $L$ = load on spring, lb; $s_s$ allowable shear stress material, lb/in$^2$. Thus, $d = [8 \times 5000 \times 1.75/(\pi \times 65{,}000)]^{1/3} = 0.7$ in

(1.8 cm). So the outside diameter $d_o$ of the spring will be $d_o = d_m + 2(d/2) = d_m + d = 1.75 + 0.70 = 2.45$ in (6.2 cm). But the spring must fit a 2-in (5.1-cm) diameter hole. Hence, a smaller value of $d_m$ must be tried.

Using $d_m = 1.5$ in (3.8 cm) and following the same procedure, we find $d = [8 \times 5000 \times 1.50/(\pi \times 65{,}000)]^{1/3} = 0.665$ in (1.7 cm). Then $d_o = 1.5 + 0.665 = 2.165$ in (5.5 cm), which is still too large.

Using $d_m = 1.25$ in (3.2 cm), we get $d = [8 \times 5000 \times 1.25/(\pi \times 65{,}000)]^{1/3} = 0.625$ in (1.6 cm). Then $d_o = 1.25 + 0.625 = 1.875$ in (4.8 cm). Since this is nearly 2 in (5.1 cm), the spring probably will be acceptable. If desired, several other $d_m$ values could be tried until a spring with a desired $d_o$ more nearly equal to 2.0 in (5.1 cm) was obtained. Use this procedure where a specific outside diameter is required for a spring.

### 2. *Compute the spring deflection*

The deflection of a helical compression spring is given by $f = 64nr_m{}^3L/(d^4Gk)$, where $f$ = spring deflection, in; $n$ = number of coils in this spring; $r_m$ = mean radius of spring coil, in; $G$ = shear modulus of elasticity of spring material, lb/in²; $k$ = spring curvature correction factor $= (4c - 1)/(4c - 4) + 0.615/c$ for heavily coiled springs, where $c = 2r_m/d$. For lightly coiled springs, $k = 1.0$.

Assuming $k = 1.0$ and $n = 10$ coils, we find $f = (64 \times 10 \times 5000)/(0.625^4 \times 5 \times 10^6 \times 10^6 \times 1.0) = 0.196$ in (0.498 cm). This is a modest deflection for a load of 5000 lb (22,241.1 N) and indicates that the spring is heavily coiled. Therefore, as a check, the value of $k$ should be determined and the deflection computed again.

Thus, $c = 2r_m/d = 2(1.25/2)/0.625 = 2.0$. Note that $r_m = d_m/2 = 1.25/2$ in this calculation for the value of $c$. Then $k = (4c - 1)/(4c - 4) + 0.615/c = (4 \times 2 - 1)/(4 \times 2 - 4) + 0.615/2 = 2.055$. Hence, the assumed value of $k = 1.0$ was inaccurate for this spring. By using the computed value of $k$, we see $f = (64 \times 10 \times 0.625^3 \times 5000)/(0.625^4 \times 4 \times 10^6 \times 2.0575) = 0.0954$ in (0.242 cm).

The number of coils $n$ assumed for this spring is based on past experience with similar springs. However, where past experience does not exist, several trial values of $n$ can be used until a spring of suitable deflection and length is obtained.

**Related Calculations:** Use this general procedure to analyze helical coil compression or tension springs. As a general guide, the outside diameter of a spring of this type is taken as (0.96)(hole diameter). The active solid height of a compression-type spring, i.e., the height of the spring when fully closed by the load, usually is $nd$, or (0.9) (final height when compressed by the design load).

## SELECTION OF HELICAL COMPRESSION AND TENSION SPRINGS

Choose a helical compression spring to carry a 90-lb (400.3-N) load with a stress of 50,000 lb/in² (344,750.0 kPa) and a deflection of about 2.0 in (5.1 cm). The spring should fit in a 3.375-in (8.6-cm) diameter hole. The spring operates at about 70°F (21.1°C). How many coils will the spring have? What will the free length of the spring be?

### Calculation Procedure:

### 1. *Determine the spring outside diameter*

Using the usual relation between spring outside diameter and hole diameter, we get $d_o = 0.96d_h$, where $d_h$ = hole diameter, in. Thus, $d_o = 0.96(3.375) = 3.24$ in, say 3.25 in (8.3 cm).

### 2. *Determine the required wire diameter*

The equations in the previous calculation procedure can be used to determine the required wire diameter, if desired. However, the usual practice is to select the wire diameter by using precomputed tabulations of spring properties, charts of spring properties, or a special slide rule available from some spring manufacturers. The tabular solution will be used here because it is one of the most popular methods.

Table 45 shows typical loads and spring rates for springs of various outside diameters and wire diameters based on a corrected shear stress of 100,000 lb/in² (689,500.0 kPa) and a shear modulus of $G = 11.5 \times 10^6$ lb/in² ($79.3 \times 10^9$ Pa).

Before Table 45 can be used, the actual load must be corrected for the tabulated stress. Do

**TABLE 45** Load and Spring Rates for Helical Compression and Tension Springs*

| Spring wire diameter | | Outside diameter of spring coil | | | | | |
|---|---|---|---|---|---|---|---|
| | | in | cm | in | cm | in | cm |
| in | cm | 3 | 7.6 | 3.25 | 8.3 | 3.5 | 8.9 |
| 0.207 | 5.258 | 113† | 502.6 | 104 | 462.6 | 97.2 | 432.4 |
| | | 121 | 211.9 | 93.6 | 163.9 | 74.1 | 129.8 |
| 0.250 | 6.350 | 198 | 880.7 | 183 | 814.0 | 170 | 756.2 |
| | | 270 | 472.8 | 208 | 364.3 | 163 | 285.5 |
| 0.283 | 7.188 | 285 | 1267.7 | 263 | 1169.9 | 247 | 1098.7 |
| | | 460 | 805.6 | 352 | 616.4 | 276 | 483.4 |

*After H. F. Ross, "Application of Tables for Helical Compression and Extension Spring Design," *Transactions ASME*, vol. 69, p. 727.
†First figure given is loads in lb at 100,000-lb/in$^2$ (in N at 689,500-kPa) stress. Second figure is spring rate in lb/in (N/cm) per coil, $G = 11.5 \times 10^6$ lb/in$^2$ ($79.3 \times 10^9$ Pa).

this by taking the product of (actual load, lb)(table stress, lb/in$^2$)/(allowable spring stress, lb/in$^2$). For this spring, tabular load, lb = (90)(100,000/50,000) = 180 lb (800.7 N). This means that a 90-lb (400.3-N) load at a 50,000-lb/in$^2$ (344,750.0-kPa) stress corresponds to a 180-lb (800.7-N) load at 100,000-lb/in$^2$ (689,500-kPa) stress.

Enter Table 45 at the spring outside diameter, 3.25 in (8.3 cm), and project vertically downward in this column until a load of approximately 180 lb (800.7 N) is intersected. At the left read the wire diameter. Thus, with a 3.25-in (8.3-cm) outside diameter and 183-lb (814.0-N) load, the required wire diameter is 0.250 in (0.635 cm).

### 3. *Determine the number of coils required*

The allowable spring deflection is 2.0 in (5.1 cm), and the spring rate per single coil, Table 45, is 208 lb/in (364.3 N/cm) at a tabular stress of 100,000 lb/in$^2$ (689,500 kPa). We use the relation, deflection $f$, in = load, lb/desired spring rate, lb/in, $S_R$; or, $2.0 = 90/S_R$; $S_R = 90/20 = 45$ lb/in (78.8 N/cm).

### 4. *Compute the number of coils in the spring*

The number of active coils in a spring is $n$ = (tabular spring rate, lb/in)/(desired spring rate, lb/in). For this spring, $n = 208/45 = 4.62$, say 5 coils.

### 5. *Determine the spring free length*

Find the approximate length of the spring in its free, expanded condition from $l$ in $= (n + i)d + f$, where $l$ = approximate free length of spring, in; $i$ = number of inactive coils in the spring; other symbols as before. Assuming two inactive coils for this spring, we get $l = (5 + 2)(0.25) + 2 = 3.75$ in (9.5 cm).

**Related Calculations:** Similar design tables are available for torsion springs, spiral springs, coned-disk (Belleville) springs, ring springs, and rubber springs. These design tables can be found in engineering handbooks and in spring manufacturers' engineering data. Likewise, spring design charts are available from many of these same sources. Spring design slide rules are generally available free of charge to design engineers from spring manufacturers.

## SIZING HELICAL SPRINGS FOR OPTIMUM DIMENSIONS AND WEIGHT

Determine the dimensions of a helical spring having the minimum material volume if the initial load on the spring is 15 lb (66.7 N), the mean coil diameter is 1.02 in (2.6 cm), the spring stroke is 1.16 in (2.9 cm), the final spring stress is 100,000 lb/in$^2$ (689,500 kPa), and the spring modulus of torsion is $11.5 \times 10^6$ lb/in$^2$ ($79.3 \times 10^9$ Pa).

## Calculation Procedure:

### 1. *Compute the minimum spring volume*

Use the relation $v_m = 8fLG/s_f^2$, where $v_m$ = minimum volume of spring, in$^3$; $f$ = spring stroke, in$^3$; $L$ = initial load on spring, lb; $G$ = modulus of torsion of spring material, lb/in$^2$; $s_f$ = final stress in spring, lb/in$^2$. For this spring, $v_m = 8(1.16)(15)(11.5 \times 10^6)/(100{,}000)^2 = 0.16$ in$^3$ (2.6 cm$^3$).

### 2. *Compute the required spring wire diameter*

Find the wire diameter from $d = [16Ld_m/\pi s_f)]^{1/3}$, where $d$ = wire diameter, in; $d_m$ = mean diameter of spring, in; other symbols as before. For this spring, $d = [16 \times 15 \times 1.02/(\pi \times 100{,}000)]^{1/3} = 0.092$ in (2.3 mm).

### 3. *Find the number of active coils in the spring*

Use the relation $n = 4v_m/(\pi^2 d^2 d_m)$, where $n$ = number of active coils; other symbols as before. Thus, $n = 4(0.16)/[\pi^2(0.092)^2(1.02)] = 7.5$ coils.

### 4. *Determine the active solid height of the spring*

The solid height $H_s = (n + 1)d$, in, or $H_s = (7.5 + 1)(0.092) = 0.782$ in (2.0 cm). For a practical design, allow 10 percent clearance between the solid height and the minimum compressed height $H_c$. Thus, $H_c = 1.1H_s = 1.1(0.782) = 0.860$ in (2.2 cm). The assembled height $H_a = H_c + f = 0.860 + 1.16 = 2.020$ in (5.13 cm).

### 5 *Compute the spring load-deflection rate*

The load-deflection rate $R = Gd^4/(8d_m{}^3 n)$, where $R$ = load-deflection rate, lb/in; other symbols as before. Thus, $R = (11.5 \times 10^6)(0.092)^4/[8(1.02)^3(7.5)] = 12.9$ lb/in (2259.1 N/m).

The initial deflection of the spring is $f_i = L/R$ in, or $f_i = 15/12.9 = 1.163$ in (3.0 cm). Since the free height of a spring $H_f = H_a + f_i$, the free height of this spring is $H_f = 2.020 + 1.163 = 3.183$ in (8.1 cm).

**Related Calculations:** The above procedure for determining the minimum spring volume can be used to find the minimum spring weight by relating the spring weight $W$ lb to the density of the spring material $\rho$ lb/in$^3$ in the following manner: For the required initial load $L_1$ lb, $W_{min} = \rho(8fL_1G/s_f^2)$. For the required energy capacity $E$ in·lb, $W_{min} = \rho(4ED/s_f^2)$. For the required final load $L_2$ lb, $W_{min} = \rho(2f_2L_2G/s_f^2)$.

The above procedure assumes the spring ends are open and not ground. For other types of end conditions, the minimum spring volume will be greater by the following amount: For squared (closed) ends, $v_m = 0.5\pi^2 d^2 d_m$. For ground ends, $v_m = 0.25\pi^2 d^2 d_m$. The methods presented here were developed by Henry Swieskowski and reported in *Product Engineering*.

## SELECTION OF SQUARE- AND RECTANGULAR-WIRE HELICAL SPRINGS

Choose a square-wire spring to support a load of 500 lb (2224.1 N) with a deflection of not more than 1.0 in (2.5 cm). The spring must fit in a 4.25-in (10.8-cm) diameter hole. The modulus of rigidity for the spring material is $G = 11.5 \times 10^6$ lb/in$^2$ ($79.3 \times 10^9$ Pa). What is the shear stress in the spring? Determine the corrected shear stress for this spring.

## Calculation Procedure:

### 1. *Determine the spring dimensions*

Assume that a 4-in (10.2-cm) diameter square-bar spring is used. Such a spring will fit the 4.25-in (10.8-cm) hole with a small amount of room to spare.

As a trial, assume that the width of the spring wire = 0.5 in (1.3 cm) = $a$. Since the spring is square, the height of the spring wire = 0.5 in (1.3 cm) = $b$.

With a 4-in (10.2-cm) outside diameter and a spring wire width of 0.5 in (1.3 cm), the mean radius of the spring coil $r_m = 1.75$ in (4.4 cm). This is the radius from the center of the spring to the center of the spring wire coil.

**2. *Compute the spring deflection***

The deflection of a square-wire tension spring is $f = 45Lr_m^3n/(Ga^4)$, where $f$ = spring deflection, in; $L$ = load on spring, lb; $n$ = number of coils in spring; other symbols as before. To solve this equation, the number of coils must be known. Assume, as a trial value, five coils. Then $f = 45(500)(1.75)^3(5)/[(11.5 \times 10^6)(0.5)^4] = 0.838$ in (2.1 cm). Since a deflection of not more than 1.0 in (2.5 cm) is permitted, this spring is probably acceptable.

**3. *Compute the shear stress in the spring***

Find the shear stress in a square-bar spring from $S_s = 4.8Lr_m/a^3$, where $S_s$ = spring shear stress, lb/in²; other symbols as before. For this spring, $S_s = (4.8)(500)(1.75)/(0.5)^3 = 33{,}600$ lb/in² (231,663.8 kPa). This is within the allowable limits for usual spring steel.

**4. *Determine the corrected shear stress***

Find the shear stress in a square-bar spring from $S_s = 4.8Lr_m/a^3$, where $s_s$ = spring correction factor $k = 1 + 1.2/c + 0.56/c^2 + 0.5/c^3$, where $c = 2r_m/a$. For this spring, $c = (2 \times 1.75)/0.5 = 7.0$. Then $k = 1 + 1.2/7 + 0.56/7^2 + 0.5/7^3 = 1.184$. Hence, the corrected shear stress is $S_s' = ks_s$, or $S_s' = (1.184)(33{,}600) = 39{,}800$ lb/in² (274,411.3 kPa). This is still within the limits for usual spring steel.

**Related Calculations:** Use a similar procedure to select rectangular-wire springs. Once the dimensions are selected, compute the spring deflection from $f = 19.6Lr_m^3n/[Gb^3(a - 0.566)]$, where all the symbols are as given earlier in this calculation procedure. Compute the uncorrected shear stress from $S_s = Lr_m(3a + 1.8b)/(a^2b^2)$. To correct the stress, use the Liesecke correction factor given in Wahl—*Mechanical Springs*. For most selection purposes, the uncorrected stress is satisfactory.

## CURVED SPRING DESIGN ANALYSIS

Find the maximum load $P$, maximum deflection $F$, and spring constant $C$ for the curved rectangular wire spring shown in Fig. 22 if the spring variables expressed in metric units are $E = 14{,}500$ kg/mm², $S_b = 55$ kg/mm², $b = 1.20$ mm, $h = 0.30$ mm, $r_1 = 0.65$ mm, $r_2 = 1.75$ mm, $L = 9.7$ mm, $u_1 = 1.7$ mm, and $u_2 = 5.6$ mm.

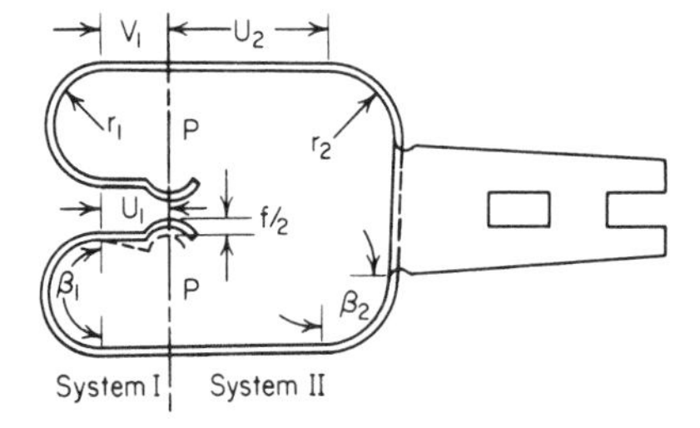


**FIG. 22** Typical curved spring. *(Product Engineering.)*

### Calculation Procedure:

**1. *Divide the spring into analyzable components***

Using Fig. 23, developed by J. Palm and K. Thomas of West Germany, as a guide, divide the spring to be analyzed into two or more analyzable components, Fig. 22. Thus, the given spring can be divided into two springs—a type D (Fig. 23), called system I, and a type A (Fig. 23), called system II.

**2. *Compute the spring force***

The spring force $P = P_I = P_{II}$. Since $(u_2 + r_2) > (u_1 + r_1)$, the spring in system II exerts a larger force. From Fig. 23 for $\beta = 90°$, $P = S\sigma_{max}/(u_2 + r_2)$, where $S$ = section modulus, mm³, of the spring wire. Since $S = bh^2/6$ for a rectangle, $P = bh^2\sigma_{max}/[6(u_2 + r_2)]$ where $b$ = spring wire width, mm; $h$ = spring wire height, mm; $\sigma_{max}$ = maximum bending stress in the spring, kg/mm³; other symbols as given in Fig. 22. Then $P = (1.20)(0.30)^2 \times (55)/[6(5.6 + 1.75)] = 0.135$ kg.

**3. *Compute the spring deflection***

The total deflection of the springs is $F = 2F_I + F_{II}$, where $F$ = spring deflection, mm, and the subscripts refer to each spring system. Taking the sum of the deflections as given in Fig. 23, we get $F = [2P/(3EI)][2K_1r_1^3(m_1 + \beta_1/2)^2 + (v_1 - u_1)^3 + K_2r_2^3(m_2 + \beta_2)^3]$, where $E$ = Young's modulus, kg/mm²; $I$ = spring wire moment of inertia, mm⁴; $K$ = correction factor for the spring from Fig. 24, where the subscripts refer to the radius being considered in the relation $u/r$; $m = u/r$; $\beta$ = angle of spring curvature, rad. Where the subscripts 1 and 2 are used in this equation,

| Spring type | Spring deflection | Spring force and bending stresses |
|---|---|---|
| 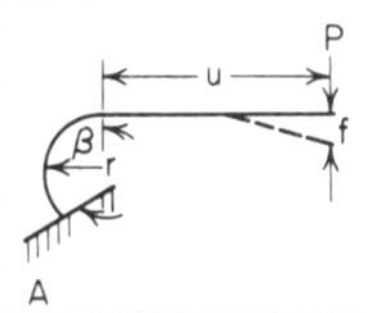 A | $F_1 = \frac{KPr^3}{3EI}(m+\beta)^3$<br>where $\alpha = \beta$ for finding K | When $\alpha = 0°$ to $90°$: $P = \frac{S\sigma}{u+\sin\beta}$, $\sigma = \frac{Pr(m+\sin\beta)}{S}$<br>When $\alpha = 90°$ to $180°$: $P = \frac{S\sigma}{u+r}$, $\sigma = \frac{Pr(m+1)}{S}$ |
| 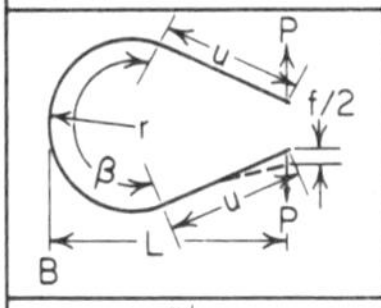 B | $F_2 = \frac{2KPr^3}{3EI}(m+\frac{\beta}{2})^3$<br>where $\alpha = \frac{\beta}{2}$ for finding K | $P = \frac{S\sigma}{L}$<br>$\sigma = \frac{PL}{S}$ |
| 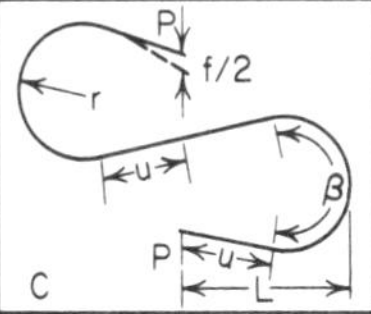 C | $F_3 = 2F_2 = \frac{4KPr^3}{3EI}(m+\frac{\beta}{2})^3$<br>where $\alpha = \frac{\beta}{2}$ for finding K | |
| D, E | $F_4 = F_5 = \frac{P}{3EI}\left[2Kr^3(m+\frac{\beta}{2})^3+(v-u)^3\right]$<br>where $\alpha = \frac{\beta}{2}$ for finding K | $P = \frac{S\sigma}{\lambda} = \frac{P\lambda}{S}$ (see table below) |

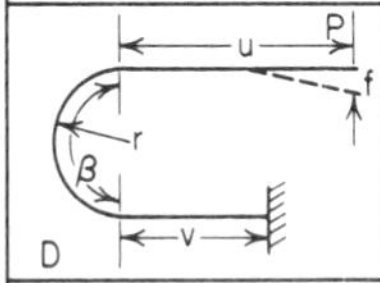

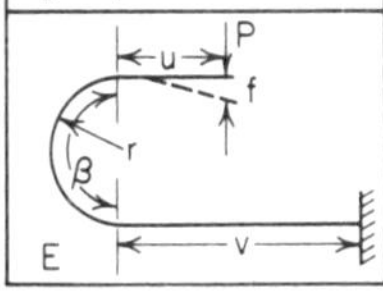

| First condition | Second condition | $\lambda$ |
|---|---|---|
| $u \geq v$ | --- | $u + r$ |
| $u < v$ | $(u-v) < (u+r)$ | $u + r$ |
| $u < v$ | $(v-u) > (u+r)$ | $v - u$ |
| $u = 0$ | $v \leq r$ | $r$ |
| $u = 0$ | $v > r$ | $v$ |

**FIG. 23** Deflection, force, and stress relations for curved springs. *(Product Engineering.)*

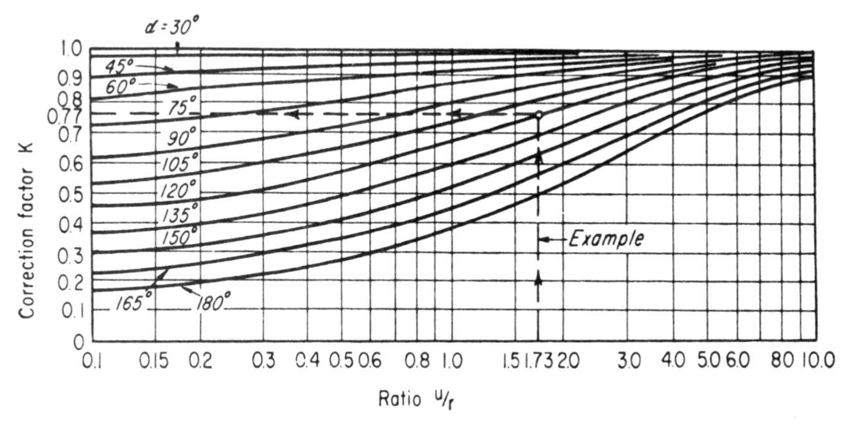


**FIG. 24** Correction factors for curved springs. *(Product Engineering.)*

they refer to the respective radius identified by this subscript. Since $I = bh^3/12$ for a rectangle, or $I = (1.20)(0.30)^3/12 = 0.0027$ mm$^4$, $F = \{[2(0.135)/[(3)(14{,}500)(0.0027)]\}[2(0.92)(0.65)(2.62 + 1.57)^3 + 0 + 0.94(1.75)^3(3.2 + 1.57)^3] = 1.34$ mm.

### 4. *Compute the spring constant*

The spring constant $C = P/F = 0.135 = 0.135/1.34 = 0.101$ kg/mm.

**Related Calculations:** The relations given here can also be used for round-wire springs. For accurate results, $h/r$ for flat springs and $d_o/r$ for round-wire springs should be less than 0.6. The various symbols used in this calculation procedure are defined in the text and illustrations. Since the equations given here analyze the springs and do not contain any empirical constants, the equations can be used, as presented, for both metric and English units. Where a round spring is analyzed, $h = b = d_o$, where $d_o$ = spring outside diameter, mm or in.

## ROUND- AND SQUARE-WIRE HELICAL TORSION-SPRING SELECTION

Choose a round-music-wire torsion spring to handle a moment load of 15.0 lb·in (1.7 N·m) through a deflection angle of 250°. The mean diameter of the spring should be about 1.0 in (2.5 cm) to satisfy the space requirements of the design. Determine the required diameter of the spring wire, the stress in the wire, and the number of turns required in the spring. What is the maximum moment and angular deflection the spring can handle? What is the maximum moment and deflection without permanent set?

## Calculation Procedure:

### 1. *Select a suitable wire diameter*

To reduce the manufacturing cost of a spring, a wire of standard diameter should be used, whenever possible, for the spring, Fig. 25. Usual torsion-spring wire diameters and the side of square-wire springs range from 0.02 to 0.60 in (0.05 to 1.52 cm), depending on the moment the spring must carry and the angular deflection.

Assume a wire diameter of 0.10 in (0.25 cm) and a bending stress of 150,000 lb/in$^2$ ($1.03 \times 10^9$ Pa) as trial values for this spring. [Typical round-wire and square-wire torsion-spring bending stresses range from 100,000 to 200,000 lb/in$^2$ ($689.5 \times 10^6$ to $1.38 \times 10^9$ Pa), depending on the material used in the spring.]

Compute the twisting moment corresponding to the assumed stress from $M_t = \pi d^3 S_b/32$, where $M_t$ = twisting moment load, lb·in; $d$ = spring wire diameter, in; $S_b$ = bending stress in spring, lb/in$^2$. Thus, $M_t = \pi(0.10)^3(150{,}000)/32 = 14.7$ lb·in (1.66 N·m). This is very close to the actual moment load of 15.0 lb·in (1.7 N·m). Therefore, the assumed spring diameter and bending stress are acceptable, thus far.

### 2. *Compute the actual spring stress*

Use the following relation to find the actual bending stress $S_b$ lb/in$^2$ in the spring: $S_b$ = (actual spring moment lb·in/computed spring moment, lb·in)(assumed stress, lb/in$^2$); $S_b = (15.0/14.7)(150{,}000) = 153{,}000$ lb/in$^2$ ($1.05 \times 10^9$ Pa).

### 3. *Check the actual vs. recommended spring stress*

Enter Fig. 26 at the wire diameter of 0.10 in (0.25 cm), and project vertically upward to the music-wire curve to read the recommended bending stress for music wire as 159,000 lb/in$^2$ (1.10

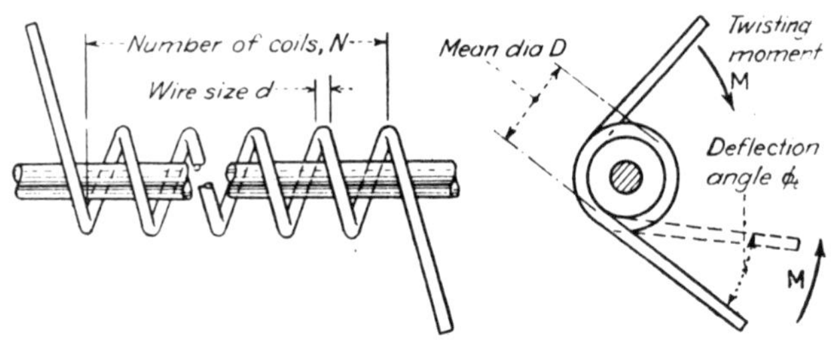


**FIG. 25** Typical torsion spring. (*Product Engineering.*)

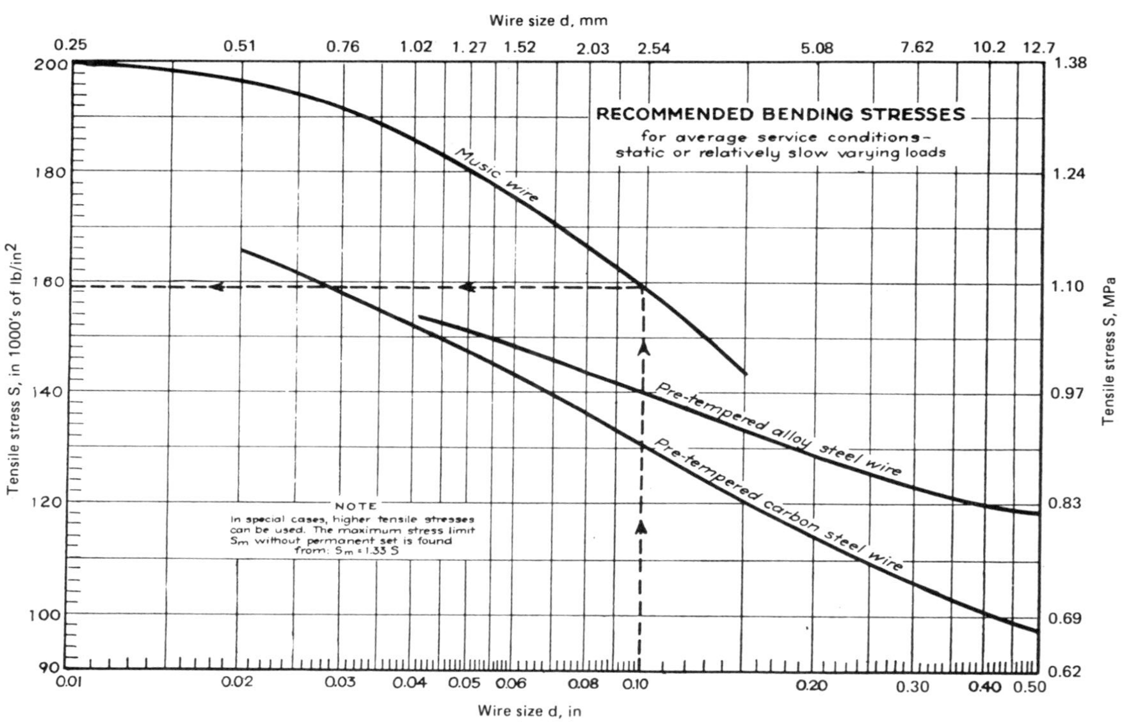


**FIG. 26** Recommended bending stresses for torsion springs. *(Product Engineering.)*

$\times 10^9$ Pa). Since the actual stress, 153,000 lb/in² (1.05 $\times 10^9$ Pa), is less than but reasonably close to the recommended stress, the selected wire diameter is acceptable for the planned load on the spring. This chart and calculation procedure were developed by H. F. Ross and reported in *Product Engineering*.

**4. *Determine the angular deflection per spring coil***

Compute the angular deflection per coil from $\phi = 360S_bd_m/(Ed)$, where $\phi$ = angular deflection per spring coil, degrees; $d_m$ = mean diameter of spring, in; $E$ = Young's modulus for spring material = 30 $\times 10^6$ lb/in² (206.9 $\times 10^9$ Pa) for spring steel; other symbols as before. Thus, by using the *assumed* bending stress in the spring, $\phi = 360(150{,}000)(1.0)/[(30 \times 10^6)(0.1)] = 18°$. This value is the maximum safe deflection per coil for the spring.

**5. *Compute the number of coils required***

The number of coils $n$ required in a helical torsion spring is $n = \phi_t$(assumed stress, lb/in²)/$\phi$ (actual stress, lb/in²), where $\phi_t$ = total angular deflection of spring, degrees; $\phi$ = maximum safe deflection per coil, degrees. Thus, $n = 250(150{,}000)/[(18)(153{,}000)] = 13.6$ coils; use 14 coils.

**6. *Determine the maximum moment the spring can handle***

On the basis of the maximum recommended stress, the moment can be increased to $M_t$ = [(maximum recommended stress, lb/in²)/(assumed stress, lb/in²)](actual moment, lb·in). Read the maximum recommended stress from Fig. 26 as 159,000 lb/in² (1.10 $\times 10^9$ Pa) for 0.1-in (0.25-cm) diameter music wire, as in step 3. Thus, $M_t = (159{,}000/150{,}000)(14.7) = 15.6$ lb·in (1.8 N·m).

**7. *Compute the maximum angular deflection***

The maximum angular deflection per coil is $\phi$ = [(maximum recommended stress, lb/in²)/(assumed stress, lb/in²)](computed angular deflection per coil, degrees) = 159,000/150,000 $\times$ 18 = 19.1° per coil.

**8. *Determine the special-case moment and deflection***

The maximum moment $M_{max}$ and deflection $\phi_{max}$, without permanent set, can be one-third greater than in steps 6 and 7, or $M_{max} = 15.6(1.33) = 20.8$ lb·in (2.4 N·m), and $\phi_{max} = 19.1(1.33) = 25.5°$ per coil. These maximum values allow for overloads on the spring.

**Related Calculations:** Use the same procedure for square-wire helical torsion springs, but substitute the length of the side of the square for $d$ in each equation where $d$ appears.

## TORSION-BAR SPRING ANALYSIS

What must the diameter of a torsion bar be if it is to have a spring rate of 2400 lb·in/rad (271.2 N·m/rad) and a total angle of twist of 0.20 rad? The bar is made of 302 stainless steel, which has a proportional limit in tension of 35,000 lb/in² (241.3 $\times 10^6$ Pa), and $G$ = torsional modulus of elasticity = $10^7$ lb/in² (68.95 $\times 10^9$ Pa). The length of the torsion bar is 26.0 in (66.0 cm), and it is solid throughout. What size square torsion bar would be required? What size equilateral triangular section would be required? What is the energy storage of each bar form?

### Calculation Procedure:

**1. *Determine the proportional limit in shear***

For stainless steel, the proportional limit $S_s$ lb/in² in shear is 0.55 times that in tension, or $S_s = 0.55(35{,}000) = 19{,}250$ lb/in² (132.7 $\times 10^6$ Pa).

**2. *Compute the required diameter of the bar***

Use the relation $d = 2S_sl/(G\theta)$, where $d$ = torsion-bar diameter, in; $l$ = torsion-bar length, in; $\theta$ = total angle of twist of torsion bar, rad; other symbols as before. Thus, $d = 2(19{,}250)(26.0)/[10^7(0.20)] = 0.50$ in (1.3 cm).

**3. *Compute the square-bar size***

Use the relation $d = 1.482S_sl/(G\theta)$, where $d$ = side of the square bar, in. Thus $d = 1.482(19{,}250)(26.0)/[10^7(0.2)] = 0.371$ in (0.9 cm).

**4. *Compute the triangular-bar size***

Use the relation $d = 2.31S_s l/(G\theta)$, where $d$ = side of the triangular bar, in. Thus, $d = 2.31(19{,}250)(26.0)/[10^7(0.2)] = 0.578$ in (1.5 cm).

**5. *Compute the energy storage of each bar***

For a solid circular torsion spring, the energy storage $e = S_s^2/(4G)$, where $e$ = energy storage in the bar, in·lb/in³. Thus, $e = (19{,}250)^2/[4(10^7)] = 9.25$ in·lb/in³ (6.4 N·cm/cm³).

For a square bar, $e = S_s^2/(6.48G)$, where the symbols are the same as before, or, $e = (19{,}250)^2/[6.48(10^7)] = 5.71$ in·lb/in³ (3.9 N·cm/cm³).

For a triangular bar, $e = S_s^2/(7.5G)$, where the symbols are the same as before. Or, $e = (19{,}250)^2/[7.5(10^7)] = 4.94$ in·lb/in³ (3.4 N·cm/cm³).

**Related Calculations:** Use this procedure for torsion-bar springs made of any metal. The energy-storage capacity of various springs in terms of the spring weight is as follows:

| | Energy storage of spring | |
|---|---|---|
| Type of spring | in·lb/lb | N·m/kg |
| Leaf | 300–450 | 74.7–112.1 |
| Helical round-wire coil | 700–1100 | 174.4–274.0 |
| Torsion-bar | 1000–1500 | 249.1–373.6 |
| Volute | 500–1000 | 124.5–249.1 |
| Rubber in shear | 2000–4000 | 498.2–996.4 |

The analyses in this calculation procedure are based on the work of Donald Bastow and D. A. Derse and are reported in *Product Engineering*.

## MULTIRATE HELICAL SPRING ANALYSIS

Determine the required spring rates, number of coils, coil clearances, and free length of two helical coil springs if spring 1 has preload of 1.2 lb (5.3 N) and spring 2 has a preload of 19.1 lb (85.0 N) in a double preload mechanism. The rod is to deflect 0.46 in (1.2 cm) before building up to the preload of 19.1 lb (85.0 N). Total deflection is to be 3.0 in (7.6 cm) with a load of 78 lb (347.0 N). The mean spring diameter $d_m$ = 1.29 in (3.28 cm) for both springs; the wire diameter is $d$ = 0.148 in (3.76 mm) for spring 1; $d$ = 0.156 in (3.96 mm) for spring 2; $G = 11.5 \times 10^6$ lb/in² ($79.3 \times 10^9$ Pa) for both springs.

### Calculation Procedure:

**1. *Determine the spring rate for each spring***

The spring rate, lb/in, is $R_s$ = (preload spring 2, lb − preload spring 1, lb)/deflection, in, before full preload. Thus, for spring 1, $R_{s1} = (19.1 - 1.2)/0.46 = 38.9$ lb/in (68.1 N/cm).

For the combination of the two springs $R_{st} = (78 - 19.1)/(3.0 - 0.46) = 23.1$ lb/in (40.5 N/cm).

For spring 2, $R_{s2} = R_{s1}R_{st}/(R_{s1} - R_{st})$, where the symbols are the same as before. Or, $R_{s2} = (38.9)(23.1)/(38.0 - 23.1) = 56.9$ lb/in (99.6 N/cm).

**2. *Check the spring rate against the spring deflection***

The deflection, in, is $f = L/R$, where $L$ = load on the spring, lb; $R$ = spring rate, lb/in. Thus for spring 1, $f_1 = (78 - 1.2)/38.9 = 1.97$ in (5.0 cm). For spring 2, $f_2 = L_2/R_{s2} = (78 - 19.1)/56.9 = 1.03$ in (2.6 cm). For the two springs, $F_t = f_1 + f_2 = 1.97 + 1.03 = 3.00$ in (7.6 cm). This agrees with the allowable deflection of 3 in (7.6 cm) at the full load of 78 lb (347.0 N). Therefore, the computed spring rates and preloads are acceptable.

### 3. Compute the number of coils for each spring

The number of coils $n = Gd^4/(8d_m^3R)$, where the symbols are as defined before. Thus, $n_1 = (11.5 \times 10^6)(0.148)^4/[8(1.29)^3(38.9)] = 8.25$ coils. And $n_2 = (11.5 \times 10^6)(0.156)^4/[8(1.29)^3(56.9)] = 7$ coils.

### 4. *Compute the solid height of each spring*

Allowing one inactive coil for each end of each spring, so that the ends may be squared and ground, we find the solid height $h_s = d$(number of coils + 2). Or $h_{s1} = (0.148)(8.25 + 2) = 1.517$ in (3.85 cm). And $h_{s2} = (0.1567)(7 + 2) + 1.404$ in (3.57 cm).

### 5. *Determine the coil clearances*

Assume a coil clearance of 3 times the spring wire diameter. Then the coil clearance *c*, in, for each spring is $c_1 = (3)(0.148) = 0.444$ in (1.128 cm) and $c_2 = (3)(0.156) = 0.468$ in (1.189 cm).

### 6. *Compute the free length of each spring*

The free length of a helical spring = $l_f$ = solid height + coil clearance + deflection + [preload, lb/(spring rate, lb/in)]. For spring 1, $l_{f1} = 1.517 + 0.444 + 1.970 + (1.2/38.9) = 3.962$ in (10.06 cm). For spring 2, $l_{f2} = 1.404 + 0.468 + 1.030 + (19.1/56.9) = 3.235$ in (8.22 cm).

**Related Calculations:** Use this procedure for springs made of any metal. This analysis is based on the work of K. A. Flesher, as reported in *Product Engineering*.

## BELLEVILLE SPRING ANALYSIS FOR SMALLEST DIAMETER

What are the minimum outside radius $r_o$ and thickness $t$ for a steel Belleville spring that carries a load of 1000 lb (4448.2 N) at a maximum compressive stress of 200,000 lb/in² ($1.38 \times 10^9$ Pa) when compressed flat?

### Calculation Procedure:

### 1. *Determine the spring radius ratio and the height-thickness ratio*

The radius ratio $r_r = r_o/r_i$ for a Belleville spring, when $r_o$ = outside radius of spring, in; $r_i$ = inside radius of spring, in = radius of hole in spring, in. Table 46 summarizes recommended values for the radius ratio for various values of the height-thickness ratio to produce the smallest diameter spring. In general, an $r_r$ value of 1.75 usually produces a spring of suitably small size. When $r_r = 1.75$, Table 46 shows that the height-thickness ratio $h/t$ with both values expressed in inches is 1.5. Assume that these two values are valid, and proceed with the calculation.

### 2. *Determine the spring outside radius*

Table 47 shows the stress constant $r_o s_c/L^{0.5}$, where $s_c$ = maximum compressive stress on the top surface at the inner edge, lb/in², Fig. 27; $L$ = total axial load on spring, lb. For $h/t = 1.5$ and $r_r = 1.75$, the stress constant $r_o s_c/L^{0.5} = -1905$. Solving gives $r_o = -1905L^{0.5}/s_c$. By substituting the given values, $r_o = -1905(1000)^{0.5}/-200{,}000 = 3.01$ in (7.65 cm). The negative sign is used for the spring stress because it is a compressive stress.

**TABLE 46** Design Constants for Belleville Springs*

| $h/t$ | $r_o/r_i$ |
|---|---|
| 1.00 | 1.25 |
| 1.25 | 1.50 |
| 1.50 | 1.75 |
| 1.75 | 2.00 |
| 2.00 | 2.50 |

*Product Engineering.

**TABLE 47** Stress Constants for Belleville Springs*

| | $r_o/r_1$ −1.75 | |
|---|---|---|
| $h/t$ | K | $r_o s_c/L^{0.5}$ |
| 1.00 | −3.2455 | −1346 |
| 1.25 | −4.3734 | −1622 |
| 1.50 | −5.6279 | −1905 |
| 1.75 | −7.0090 | −2197 |

*Product Engineering.

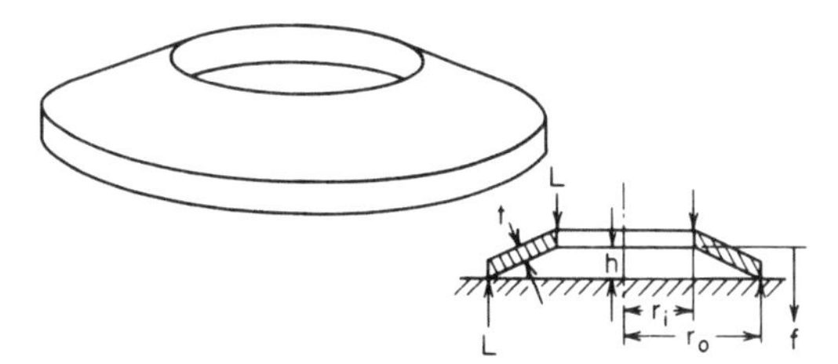


**FIG. 27** Belleville spring: appearance and dimensions. *(Product Engineering.)*

### 3. *Determine the radius of the hole in the spring*

For this Belleville spring, $r_i = r_o/r_r = 3.01/1.75 = 1.72$ in (4.37 cm).

### 4. *Compute the spring thickness*

The thickness of a Belleville spring is given by $t = [s_c r_o^2/(KE)]^{0.5}$, where $K$ = a stress constant from Table 47; $E$ = modulus of elasticity of the spring material, lb/in$^2$; other symbols as before. Thus, with $E = 30 \times 10^6$ ($206.9 \times 10^9$ Pa), $t = [200{,}000 \times 3.01^2/(5.6279 \times 30 \times 10^6)]^{0.5} =$ 0.1037 in (2.63 mm).

### 5. *Compute the spring height*

Since $h/t = 1.5$ for this spring, $h = 1.5(0.1037) = 0.156$ in (3.96 mm).

**Related Calculations:** Professor M. F. Spotts developed the analytical procedure and data presented here. His studies show that space is usually the limiting factor in spring selection, and the designer generally must determine the minimum permissible outside diameter of the spring to carry a given load at a specified stress. Further, the ratio of the outside to the inside diameter for the smallest spring is about 1.75, assuming that the load spring is compressed nearly flat, which is the usual design assumption. A value of $h/t$ of 1.5 is recommended for most spring applications. Belleville springs are used in disk brakes, the preloading of bolted assemblies, ball bearings, etc. The analysis presented here is useful for all usual applications of Belleville springs.

## RING-SPRING DESIGN ANALYSIS

Determine the major dimensions of a ring spring made of material having an allowable stress of 175,000 lb/in$^2$ ($1.21 \times 10^9$ Pa), $E = 20 \times 10^6$ lb/in$^2$ ($137.9 \times 10^9$ Pa), a coefficient of friction of 0.12, an inside diameter of 7.0 in (17.8 cm), an outside diameter of 9.0 in (22.9 cm) or less, a taper angle of 14°, an axial load of 56 tons (50.8 t), and a deflection of not more than 8.0 in (20.3 cm).

## Calculation Procedure:

### 1. *Determine the inner-ring dimensions*

For the usual ring spring, the ring height $h$ is 15 percent of the allowable outside diameter, or (0.15)(9.0) = 1.35 in (3.4 cm), Fig. 28. The axial gap between the rings $g$ is usually 25 percent of the ring height.

Compute the area of the internal ring from $A_i = L/(\pi K_c s_i)$, where $A_i$ = area of internal ring, in$^2$; $L$ = axial load on spring, lb; $K_c$ = spring constant from Fig. 29; $s_i$ = allowable stress in the inner ring of the spring, lb/in$^2$. With a coefficient of friction $\mu = 0.12$ and a taper angle of 14°, $K_c = 0.38$. Then $A_i = 56 \times 2000/[\pi(0.38)(175{,}000)] = 0.537$ in$^2$ (3.47 cm$^2$).

The width $w_i$ of the inner ring is $w_i = [A_i - (h_i^2 \tan\theta)/4]/h_i$, where $\tan\theta$ = tangent of taper angle; $h_i$ = height of inner ring, in. Thus, $w_i = [0.537 - (1.35^2 \tan 14°)/4]/1.35 = 0.314$ (7.98 mm).

Use a trial-and-error process to determine the dimensions of the outer ring. Do this by assuming a cross-sectional area for the outer ring; then compute whether the outside diameter and stress meet the specifications for the spring.

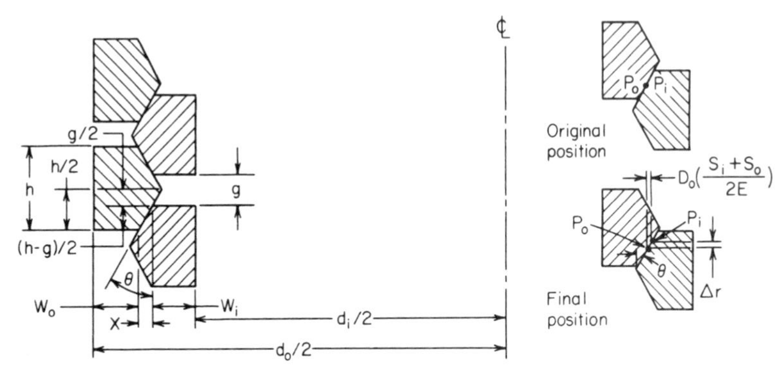


**FIG. 28** Ring-spring positions and dimensions. *(Product Engineering.)*

Assume that $A_o$ = 0.609 in$^2$ (3.93 cm$^2$). Then $s_o = L/(\pi A_o K_c)$, where $s_o$ = stress in outer ring, lb/in$^2$; other symbols as before. So $s_o$ = 56 × 2000/[$\pi$(0.609)(0.38)] = 154,200 lb/in$^2$ (1.06 × 10$^9$ Pa). This stress is within the allowable limits.

In the usual ring spring, $h_o = h_i$ = 1.35 in (3.4 cm) for this spring. Then, by using a relation similar to that for the inner ring, $w_o = [A_o - (h_o^2 \tan\theta)/4]/h_o$ = [0.609−(1.35$^2$tan14°)/4]/1.35 = 0.366 in (9.3 mm).

Find the outside diameter of the ring from $d_o = d_i + 2w_i + 2w_o + (h - g) \tan \theta$, where $d_o$ = outside diameter of outer ring, in; $d_i$ = inside diameter of inner ring, in; $g$ = axial gap of rings, in = 25 percent of ring height for this spring, or 0.25(1.35) = 0.3375 in (8.57 mm). Hence, $d_o$ = 7.0 + 2(0.314) + 2(0.366) + (1.35 − 0.3375)tan 14° = 8.613 in (21.9 cm). This is close enough to the maximum allowable outside diameter of 9 in (22.9 cm) to be acceptable. Were the value of $d_o$ unacceptable, another value of $A_o$ would be assumed and the calculation repeated until the stress and $d_o$ values were acceptable.

## 2. *Compute the number of rings required*

Find the axial deflection per ring $f$, in, from $f = d_a[(s_i + s_o)/(2E)] \cot \theta$, where $d_a$ = mean diameter of the spring, in; $E$ = modulus of elasticity of the spring material, lb/in$^2$. Compute $d_a$ = $[(d_o - 2w_o) + (d_i + 2w_i)]/2$ = [(8.613 − 2 × 0.366) + (7.0 + 2 × 0.314)]/2 = 7.755 in (19.7 cm). Then $f$ = 7.755[(175,000 + 154,200)/(2 × 29 × 10$^6$)] cot 14° = 0.176 in (4.47 mm). Since the axial deflection must not exceed 8 in (20.3 cm), the number of rings required = axial deflection, in/deflection per ring, in = 8.0/0.176 = 45.5, or 46 rings. Figure 30 shows the spring dimensions.

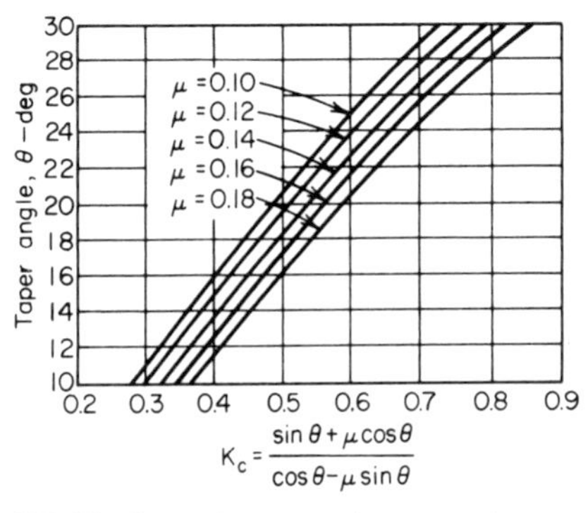


**FIG. 29** Ring-spring compression constant in terms of the taper angle for various values of the coefficient of friction. *(Product Engineering.)*

**Related Calculations:** Ring springs are suitable for pipe-vibration isolation, shock absorbers, plows, trench diggers, railroad couplers, etc. The recommended approximate proportions of ring springs are as follows: (1) Compressed height should be at least 4 times the deflection of the spring. (2) Ring height should be 15 to 20 percent of the ring outside diameter. (3) Spring outside diameter and height are usually as large as space permits.(4) Thin ring sections are preferred to thick ones. (5) Ring taper should be 1:4. (6) Coefficient of friction for ring springs varies from 0.10 to 0.18. (7) Allowable spring stresses are 160,000 lb/in$^2$ (1.10 × 10$^9$ Pa) for nonmachined steel, 200,000 lb/in$^2$ (1.38 × 10$^9$ Pa) for machined steel. For vibratory loads, the allowable stress is about one-half these values. (8) Load capacities

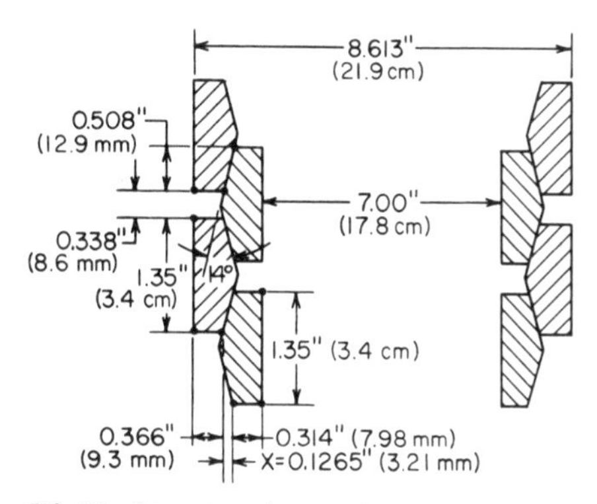


**FIG. 30** Dimensions of a typical ring spring. *(Product Engineering.)*

of ring springs vary between 2 and 150 tons (1.8 and 136.1 t). (9) Spring deflections vary between 1 in (2.5 cm) and 1 ft (0.3 m). (10) The equations given above can be used for spring design or for analysis of an existing spring.

The design method given here was developed by Tyler G. Hicks and reported in *Product Engineering.*

## LIQUID-SPRING SELECTION

Select a liquid spring to absorb a 50,000-lb (222,411.1-N) load with a 5-in (12.7-cm) stroke. The rod diameter is 1 in (2.5 cm). What is the probable temperature rise per stroke? Compare this spring with metal-coil, Belleville, and ring springs.

### Calculation Procedure:

#### 1. *Compute the liquid volume required*

Assume that the final pressure of the compressed liquid is 50,000 lb/in$^2$ (344,750 kPa) and that the liquid is compressed 18 percent on application of full load on the spring. This means that 82 percent (100 − 18) of the original volume remains after application of the load.

Compute the liquid volume required from $v = \pi S d^2/(4c)$, where $v$ = liquid volume required, in$^3$; $S$ = stroke length, in; $d$ = rod diameter, in; $c$ = liquid compressibility, expressed as a decimal. Thus $v = \pi(5)(1)^2/[4(0.18)] = 21.8$ in$^3$ (357.2 cm$^3$).

#### 2. *Determine the cylinder length*

In a liquid spring, the cylinder inside diameter $d_i$ is usually greater than that of the rod. Assuming an inside diameter of 1.8 in (4.6 cm) for the cylinder, we find length $= 4v/(\pi d_i^2)$, where $d_i$ = cylinder inside diameter, in; other symbols as before. For this cylinder, length $= 4(21.8)/[\pi(1.8)^2] = 8.56$, say 8.6, in (21.8 cm).

#### 3. *Determine the cylinder dimensions*

With a 1.8-in (4.6-cm) inside diameter, a 3-in (7.6-cm) outside diameter will be required, based on the usual cylinder proportions. Allowing 3.4 in (8.6 cm) for the cylinder ends and seals and 5 in (12.7 cm) for the stroke, we find that the total length of the cylinder will be 8.6 + 3.4 + 5.0 = 17.0 in (43.2 cm).

#### 4. *Compute the cylinder temperature rise*

Assume that the average friction load is 10 percent of the load on the spring, or 0.1 × 50,000 = 5000 lb (22,241.1 N). A friction load of 10 percent is typical for liquid springs.

The energy absorbed per stroke of the spring is $e = Fl$, where $e$ = energy absorbed, ft·lb; $F$ = friction force, lb; $l$ = stroke length, ft. For this spring, $e = 5000(5/12) = 2085$ ft·lb (2826.9 N·m). Since 778.2 ft·lb = 1 Btu = 1.1 kJ, $e = 2085/778.2 = 2.68$ Btu (2827.6 J).

**TABLE 48** Performance of Four Typical Spring Types*

| | Coil | Nested: Belleville washers | Nested: Tapered rings | Liquid |
|---|---|---|---|---|
| Useful range: | | | | |
| Low load | 1 oz (28.3 g) | 20 lb (9.1 kg) | 2 tons (1.8 t) | 100 lb (45.4 kg) |
| High load | 10 ton (9.1 t) | 100 ton (90.7 t) | 150 ton (136.1 t) | 200 ton (181.4 t) |
| Force vs. deflection | Low to high | High | High | Medium to high |
| Stroke | Short to long | Short | Short to medium | Short to long |
| Damping ability | Low | Low | Low | Low to high |
| Relative cost | Low | Low | Medium | High |

*Product Engineering.*
Note: An example: For 50,000-lb (222,411.1-N) load, 5-in (12.7-cm) stroke:

| | | | | | |
|---|---|---|---|---|---|
| Size | Length, in (cm) | 68 (172.7) | 37 (94.0) | 24 (61.0) | 17 (43.2) |
| | Diameter, in (cm) | 11.5 (29.5) | 8 (20.3) | 5 (12.7) | 3 (7.6) |

An assembly of the dimensions computed in step 3 will weigh about 35 lb (15.9 kg) and will have an average overall specific heat of 0.15 Btu/(lb·°F) [628.0 J/(kg·°C)]. Hence, the temperature rise per stroke will be: Btu of heat generated per stroke/[(specific heat)(cylinder weight, lb)] = 0.51°F (0.28°C) per stroke. A temperature rise of this magnitude is easily dissipated by the external surfaces of the cylinder. But a smaller liquid spring under rapidly fluctuating loads may have an excessive temperature rise. Each spring must be analyzed separately.

**5. *Compare the various types of springs***

By using previously presented calculation procedures, Table 48 can be constructed. This table and the spring analysis given above are based on the work of Lloyd M. Polentz, Consulting Engineer. The tabulation shows that the liquid spring is the shortest and has the smallest diameter for the load in question. Figure 31 shows typical liquid springs; Fig. 32 shows the compressibility of liquids used in various liquid springs.

**Related Calculations:** Use the method given here to select liquid springs for applications in any of a variety of services where a large load must be absorbed. The seals at the cylinder ends must be absolutely tight. Liquid springs are best applied in atmospheres where the temperature variation is minimal.

## SELECTION OF AIR-SNUBBER DASHPOT DIMENSIONS

Determine the required orifice area, peak actuator pressure, peak negative acceleration, and the time required for the stroke of a 3-in$^3$ (49.2-cm$^3$) capacity air snubber if the total load mass $M$ = 0.1 lbs·s$^2$/in (17.9 g·s$^2$/cm); the snubber pressure $P_i$ = 100 lb/in$^2$ (689.5 kPa); piston area $A_p$ = 3 in$^2$ (19.4 cm$^2$); initial snubber active length $S$ = 1.0 in (2.5 cm); initial piston velocity $v_i$ = 100 in/s (254 cm/s); piston velocity at the end of travel $v$ = 29 in/s (73.7 cm/s); constant external force on snubber $F$ = 150 lb (667.2 N); initial gas temperature $T_i$ = 530 R (294.1 K); gas constant $R$ = [639.6 in·lb/(lb·°R)] (air); $C_D$ = orifice discharge coefficient = 0.9 dimensionless.

### Calculation Procedure:

**1. *Compute the snubber dimensionless parameters***

The first dimensionless parameter $K_E$ = stored energy/kinetic energy = $P_iV_i(Mv_i^2)$ = (100)(3)/[(0.1)(100)$^2$] = 0.3. The next parameter $K_F$ = constant external force/initial pressure force = $F$/

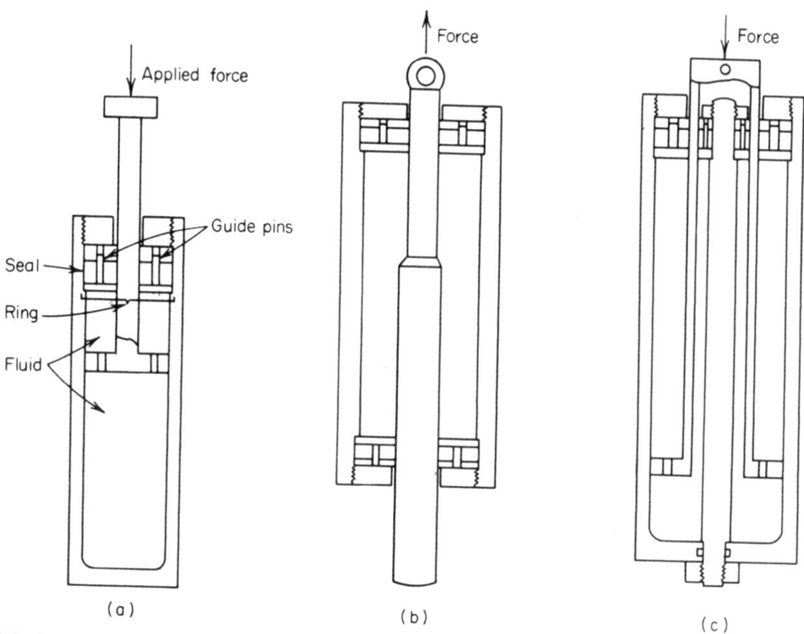


**FIG. 31** Typical liquid springs. (*a*) General design; (*b*) tension type; (*c*) long-stroke type. *(Product Engineering.)*

$(P_iA_p)$ = 150/[(100)(3)] = 0.5. The third parameter $K_v$ = piston velocity at end of stroke/initial piston velocity = $v/v_i$ = 20/100 = 0.20.

## 2. *Determine the actual value of the orifice parameter*

The parameter $K_w$ = initial orifice flow/initial displacement flow = $w_i/(\rho A_p v_i)$, where $\rho$ = gas density, lb/in$^3$. Figure 33 gives values of $K_w$ for $K_F = 0$ and $K_F = 1.0$. However, $K_F$ for this snubber = 0.5. Therefore, it is necessary to interpolate between the charts for $K_F = 0$ and $K_F = 1.0$.

Interpolate by constructing a chart, Fig. 34, using values of $K_w$ read from each chart in Fig. 33. Thus, when $K_F = 1$, $K_v = 0.2$, $K_E = 0.3$, $K_w = 0.295$. After the curve is constructed, read $K_w = 0.375$ for $K_f = 0.5$.

## 3. *Compute the true flow through the orifice*

The true initial flow rate $w_i$, lb/s = $K_w[P_i/(RT_i)]A_p v_i$, where all the symbols are as defined earlier. Thus, $w_i$ = (0.375){100/[(639.6)(530)]}(3)(100) = 0.0332 lb/s (15.1 g/s).

## 4. *Compute the required orifice area*

Use the equation $A_o = w_i/P_iC_D\{(kg/RT_i)[2/(k+1)](k+1)/(k-1)\}^{0.5}$, where $k = 1.4$; $g$ =

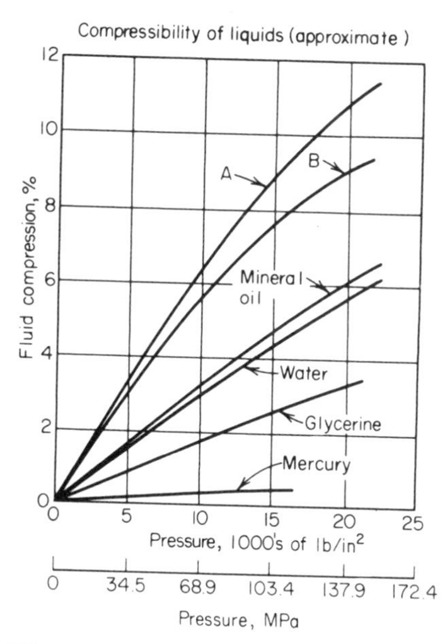


**FIG. 32** Common fluids for liquid springs are Dow-Corning type F-4029, curve *A*, and type 200, curve *B*. *(Product Engineering.)*

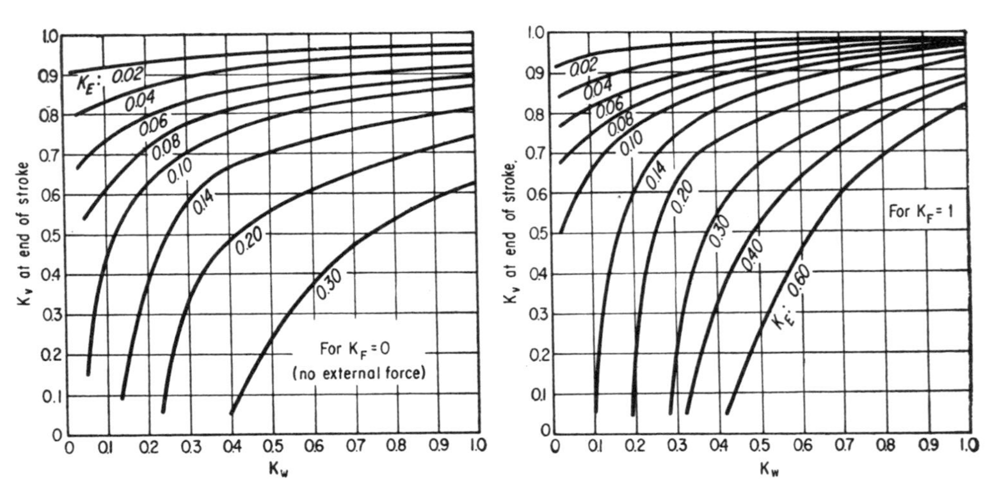


**FIG. 33** Impact velocity vs. orifice flow (dimensionless) for air snubber. *(Product Engineering.)*

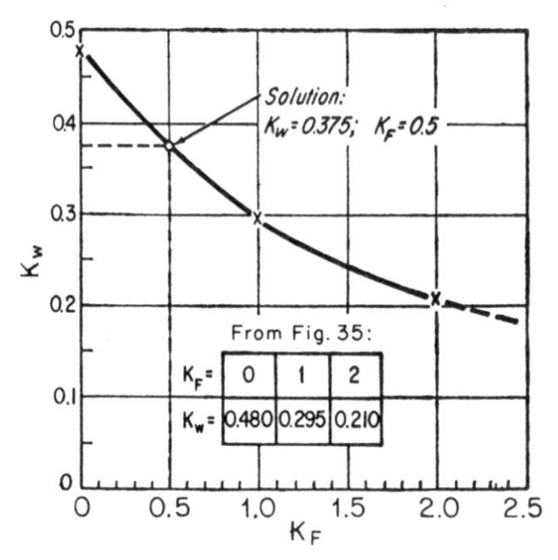


**FIG. 34** Cross plot for an air snubber. *(Product Engineering.)*

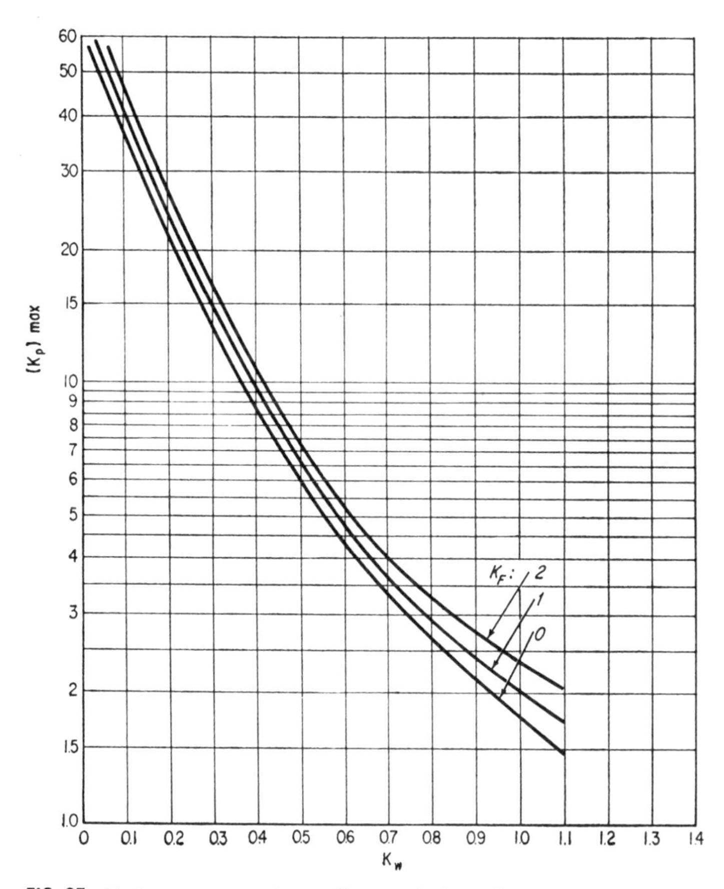


**FIG. 35** Maximum pressure at impact (dimensionless). *(Product Engineering.)*

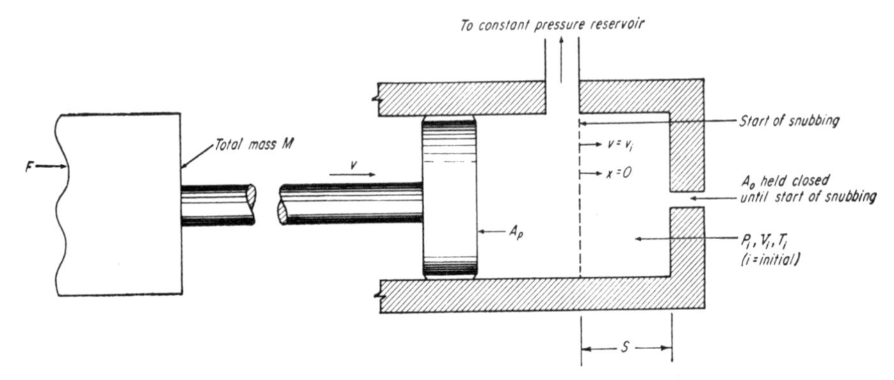


**FIG. 36** Simplified air-snubber design. *(Product Engineering.)*

32.2 ft/s² (9.8 m/s²); other symbols as before. Thus, $A_o = 0.0332/[100(0.9)]\{(1.4 \times 32.2/639.6 \times 530)[(2/(1.4 + 1)](1.4 + 1)/(1.4 - 1)\}^{0.5} = 0.016$ in² (0.103 cm²).

### 5. *Determine the maximum pressure at the end of the stroke*

Read, from Fig. 35, $K_{p,\max} = 10.2$ for $K_F = 0.5$, $K_w = 0.375$. Then the true $P_{\max}$ at end of stroke $= K_pP_i = 10.2(100) = 1020$ lb/in² (7032.9 kPa).

### 6. *Determine the maximum acceleration of the piston*

For an air snubber, $K_{a,\max} = K_F - K_p = 0.5 - 10.2 = -9.7$. Also, the maximum acceleration $a_{\max} = K_aP_iA_p/M = (-9.7)(100)(3.0)/0.1 = -29{,}100$ in/s² ($-739.1$ m/s²).

### 7. *Determine the approximate travel time of the piston*

The travel time for the piston is $t = K_tS/v_i$, where $t$ = travel time, s. Or, $t = 0.95 \times (1.0)/100 = 0.0095$ s, assuming $K_t = 0.95$.

**Related Calculations:** The equations in this procedure were developed by Tom Carey and T. T. Hadeler, and are based on these assumptions: (1) They apply only to a piston-orifice-type dashpot; (2) the piston is firmly stopped at the end of the stroke and does not rebound, oscillate, or bounce; (3) friction is zero; (4) the external force is constant; (5) $k = 1.4$, which means that the equations are valid for air, hydrogen, nitrogen, oxygen, and any other gas having a specific-heat ratio of about 1.4; (6) the contained air or gas is ideal; (7) compression is adiabatic; (8) flow through the bleed orifice is critical (a valid assumption except when the dashpot initial pressure is atmospheric, as in screendoor snubbers). When actual friction exists, there is a slight increase in the value of $K_t$. Figure 36 shows a simplified design of a typical air snubber.

## DESIGN ANALYSIS OF FLAT REINFORCED-PLASTIC SPRINGS

A large shaker unit in a vibrating screen system is supported on a series of six steel leaf springs, each a cantilever 6 in (15.2 cm) wide by 0.125 in (3.18 mm) thick. How thick should a single epoxy-glass leaf spring of the same width be if it is to replace the composite steel spring? The cantilever is 30 in (76.2 cm) long with a 24-in (60.9-cm) free length; maximum deflection = 0.375 in (9.53 mm); axial load per spring = 2500 lb (11,120.6 N); safety factor = 8; $E = 4.5 \times 10^6$ lb/in² ($31.0 \times 10^9$ Pa) for the plastic spring; ultimate flexure strength = 100,000 lb/in² (689,475.7 kPa).

### Calculation Procedure:

### 1. *Compute the spring thickness for minimum bending stress*

The equation for the thickness giving the minimum bending stress is $t = [4Ll^2/(wE)]^{1/3}$, where $t$ = spring thickness, in; $L$ = axial load on spring, lb; $l$ = spring free length, in; $w$ = spring width, in; $E$ = modulus of elasticity of spring material, lb/in². For this spring, $t = [4 \times 2500 \times 24^2/(6 \times 4.5 \times 10^6)]^{1/3} = 0.598$, say 0.6 in (1.5 cm).

**2. Determine the total combined stress in the beam**

The maximum combined stress $s_B = (3Et/l^2 + 6L/wt^2)f$, where $f$ = spring deflection, in; other symbols as before. Thus, $s_B = \{e \times 4.5 \times 10^6 \times 0.6/[24^2 + 6 \times 2500/(6 \times 0.6^2)]\} \times 0.375 =$ 7875 lb/in² (54,290 kPa).

The total stress in the spring $s_T = s_B + L/A$, where $A$ = spring cross-sectional area. Thus, $s_T = 7875 + 2500/(6 \times 0.6) = 8570$ lb/in² (59,082 kPa). The allowable stress = ultimate flexure strength, lb/in²/factor of safety = 100,000/8 = 12,500 lb/in² (86,175.5 kPa). Since $s_T < 12{,}500$ lb/in² (86,175.5 kPa), the dimensions of the spring are satisfactory.

**3. Check the critical buckling stress of the spring**

To prevent buckling of the spring, the following must hold: $L/A \leq \pi^2Et^2/36^2 = s_{CR}$, or $2500/(6 \times 0.6) \leq \pi^2 \times 4.5 \times 10^6 \times 0.6^2/(3 \times 24^2) = 694$ lb/in² (4784 kPa) $<$ 9240 lb/in² (63,701 kPa). Hence, the spring dimensions are satisfactory.

**4. Determine whether the computed thickness gives adequate stiffness**

For a plastic spring to have a stiffness equal to a steel spring having $n$ leaves, the plastic spring should have a thickness of $t = t_s(nE_s/E)^{1/3}$, where $t_s$ and $E_s$ refer to the thickness, in, and modulus of elasticity, lb/in², respectively, of the steel spring; other symbols as before. Thus, $t = 0.125[6 \times 30 \times 10^6/(4.5 \times 10^6)]^{1/3} = 0.43$ in (1.1 cm). Since $t = 0.60$ in (1.5 cm), as computed in step 1, the plastic spring is slightly too stiff.

**5. Check the thickness required for equivalent thickness**

Using the equations for $s_B$ and $s_T$ from step 2, and the equation for $s_{CR}$ from step 3, compute the respective stresses for values of $t$ less than, and greater than, 0.43. Thus

| $t$ | | $s_B$ | | $Q/A$ | | $s_T$ | | $s_{CR}$ | |
|---|---|---|---|---|---|---|---|---|---|
| in | cm | lb/in² | kPa | lb/in² | kPa | lb/in² | kPa | lb/in² | kPa |
| 0.500 | 1.270 | 8,143 | 56,144.0 | 833 | 5,743.5 | 8,976 | 61,889.5 | 6,400 | 44,128.0 |
| 0.600 | 1.524 | 7,875 | 54,296.2 | 695 | 4,792.0 | 8,570 | 59,090.2 | 9,240 | 63,709.8 |
| 0.625 | 1.588 | 7,900 | 54,468.6 | 666 | 4,592.1 | 8,566 | 59,062.6 | 10,000 | 68,947.6 |

Plot the results as in Fig. 37. This plot clearly shows that $t = 0.43$ in (1.09 cm) gives $s_T < 12{,}500$ lb/in2 (86,175 kPa), and $Q/A < s_{CR}$. Hence, this thickness is satisfactory.

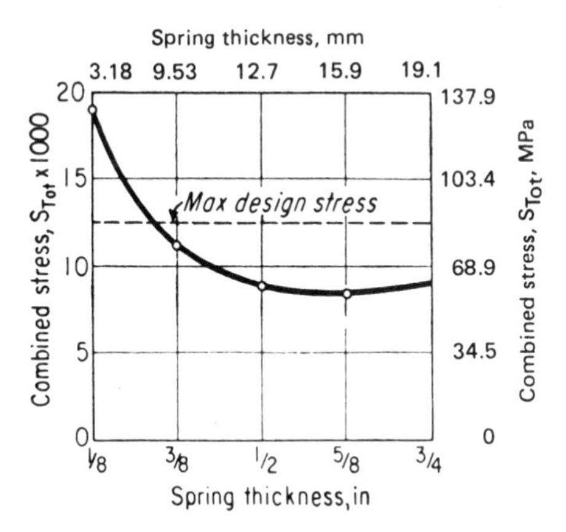


**FIG. 37** Combined stress in a plastic spring. *(Product Engineering.)*

**6. *Determine whether a thinner spring can be used***

A thinner spring will save money. From Fig. 37, $t$ = 0.375 in (9.53 mm) gives an $s_T$ value well below the maximum design stress, and the actual spring stress is one-third the critical buckling stress. If tests on a 0.375-in (9.53-mm) thick spring show no serious disruption of harmonic operation, then specify the thinner material to lower the cost. Otherwise, use the thicker 0.43-in (1.09-cm) spring.

**Related Calculations:** Use this procedure for unidirectional, cross-plied, or isotropic-ply plastic springs. Obtain the allowable stress for the spring from the plastic manufacturer. The method given here is the work of L. A. Heggernes, reported in *Product Engineering*.

## LIFE OF CYCLICALLY LOADED MECHANICAL SPRINGS

What is the probable life in cycles of a Belleville spring under a bending load if it is made of carbon steel having a Rockwell hardness of C48?

### Calculation Procedure:

**1. *Determine the spring material tensile strength***

Enter Fig. 38 at the Rockwell hardness C48, and project vertically upward to read the tensile strength of the carbon-steel spring material as 235,00 lb/in$^2$ (1620 MPa).

**2. *Compute the actual stress in the spring***

Using the spring dimensions and the equations presented in the Belleville spring calculation procedure, compute the actual stress in the spring. For the spring in question, the actual stress is found to be 150,000 lb/in$^2$ (1034 MPa). This is 150,000(100)/235,000 = 63.8, say 64, percent of the spring material tensile strength.

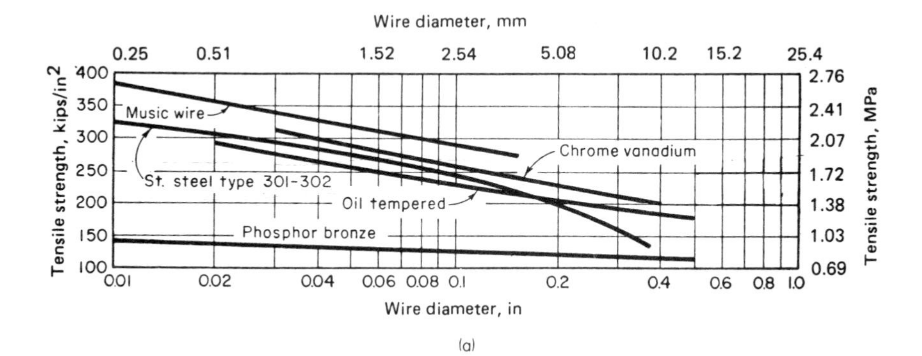


(a)

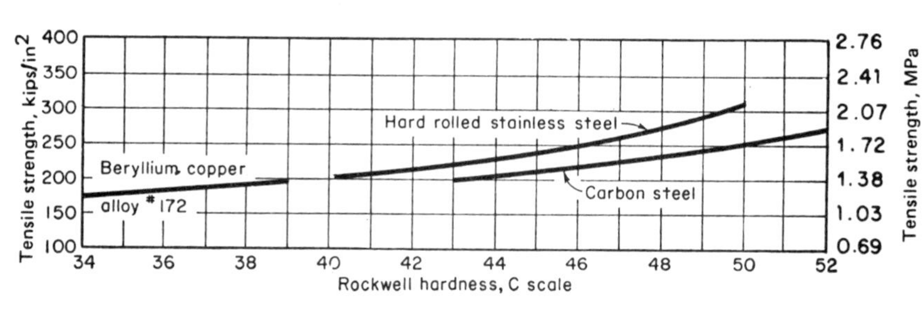


(b)

**FIG. 38** (*a*) Tensile strength of spring wire; (*b*) tensile strength of spring strip. *(Product Engineering.)*

### 3. *Estimate the spring cycle life*

Enter the upper part of Table 49 for springs in bending. This tabulation shows that at a stress of 65 percent of the tensile strength, the spring will have a life between 10,000 and 100,000 stress cycles. Actual test of the spring caused failure at about 100,000 cycles.

**Related Calculations:** Use this procedure for helical torsion springs, cantilever springs, wave washers, flat springs, motor springs, helical compression and extension springs, torsion bars, and Belleville springs. Be sure to enter the proper portion of Table 49 when finding the approximate number of repetitive stress cycles. The method presented here is the work of George W. Kuasz and William R. Johnson and is reported in *Product Engineering*.

## SHOCK-MOUNT DEFLECTION AND SPRING RATE

Determine the maximum probable acceleration, the shock isolator deflection, and the isolator spring rate for a 25-lb (11.3-kg) piece of electronic equipment which drops from a 24-in (61.0-cm) high tailgate of a truck onto a concrete road. The product lands on one corner point and should be considered as rigid steel for analysis purposes. In its carton, the load will be supported by 16 shock isolators.

### Calculation Procedure:

### 1. *Compute the acceleration of the load*

Use the relation $g = (72/t)(h)^{0.5}$, where $g$ = load acceleration in $g$ [1 $g$ = 32.2 ft/s$^2$ (9.8 m/s$^2$) at sea level]; $t$ = shock-rise time, ms, from Table 50; $h$ = drop height, in. From Table 50, $t$ = 2 ms for rigid steel making point contact with concrete. Then $g = 72/2 \times (24)^{0.5} = 176.5g$.

### 2. *Compute the isolator deflection*

Use the relation $d = 2h/(g - 1)$, where $d$ = isolator deflection, in; other symbols as before. For this load, $d = 2 \times 24/(176.5 - 1) = 0.273$ in (6.93 mm).

### 3. *Compute the required specific spring rate for the isolator*

Use the relation $K = g/d$, where $D$ = isolator specific spring rate, lb/(in·lb). Thus $K = 176.5/0.273 = 646$ lb/(in·lb) [254.3 N/(N·cm)].

**TABLE 49** Design Stresses for Springs*

| No. of repetitive stress cycles | Maximum design stress (percent of the tensile strength shown in charts) |
|---|---|
| Design stress for springs in bending† | |
| 10,000 | 80 |
| | 65‡ |
| 100,000 | 53 |
| 1,000,000 | 50 |
| 10,000,000 | 48 |
| Design stress for springs in torsion§ | |
| 10,000 | 45 |
| | 35‡ |
| 100,000 | 35 |
| 1,000,000 | 33 |
| 10,000,000 | 30 |

* *Product Engineering.*
† For example, helical torsion springs, Bellevilles, cantilever springs, wave washers, flat springs, and motor springs.
‡ For stainless-steel and phosphor-bronze materials. Tests show that such materials have low yield points.
§ For example, helical compression springs, helical extension springs, and torsion bars.

**TABLE 50** Typical Value for Shock-Time Rise

| Condition | Shock-time rise, ms | |
|---|---|---|
| | Flat face | Point |
| Rigid steel against concrete | 1 | |
| Rigid steel against wood or mastic | 2–3 | 5–6 |
| Steel or aluminum against compact earth | 2–4 | 6–8 |
| Steel or aluminum against sand | 5–6 | 15 |
| Product case against mud | 15 | 20 |
| Product case against 1-in (2.5-cm) felt | 20 | 30 |

*Note:* Mass of struck surface is assumed to be at least 10 times the striking mass. Point contact with spherical radius of 1 in (2.5 cm).

**4. *Determine the required spring rate per isolator***

With $n$ shock isolators, the required spring rate, lb/in per isolator is $k = KW/n$, where $W$ = weight of part, lb; $n$ = number of isolators used in the carton. Thus, $k = (646)(25)/16 = 1020$ lb/in (182.2 kg/cm).

**Related Calculations:** Some of the largest stock loads encountered by equipment occur during transportation. Thus, vertical accelerations on the body of a 2-ton (1.8-t) truck traveling at 30 mi/h (13.4 m/s) on good pavement range from 1 to 2 $g$, with a rise time of 10 to 15 ms. Higher speeds, rougher roads, stiffer truck springs, and careless driving all decrease the rise time and thus double or triple the acceleration loads.

The highest acceleration forces in railroad freight cars occur during humping, when the impact loads on a product container may range from 4.5 to 28 $g$.

For most components that are sensitive to shock, suppliers include maximum safe acceleration loads in engineering data. Maximum allowable loads on vacuum tubes are 2 to 5 $g$; relays may withstand higher accelerations, depending on the type and direction of the acceleration. Transistors have low mass and good rigidity and are highly resistant to shock when properly supported. Ball-bearing races may be indented by the balls; sleeve bearings are usually much more resistant to shock.

The function of a shock mount is to provide enough protection to avoid damage under expected conditions. But overdesign can be costly, both in the design of the product and in the shock-mount components. Underdesign can lead to failures of the shock mount in service and possible damage to the product. Therefore, careful design of shock mounts is important. The method presented here is the work of Raymond T. Magner, reported in *Product Engineering*.

## CLUTCH SELECTION FOR SHAFT DRIVE

Choose a clutch to connect a 50-hp (37.3-kW) internal-combustion engine to a 300-r/min single-acting reciprocating pump. Determine the general dimensions of the clutch.

**TABLE 51** Clutch Characteristics

| Type of clutch | Typical applications* |
|---|---|
| Friction: | |
| Cone | Varying loads; 0 to 200 hp (0 to 149.1 kW); losing popularity for many applications, particularly in the higher hp ranges |
| Disk or plate | Varying loads; 0 to 500 hp (0 to 372.9 kW); widely used; more popular than the cone clutch |
| Rim: | |
| Band | Varying loads; 0 to 100 hp (0 to 74.6 kW); not too widely used |
| Overrunning | Constant or moderately varying loads; 0 to 200 hp (0 to 149.1 kW); engages in one direction; freewheels in the opposite direction |
| Centrifugal | Constant loads; 0 to 50 hp (0 to 37.3 kW) |
| Inflatable | Varying loads; 0 to 5000 hp (0 to 3728.5 kW); compressed air inflates clutch; have 360° friction surface |
| Magnetic | Varying loads; 0 to 10,000 hp (0 to 7457.0 kW); high speeds; also used where disk clutch would be overloaded |
| Positive-engagement | Nonslip operation; low-speed (10 to 150 r/min) engagement; has sudden starting action |
| Fluid | Large, varying loads; 0 to 10,000 hp (0 to 7457.0 kW); variable-speed output; can produce a desired slip |
| Electromagnetic | Large, varying loads; 0 to 10,000 hp (0 to 7457.0 kW); variable-speed output; characteristics similar to fluid clutches |

*Clutch capacity depends on the design, materials of constructions, type of load, shaft speed, and operating conditions. The applications and capacity ranges given here are typical but should not be taken as the only uses for which the listed clutches are suitable.

**TABLE 52** Clutch Service Factors

| Type of service | Service factor |
|---|---|
| Driver: | |
| Electric motor: | |
| Steady load | 1.0 |
| Fluctuating load | 1.5 |
| Gas engine: | |
| Single cylinder | 1.5 |
| Multiple cylinder | 1.0 |
| Diesel engine: | |
| High-speed | 1.5 |
| Large, slow-speed | 2.0 |
| Driven machine: | |
| Generator: | |
| Steady load | 1.0 |
| Fluctuating load | 1.5 |
| Blower | 1.0 |
| Compressor, depending on number of cylinders | 2.0–2.5 |
| Pumps: | |
| Centrifugal | 1.0 |
| Reciprocating, single-acting | 2.0 |
| Reciprocating, double-acting | 1.5 |
| Lineshaft | 1.5 |
| Woodworking machinery | 1.75 |
| Hoists, elevators, cranes, and shovels | 2.0 |
| Hammer mills, ball mills, and crushers | 2.0 |
| Brick machinery | 3.0 |
| Rock crushers | 3.0 |

## Calculation Procedure:

### 1. *Choose the type of clutch for the load*

Table 51 shows typical applications for the major types of clutches. Where economy is the prime consideration, a positive-engagement or a cone-type friction clutch would be chosen. Since a reciprocating pump runs at a slightly varying speed, a centrifugal clutch is not suitable. For greater dependability, a disk or plate friction clutch is more desirable than a cone clutch. Assume that dependability is more important than economy, and choose a disk-type friction clutch.

### 2. *Determine the required clutch torque starting capacity*

A clutch must start its load from a stopped condition. Under these circumstances the instantaneous torque may be two, three, or four times the running torque. Therefore, the usual clutch is chosen so it has a torque capacity of at least twice the running torque. For internal-combustion engine drives, a starting torque of three to four times the running torque is generally used. Assume 3.5 times is used for this engine and pump combination. This is termed the *clutch starting factor*.

Since $T = 63{,}000hp/R$, where $T$ = torque, lb·in; $hp$ = horsepower transmitted; $R$ = shaft rpm; $T = 63{,}000(50)/300 = 10{,}500$ lb·in (1186.3 N·m). This is the required starting torque capacity of the clutch.

### 3. *Determine the total required clutch torque capacity*

In addition to the clutch starting factor, a service factor is also usually applied. Table 52 lists typical clutch service factors. This tabulation shows that the service factor for a single-reciprocating pump is 2.0. Hence, the total required clutch torque capacity = required starting torque capacity × service factor = 10,500 × 2.0 = 21,000-lb·in (2372.7-N·m) torque capacity.

**4. *Choose a suitable clutch for the load***

Consult a manufacturer's engineering data sheet listing clutch torque capacities for clutches of the type chosen in step 1 of this procedure. Choose a clutch having a rated torque equal to or greater than that computed in step 3. Table 53 shows a portion of a typical engineering data sheet. A size 6 clutch would be chosen for this drive.

**TABLE 53** Clutch Ratings

| Clutch number | Torque rating | | Power (100 r/min) | |
|---|---|---|---|---|
| | lb·in | N·m | hp | kW |
| 1 | 2,040 | 230.5 | 3 | 2.2 |
| 2 | 4,290 | 484.7 | 6 | 4.5 |
| 3 | 8,150 | 920.8 | 12 | 8.9 |
| 4 | 13,300 | 1,502.7 | 21 | 15.7 |
| 5 | 19,700 | 2,225.8 | 31 | 23.1 |
| 6 | 35,200 | 3,977.1 | 55 | 41.0 |
| 7 | 44,000 | 4,971.3 | 69 | 51.5 |

**Related Calculations:** Use the general method given here to select clutches for industrial, commercial, marine, automotive, tractor, and similar applications. Note that engineering data sheets often list the clutch rating in terms of torque, lb·in, and hp/(100 r/min).

Friction clutches depend, for their load-carrying ability, on the friction and pressure between two mating surfaces. Usual coefficients of friction for friction clutches range between 0.15 and 0.50 for dry surfaces, 0.05 and 0.30 for greasy surfaces, and 0.05 and 0.25 for lubricated surfaces. The allowable pressure between the surfaces ranges from a low of 8 lb/in$^2$ (55.2 kPa) to a high of 300 lb/in$^2$ (2068.5 kPa).

## BRAKE SELECTION FOR A KNOWN LOAD

Choose a suitable brake to stop a 50-hp (37.3-kW) motor automatically when power is cut off. The motor must be brought to rest within 40 s after power is shut off. The load inertia, including the brake rotating member, will be about 200 lb·ft$^2$ (82.7 N·m$^2$); the shaft being braked turns at 1800 r/min. How many revolutions will the shaft turn before stopping? How much heat must the brake dissipate? The brake operates once per minute.

### Calculation Procedure:

**1. *Choose the type of brake to use***

Table 54 shows that a shoe-type electric brake is probably the best choice for stopping a load when the braking force must be applied automatically. The only other possible choice—the eddy-current brake—is generally used for larger loads than this brake will handle.

**2. *Compute the average brake torque required to stop the load***

Use the relation $T_a = Wk^2n/(308t)$, where $T_a$ = average torque required to stop the load, lb·ft; $Wk^2$ = load inertia, including brake rotating member, lb·ft$^2$, $n$ = shaft speed prior to braking, r/min; $t$ = required or desired stopping time, s. For this brake, $T_a$ = (200)(1800)/[308(40)] = 29.2 lb·ft, or 351 lb·in (39.7 N·m).

**3. *Apply a service factor to the average torque***

A service factor varying from 1.0 to 4.0 is usually applied to the average torque to ensure that the brake is of sufficient size for the load. Applying a service factor of 1.5 for this brake yields the required capacity = 1.5(351) = 526 in·lb (59.4 N·m).

**4. *Choose the brake size***

Use an engineering data sheet from the selected manufacturer to choose the brake size. Thus, one manufacturer's data show that a 16-in (40.6-cm) diameter brake will adequately handle the load.

**5. *Compute the revolutions prior to stopping***

Use the relation $R_s = tn/120$, where $R$ = number of revolutions prior to stopping; other symbols as before. Thus, $R_s$ = (40)(1800)/120 = 600 r.

**6. *Compute the heat the brake must dissipate***

Use the relation $H = 1.7\ FWk^2(n/100)^2$, where $H$ = heat generated at friction surfaces, ft·lb/min; $F$ = number of duty cycles per minute; other symbols as before. Thus, $H$ = 1.7(1)(200)(1800/100)$^2$ = 110,200 ft·lb/min (2490.2 N·m/s).

**TABLE 54** Mechanical and Electrical Brake Characteristics

| Type of brake | Typical characteristics |
|---|---|
| Block | Wooden or cast-iron shoe bearing on iron or steel wheel; double blocks prevent bending of shaft; used where economy is prime consideration; leverage 5:1 |
| Band | Asbestos fabric bearing on metal wheels; fabric may be reinforced with copper wire and impregnated with asphalt; bands are faced with wooden blocks; used where economy is a major consideration; leverage 10:1 |
| Cone | Friction surface attached to metal cone; popular for cranes; coefficient of friction = 0.08 to 0.10; useful for intermittent braking applications |
| Disk | Have one or more flat braking surfaces; effective for large loads; continuous application |
| Internal-shoe | Popular for vehicles where shaft rotation occurs in both directions; self-energizing, i.e., friction makes shoe follow rotating brake drum; capable of large braking power |
| Eddy-current | Used for flywheels requiring quick braking and where large kinetic energy of rotating masses precludes use of block brakes because of excessive heating |
| Electric, shoe-type | Used where automatic application of brake is required as soon as power is turned off; spring-activated brake shoes apply the braking action |
| Electric, friction-disk type | Best for duty cycles requiring a number of stops and starts per minute; may have one or multiple disks |

**7.** ***Determine whether the brake temperature will rise***

From the manufacturer's data sheet, find the heat dissipation capacity of the brake while operating and while at rest. For a 16-in (40.6-cm) shoe-type brake, one manufacturer gives an operating heat dissipation $H_o$ = 150,000 ft · lb/min (3389.5 N · m/s) and an at-rest heat dissipation of $H_v$ = 35,000 ft · lb/min (790.9 N · m/s).

Apply the cycle time for the event; i.e., the brake operates for 40 s, or 40/60 of the time, and is at rest for 20 s, or 20/60 of the time. Hence, the heat dissipation of the brake is (150,000)(40/60) + (35,000)(20/60) = 111,680 ft · lb/min (2523.6 N · m/s). Since the heat dissipation, 111,680 ft · lb/min (2523.6 N · m/s), exceeds the heat generated, 110,200 ft · lb/min (2490.2 N · m/s), the temperature of the brake will remain constant. If the heat generated exceeded the heat dissipated, the brake temperature would rise constantly during the operation.

Brake temperatures higher than 250°F (121.1°C) can reduce brake life. In the 250 to 300°F (121.1 to 148.9°C) range, periodic replacement of the brake friction surfaces may be necessary. Above 300°F (148.9°C), forced-air cooling of the brake is usually necessary.

**Related Calculations:** Because electric brakes are finding wider industrial use, Tables 55 and 56, summarizing their performance characteristics and ratings, are presented here for easy reference.

The coefficient of friction for brakes must be carefully chosen; otherwise, the brake may "grab," i.e., attempt to stop the load instantly instead of slowly. Usual values for the coefficient of friction range between 0.08 and 0.50.

The methods given above can be used to analyze brakes applied to hoists, elevators, vehicles, etc. Where $Wk^2$ is not given, estimate it, using the moving parts of the brake and load as a guide to the relative magnitude of load inertia. The method presented is the work of Joseph F. Peck, reported in *Product Engineering*.

## MECHANICAL BRAKE SURFACE AREA AND COOLING TIME

How much radiating surface must a brake drum have if it absorbs 20 hp (14.9 kW), operates for half the use cycle, and cannot have a temperature rise greater than 300°F (166.7°C)? How long will it take this brake to cool to a room temperature of 75°F (23.9°C) if the brake drum is made of cast iron and weighs 100 lb (45.4 kg)?

**TABLE 55** Performance Characteristics of Electric Brakes

| Brake type | Operational mode | | Design characteristics | | | | | Brake functions performed | | | | | On-off duty-cycling capability |
|---|---|---|---|---|---|---|---|---|---|---|---|---|---|
| | On-off | Continuous | Torque adjustment | Torque-control range | Wear adjustment | Residual drag | Heat dissipation | Instant stop | Cushioned stop | Retard (drag) | Hold | Failsafe brake | |
| Magnetic particle | Yes | Yes | Electrical | Wide | Nonwearing | High | Limited | No | Yes | Yes | Yes | No | Limited by heat-dissipation capability to low-inertia loads |
| Eddy-current, air-cooled | Yes | Yes | Electrical | Wide | Nonwearing | Moderate to low | Good | No | Yes | Yes | No | No | Limited to long time cycles |
| Eddy-current, water-cooled | Yes | Yes | Electrical | Wide | Nonwearing | High to moderate | Excellent | No | Yes | Yes | No | No | Limited to long time cycles |
| Single-disk friction, electrically actuated | Yes | Yes | Electrical | Wide | Self-compensating | None | Excellent | Yes | Yes | Yes | Yes | No | Excellent—up to several hundred stops per minute |
| Multidisk friction, electrically actuated, direct-acting | Yes | No | Electrical | Moderate | | Low | Limited | Yes | Yes | No | Yes | No | Same as comparable size electric motor: 12 stops per minute (maximum) |
| Multidisk friction, electrically actuated, indirect-acting | Yes | No | Mechanical | Limited | Mechanical | | | Yes | Semisoft | No | Yes | No | Same as comparable size electric motor: 12 stops per minute (maximum) |
| Multidisk friction, spring-actuated | Yes | No | Mechanical | Limited | Mechanical | Low | Limited | Yes | Semisoft | No | Yes | Yes | Same as comparable size electric motor: 12 stops per minute (maximum) |
| Shoe brake, spring-actuated | Yes | No | Mechanical | Limited | Mechanical | None | Good | Yes | Semisoft | No | Yes | Yes | Generally not over 3 stops per minute without derating |

**TABLE 56** Representative Range of Ratings and Dimensions for Electric Brakes

| Brake type | hp (W) | Torque, maximum, lb·ft (N·m) | Shaft speed, maximum, r/min | Diameter, in (cm) | Length, in (cm) | Inertia of rotating member, lb·ft$^2$ (N·m$^2$) |
|---|---|---|---|---|---|---|
| Magnetic particle brakes | 1/50–25 (14.9–18,643) | 0.6–150 (0.8–203.4) | 1,000–2,000 | 2–10 (5.1–25.4) | 2–6 (5.1–15.2) | $1.5 \times 10^{-4}$–0.27 ($6.2 \times 10^{-5}$–0.11) |
| Eddy-current brakes: | | | | | | |
| Air cooled | 3/4–75 (559.3–55,928) | 5–1,740 (6.8–2,359) | 2,000–900 | 6½–24¾ (16.5–62.9) | 9½–43½ (24.1–110.5) | 0.12–100 (0.05–41.3) |
| Water-cooled | 40–800 (29,828–596,560) | 130–4,600 (176.3–6,237) | 1,800–1,200 | 14¾–36½ (37.5–92.7) | 18½–43 (47.0–109.2) | 8.5–725 (3.5–299.6) |
| Friction disk brakes: | | | | | | |
| Single-disk, electrically actuated | 1/50–200 (14.9–149,140) | 0.17–700 (0.23–949.1) | 10,000–1,800 | 1½–15¼ (3.8–38.7) | 1½–4½ (3.8–11.4) | 0.000125–3 (0.000052–1.2) |
| Multiple-disk, electrically actuated | ¼–2,000 (186.4–1,491,400) | 3–15,000 (4.1–20,337) | 5,000–750 | 2¼–21 (5.7–53.3) | 2–8 (5.1–20.3) | Up to 90 (Up to 37.2) |
| Multiple-disk, spring actuated | ¼–2,000 (186.4–1,491,400) | 4–7,500 (5.4–10,169) | 5,000–1,200 | 4–29 (10.2–73.7) | 2½–16½ (6.4–41.9) | |
| Shoe brakes, spring-actuated | 1–2,500 (745.7–1,864,250) | 3–10,000 (4.1–13,558) | 10,000–1,200 | 2–28 (5.1–71.1) | 4½–12 (11.4–30.5) | 0.023–485 (0.010–200.4) |

## Calculation Procedure:

### 1. *Compute the required radiating area of the brake*

Use the relation $A = 42.4hpF/K$, where $A$ = required brake radiating area, in$^2$; hp = power absorbed by the brakes; $F$ = brake load factor = operating portion of use cycle; $K$ = constant = $Ct_r$, where $C$ = radiating factor from Table 57, $t_r$ = brake temperature rise, °F. For this brake, assuming a full 300°F (166.7°C) temperature rise and using data from Table 57, we get $A = 42.4(20)(0.5)/[(0.00083)(300)] = 1702$ in$^2$ (10,980.6 cm$^2$).

**TABLE 57** Brake Radiating Factors

| Temperature rise of brake | | |
|---|---|---|
| °F | °C | Radiating factor $C$ |
| 100 | 55.6 | 0.00060 |
| 200 | 111.1 | 0.00075 |
| 300 | 166.7 | 0.00083 |
| 400 | 222.2 | 0.00090 |

### 2. *Compute the brake cooling time*

Use the relation $t = (cW \ln t_r)/(K_cA)$, where $t$ = brake cooling time, min; $c$ = specific heat of brake-drum material, Btu/(lb·°F); $W$ = weight of brake drum, lb; $t_r$ = drum temperature rise, °F; ln = log to base $e$ = 2.71828; $K_c$ = a constant varying from 0.4 to 0.8; other symbols as before. Using $K_c = 0.4$, $c = 0.13$, $t = (0.13 \times 100 \ln 300)/[(0.4)(1702)] = 0.1088$ min.

**Related Calculations:** Use this procedure for friction brakes used to stop loads that are lifted or lowered, as in cranes, moving vehicles, rotating cylinders, and similar loads.

# INVOLUTE SPLINE SIZE FOR KNOWN LOAD

Choose the type and size of involute spline to transmit a torque of 10,000 lb·in (1129.8 N·m) from an electric motor to a centrifugal pump. What are the required face width and number of teeth for this spline?

## Calculation Procedure:

### 1. *Select the type of spline to use*

Involute splines are usually chosen for industrial drives because this type transmits more torque for its size than a parallel-side spline does. The involute spline has almost no speed limitation, being used at speeds of 10,000 r/min and higher. Further, an involute spline can be cut and measured by the same machines that cut and measure gear teeth. A spline, however, differs from a gear in that the spline has no rolling action and all teeth are in contact at once.

Involute splines may be either *flexible* or *fixed*. Flexible splines allow some rocking motion; and under torque, the teeth slip axially to accommodate axial expansion or runout. Fixed splines allow no relative or rocking motion between the internal and external teeth. The fixed-type spline can be either shrink-fitted or loosely fitted together. For a centrifugal-pump drive, the flexible-type spline is generally preferred. Therefore, a flexible involute spline will be chosen for this drive. A standard commercial grade will be acceptable.

### 2. *Determine the pitch diameter of the spline*

Enter Fig. 39 at a torque of 10,000 lb·in (1129.8 N·m), and project vertically upward to the curve marked *Commercial flexible*. From the intersection with this curve, project horizontally to the left to find the required spline pitch diameter as 3.75 in (9.5 cm). This is also the required outside diameter of a keyed shaft to transmit the same torque.

### 3. *Determine the maximum effective face width*

Enter Fig. 40 at the pitch diameter of 3.75 in (9.5 cm), and project horizontally to the curve marked *For flexible splines*. From the intersection, project vertically downward to read the maximum effective face width as 1.75 in (4.4 cm).

### 4. *Choose the number of teeth for the spline*

Table 58 lists the recommended minimum number of teeth for an involute spline. Cost and manufacturing considerations determine the number of teeth to use, because the number of teeth

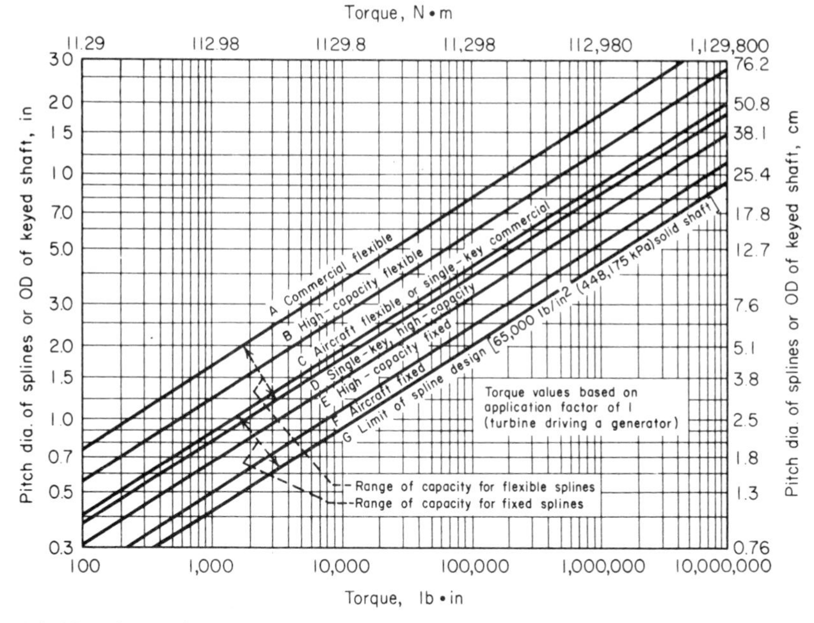


**FIG. 39** Spline size based on diameter-torque relationships. *(Product Engineering.)*

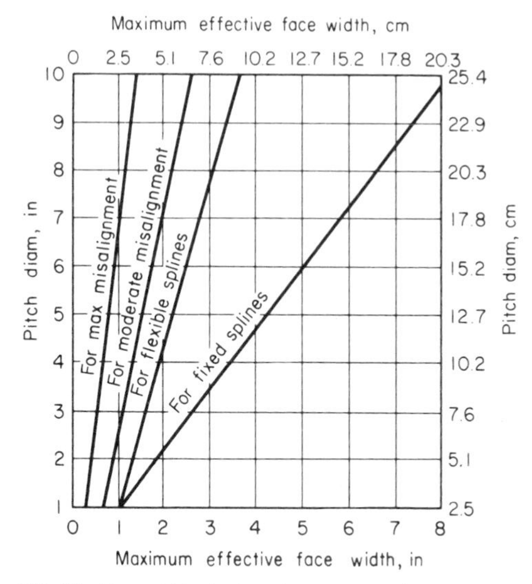


**FIG. 40** Face width of splines for various applications. *(Product Engineering.)*

**TABLE 58** Recommended Minimum Numbers of Spline Teeth*

| Pitch diameter | | Broaching | | | Shaping | | | | Shaving or grinding | |
|---|---|---|---|---|---|---|---|---|---|---|
| | | Angle: 30° | 20° or 14½° | | 30° | 25° | 25° | 14½° | 25% | 20° |
| in | cm | Depth: 50% | 50% | 30% | 50% | 70% | 75% | 30% | 70% | 75% |
| 0.5 | 1.3 | 6 | 10 | 10 | 12 | 18 | | | | |
| 0.8 | 2.0 | 8 | 12 | 10 | 14 | 20 | 22 | 16 | | |
| 1.0 | 2.5 | 8 | 12 | 10 | 16 | 20 | 24 | 16 | | |
| 2.0 | 5.1 | 8 | 12 | 12 | 20 | 20 | 24 | 24 | | |
| 4.0 | 10.2 | 10 | 16 | 16 | 24 | 24 | 32 | 24 | 48 | 48 |
| 8.0 | 20.3 | 20 | 20 | 24 | 32 | 32 | 40 | 32 | 56 | 56 |
| 12.0 | 30.5 | 30 | 30 | 36 | 36 | 36 | 48 | 36 | 60 | 60 |

**Product Engineering.*

chosen has no effect on tooth stress. An even number of teeth should be used whenever possible. When a large number of teeth are used on a spline, the root diameter of the external member is greater, tool design is easier, and lubrication is improved. Generally, however, the cost of the spline increases with a larger number of teeth.

For industrial drives, where the spline cost is usually more important than the weight of the spline or the space it occupies, a tooth with a 20° pressure angle is generally chosen. The nominal tooth depth, compared with gear teeth, is 75 percent. Using these data and a pitch diameter of 3.75 in (9.5 cm), as determined in step 2, shows that 32 teeth should be used.

**Related Calculations:** Involute splines for use in aircraft applications generally have a 30° pressure angle and 50 percent depth. In automotive service, shaved splines having the same proportions as the industrial splines mentioned above are often used. Rolled splines having 30 or 40° pressure angles and 50 and 40 percent depth, respectively, are also used. ANSA standards covering involute and straight-sided splines are available.

The method presented here is the work of Darle W. Dudley, reported in *Product Engineering*.

## FRICTION DAMPING FOR SHAFT VIBRATION

Design a *dry-friction* (also termed *coulomb friction*) sleeve for a shaft transmitting power to an air compressor. The shaft has an outside diameter of 7.5 in (19.1 cm) and a length of 8 ft (2.4 m), and it drives the compressor as shown in Fig. 41. The angular value of torsional vibration should be limited to 10 percent of the steady displacement caused by the mean torque in the shaft. The compressor torque is 800,000 lb·in (90,387.8 N·m).

### Calculation Procedure:

#### 1. *Compute the required damping ratio*

To apply the friction-damping technique to a shaft, a sleeve (Fig. 42) is added which is attached to the shaft at one end, A. The sleeve is extended along the shaft and makes contact with some point on the shaft through the disk. This disk may be welded to or tightly pressed on the shaft and snugly fits into the sleeve.

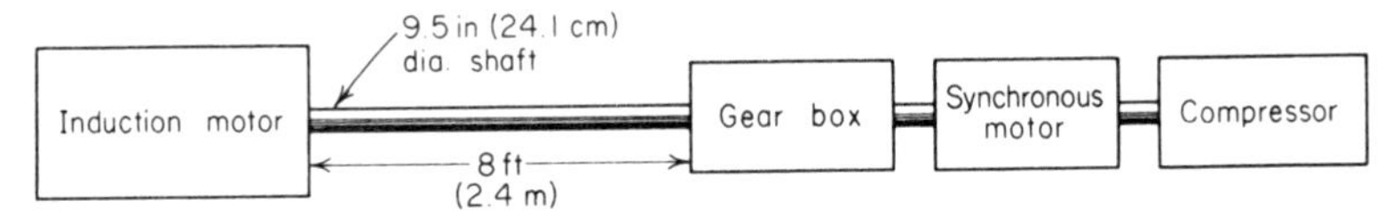


**FIG. 41** Transmission system designed for friction damping. *(Product Engineering.)*

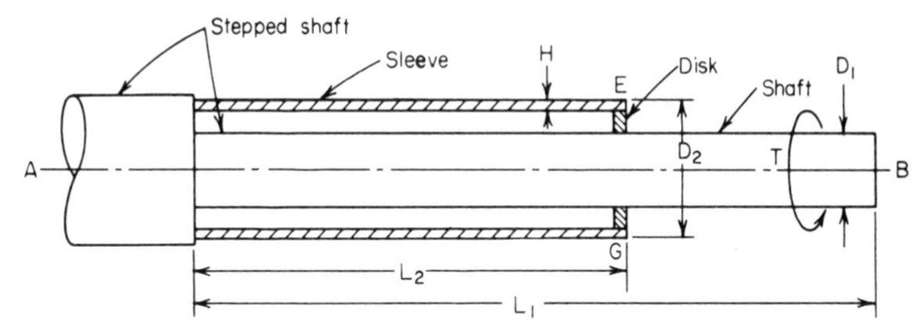


**FIG. 42** Thin sleeve added to rotating shaft reduces torsional vibrations. *(Product Engineering.)*

In most dry-friction damping, about 3 percent of the damping takes place per cycle. If the forcing torque were reduced for one cycle, the strain energy would drop to 97 percent of its maximum value and the angular displacement $\theta$ of the shaft would drop to $0.97\theta$. Hence, the forcing torque must be such as to increase the angular displacement by an amount, or (in the absence of damping) $\Delta\theta = 0.03\theta$ per cycle $= 0.015\theta$ for a half-cycle.

Compute the damping ratio for the system from $R = 1 - \Delta\theta/\theta$, where $R$ = damping ratio. Thus, $R = 1 - 0.015\theta/(0.1\theta) = 0.85$. The value $0.1\theta$ is used in the denominator because the design requires that $\Delta\theta$ be limited to 10 percent of the steady displacement $\theta$, which results from the mean torque in the shaft.

## 2. *Determine the shaft damping/critical damping value*

With $R = 0.85$, enter Fig. 43, and find $m = 5.2$, where $m = D_1/(8HC^3)$ = ratio of the torsional stiffness of the shaft to that of the sleeve, a dimensional constant; $D_1$ = shaft diameter, in; $H$ = thickness of the sleeve wall, in; $C = D_2/D_1$, where $D_2$ = outside diameter of sleeve, in. Thus, damping/critical damping value = 0.026, or 2.6 percent, from Fig. 43, assuming $L_1/L_2 = 1.0$.

## 3. *Select the sleeve outside diameter*

Since $m = D_1/(8HC^3) = 5.2$, and $D_1 = 7.5$ in (19.1 cm), $HC^3 = 0.1802$, by substitution of the value of $D_1$ and $m$ in this relation.

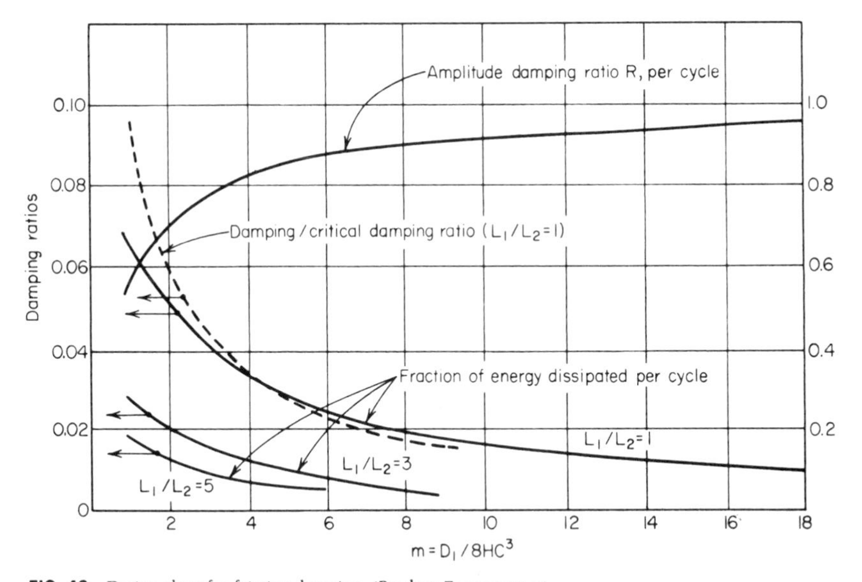


**FIG. 43** Design chart for friction damping. *(Product Engineering.)*

Choose how $HC^3$ is to be made up. Assuming $D_2 = 2D_1$, $C = D_2/D_1 = 2.0$. Since $HC^3 = 0.1802$, $H = -0.1802/2^3 = 0.0225$ in (0.572 mm). This provides a sleeve thickness of about 24 gage. The sleeve will weigh only about 2.7 percent of the shaft weight. Thus, a 10:1 reduction in the vibration is obtained with very little extra weight.

**4. *Compute the resisting torque of the system***

The ratio of resisting frictional torque applied by the sleeve $T_r$ lb·in to the applied torque on the shaft $T$ lb·in is $T_r/T = 1[1 - (1 - 1/ar)^{0.5}]$, where $a = L_1/L_2$; $r = 1 + m$. Or, $T_r/T = 1\{1 - [1/(1 \times 6.2)]^{0.5}\} \doteq 0.09$. Since the compressor torque $T = 800{,}000$ lb·in (90,387.8 N·m), $T_r = 0.09(800{,}000) = 72{,}000$ lb·in (8134.9 N·m).

**5. *Compute the friction force on the sleeve***

In step 4 the sleeve diameter $D_2$ was chosen as $2D_1$. Since $D_1 = 7.5$ in (19.1 cm), $2D_1 = 15$ in (38.10 cm). The frictional torque acts, through the disk, over the circumference of the inner surface of the sleeve. The diameter of the inner surface of the sleeve is $15.0 - 2(0.0225) = 14.955$ in (38.09 cm), using the sleeve thickness obtained in step 3. The circumference of the inner surface of the sleeve is $14.955\pi = 47.0$ in (119.4 cm). Hence, the friction force acting on the sleeve $F_f = T_r$/circumference, in $= 72{,}000/47.0 = 1532$ lb (6814.7 N).

**6. *Determine the disk normal force***

Assume that the disk has a coefficient of friction of 0.6. Then the normal force acting on the sleeve is $F_n = F_f/f$, where $f$ = coefficient of friction, or $F_n = 1532/0.6 = 2550$ lb (11,343.0 N).

**Related Calculations:** Dry-friction damping can be applied to industrial machines of many types, military equipment (submarines, missiles, aircraft), internal-combustion engines, and similar machinery. Vibration amplitudes in a shaft become a problem when the shaft length-to-thickness ratio $L_1/D_1$ becomes large. Although the shaft diameter can be increased to reduce the ratio, this adds to the weight and cost of the machine.

Here are several useful design pointers: (1) If weight is a primary objective, make the damping-sleeve diameter as large as possible. (2) If weight is not important, use a sleeve diameter only slightly larger than the shaft diameter. (3) Sleeve length can vary from 0.1 to 1.0 shaft length. With short sleeves, be sure the sleeve has sufficient rigidity and stiffness. (4) Reduce the sleeve wall thickness at the end of the sleeve in contact with the disk so that the contact pressure will not induce large stresses in the shaft. The method presented here was developed by Burt Zimmerman and reported in *Product Engineering*.

## DESIGNING PARTS FOR EXPECTED LIFE

A machined and ground rod has an ultimate strength of $s_u = 90{,}000$ lb/in² (620,350 kPa) and a yield strength $s_y = 60{,}000$ lb/in² (413,700 kPa). It is grooved by grinding and has a stress concentration factor of $K_f = 1.5$. The expected loading in bending is 10,000 to 60,000 lb/in² (68,950

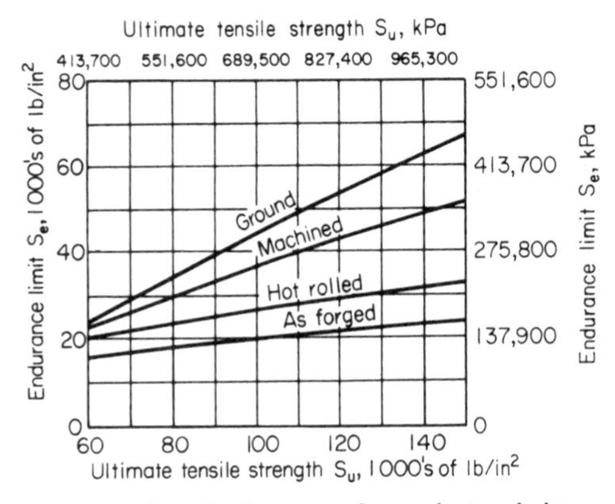


**FIG. 44** Relationship between endurance limit and ultimate tensile strength. *(Product Engineering.)*

to 413,700 kPa) for 0.5 percent of the time; 20,000 to 50,000 lb/in$^2$ (137,900 to 344,750 kPa) for 9.5 percent of the time; 20,000 to 45,000 lb/in$^2$ (137,900 to 310,275 kPa) for 20 percent of the time; 30,000 to 40,000 lb/in$^2$ (206,850 to 275,800 kPa) for 30 percent of the time; 30,000 to 35,000 lb/in$^2$ (206,850 to 241,325 kPa) for 40 percent of the time. What is the expected fatigue life of this part in cycles?

## Calculation Procedure:

### 1. *Determine the material endurance limit*

For $s_u$ = 90,000 lb/in$^2$ (620,550 kPa), the endurance limit of the material $s_e$ = 40,000 lb/in$^2$ (275,800 kPa), closely, from Fig. 44.

### 2. *Compute the equivalent completely reversed stress*

The largest equivalent completely reversed stress for each load in bending is $s_F = (s_e/s_y)s_a + K_f s_a$ lb/in$^2$; where $s_a$ = average or steady stress, lb/in$^2$; other symbols as before. Since $s_e/s_y$ = 40,000/60,000 = ⅔ and $K_f$ = 1.5, $s_v = ⅔s_a + 1.5s_a$. Then

$$s_{v1} = (⅔)(35{,}000) + (1.5)(25{,}000) = 60{,}830 \text{ lb/in}^2 \text{ (419,422.9 kPa)}$$

$$s_{v2} = (⅔)(35{,}000) + (1.5)(15{,}000) = 45{,}830 \text{ lb/in}^2 \text{ (315,997.9 kPa)}$$

$$s_{v3} = (⅔)(32{,}500) + (1.5)(12{,}500) = 40{,}420 \text{ lb/in}^2 \text{ (278,695.9 kPa)}$$

$$s_{v4} = (⅔)(35{,}000) + (1.5)(5000) = 30{,}830 \text{ lb/in}^2 \text{ (212,572.9 kPa)}$$

$$s_{v5} = (⅔)(32{,}500) + (1.5)(2500) = 25{,}420 \text{ lb/in}^2 \text{ (175,270.9 kPa)}$$

### 3. *Compute the fatigue life for the initial stress*

The initial stress is $s_{v1}$; the fatigue life at this stress is, in cycles, $N_1 = 1000(s_u/s_{vi})^{3/\log 1(s_u/s_e)}$ where $s_{vi}$ = equivalent completely reversed stress, lb/in$^2$; other symbols as before. By taking the first value of $s_{vi} = s_{v1}$ = 60,830 lb/in$^2$ (419,422.9 kPa), $N_1 = 1000(90{,}000/60{,}830)^{3/\log 2.25}$ = 28,100 cycles.

### 4. *Compute the exponent for the fatigue-life equation*

The exponent for the fatigue-life equation is $2.55/\log(s_u/s_e) = 2.55/\log(90{,}000/40{,}000)$ = 7.2406.

### 5. *Compute the factors for the fatigue-life equation*

The factors needed for the fatigue-life equation are $s_{v1}/s_{vi}$, $(s_{vi}/s_{v1})^{2.55/\log(s_u/s_e)}$, and $\alpha_i(s_{vi}/s_{v1})^{2.55/\log(s_u/s_e)}$. In these factors, the value of $s_{vi} = s_{v2}$, $s_{v3}$, and so forth, as summarized in Table 59. The value $\alpha_i$ = percent-time duration of a stress, expressed as a decimal. The numerical values computed are summarized in Table 59.

### 6. *Compute the part fatigue life in cycles*

The part fatigue life, in cycles, is $N = N_1/\alpha_1 + \alpha_2[1/(s_{v1}/s_{v2})]^{2.55\log(s_u/s_e)} + 3[1/(s_{v1}/s_{v3})]^{2.55\log(s_u/s_e)} + \ldots$ for each bending load. In this equation, $\alpha_1, \alpha_2, \ldots$ = percent-time

**TABLE 59** Values for Cycles-to-Failure Analysis

| $s_{vi}$ | | | | | |
|---|---|---|---|---|---|
| lb/in$^2$ | kPa | $s_{v1}/s_{vi}$ | $(s_{vi}/s_{v1})^{7.2406}$ | $i$ | $i(s_{vi}/s_{v1})^{7.2406}$ |
| 60,830 | 419,422.9 | 1.00 | 1.00 | 0.005 | 0.00500 |
| 45,830 | 315,997.9 | 1.3273 | 0.12873 | 0.095 | 0.01223 |
| 40,420 | 278,695.9 | 1.5051 | 0.05179 | 0.200 | 0.01036 |
| 30,830 | 212,572.9 | 1.9730 | 0.007296 | 0.300 | 0.00219 |
| 25,420 | 175,270.9 | 2.3934 | 0.001802 | 0.400 | 0.00072 |
| | | | | | 0.03050 |

duration of a stress, expressed as a decimal, the subscript referring to the stress mentioned in step 2, above.

Since Table 59 summarizes the denominator of the fatigue-life equation, $N = N_1/0.03050 = 28{,}100/0.03050 = 922{,}000$ cycles.

**Related Calculations:** Data on the endurance limit, yield point, and ultimate strength of ferrous materials are tabulated in Baumeister and Marks—*Standard Handbook for Mechanical Engineers*. The equations presented in this calculation procedure hold for both simple and complex loading. These equations can be used for analysis of an existing part or for design of a part to fail after a selected number of cycles. The latter procedure is sometimes used for components in an assembly in which the principal part has an accurately known life.

The method presented here is the work of Professor M. F. Spotts, reported in *Product Engineering*.

## WEAR LIFE OF ROLLING SURFACES

Determine the maximum allowable bearing load in various bearing materials to avoid pronounced wear before $40 \times 10^6$ stress cycles if the roller bearing made of these materials has these dimensions: outside diameter = 4.3307 in (11.0 cm), bore = 2.3622 in (6.0 cm), width = 0.866 in (2.2

**TABLE 60** Load-Stress *K* Factors for Various Materials*

| Roller 1 | | Roller 2 | *K* of roller 1 at number of cycles | | | |
|---|---|---|---|---|---|---|
| Material | Hardness | Material | $10^6$ | $10^7$ | $4 \times 10^7$ | $10^8$ |
| Gray cast iron | 130–180 BHN | Same as roller 1 | 4,000 | 2,000 | 1,300 | |
| GM Meehanite | 190–240 BHN | | 4,000 | 2,500 | 1,950 | |
| Nodular cast iron | 207–241 BHN | | 10,000 | 5,600 | . . . . | 3,400 |
| Gray cast iron | 270–290 BHN | | 7,500 | 5,300 | 4,200 | |
| Gray cast iron, phosphate-coated | 140–160 BHN | Carbon tool steel 60–62 Rc | 2,600 | 1,400 | 1,000 | |
| | 160–190 BHN | | 3,200 | 1,900 | 1,300 | |
| | 270–290 BHN | | 5,500 | 4,000 | 3,100 | |
| SAE 1020 steel, phosphate-coated | 130–150 BHN | | 4,500 | 2,700 | 1,700 | |
| SAE 4150 steel, chromium-plated | 270–300 BHN | | 13,500 | 11,000 | . . . . | 9,000 |
| SAE 6150 steel | 270–300 BHN | | 2,600 | 1,300 | | |
| SAE 1020 steel, induction-hardened | 45–55 RC | | 21,000 | 14,500 | . . . . | 10,000 |
| SAE 1340 steel, case-hardened | 50–58 RC | | 26,000 | 20,000 | . . . . | 15,000 |
| Phosphor bronze | 67–77 BHN | | 3,600 | 1,600 | 1,000 | |
| Yellow brass | Drawn | | 5,600 | 3,000 | 2,000 | |
| | Extruded | | 4,500 | 2,400 | 1,700 | |
| Zinc diecasting | . . . . . . . . . . . . | | 1,100 | 500 | 320 | |
| Laminated-graphitized phenolic | . . . . . . . . . . . . | | 1,700 | 1,300 | 1,000 | |
| Cast aluminum SAE 39 | 60–65 BHN | Gray cast iron 340–360 BHN | 1,200 | 500 | 300 | |

* *Product Engineering*. Based on data presented by W. C. Cram at a University of Michigan symposium on surface damage.

cm), inner-race radius $r_2$ = 1.439 in (3.655 cm), roller diameter = 0.468 in (1.19 cm), roller width $f$ = 0.468 in (1.19 cm), number of rollers $n$ = 16. Materials being considered are gray cast iron, Meehanite; and hardened-steel rollers on cast-iron races, on heat-treated cast-iron races, on heat-treated and medium-steel races, and on carburized low-carbon steel races.

## Calculation Procedure:

### 1. *Determine the load-stress factor for each material*

Table 60 lists the load-stress factor $K$ for various materials at varying load cycles. Thus for gray cast iron, $K = 1300$ at $40 \times 10^6$ cycles.

### 2. *Compute the maximum allowable bearing load*

Use the relation $F = nfK/[5(1/r_1 + 1/r_2)]$ where all symbols are defined in the problem statement above. Thus, $F = (16)(0.468)(1300)/[5(1/0.234 + 1/1.439)] = 391$ lb (1739.3 N) for gray cast iron.

For the other materials, by using the appropriate $K$ value from Table 60 and the above procedure, the allowable load is as follows:

| Bearing materials | Allowable load, lb (N) |
|---|---|
| Meehanite rollers and races | 587 (2,611.1) |
| Hardened-steel rollers, cast-iron races | 300 (1,334.5) |
| Hardened-steel rollers, heat-treated cast-iron races | 933 (4,150.2) |
| Hardened-steel rollers, heat-treated medium-steel races | 2,700 (12,010.2) |
| Hardened-steel rollers, carburized low-carbon steel races | 3,912 (17,401.4) |

**Related Calculations:** The same, or similar, procedures can be used for computing the wear life of gears, cams, bearings, clutches, chains, and other devices having rolling surfaces. Thus, in a joint composed of a pin having radius $r_1$ in a hole of radius $r_2$, $F = fK/(1/r_1 - 1/r_2)$. This relation also applies to a roller chain. For a cam, $F = fK \cos \infty/(1/\rho - 1/r)$, where $\infty$ = cam pressure angle; $\rho$ = cam radius of contact, in; $r$ = radius of contact, in.

The method presented here is the work of Professor Donald J. Myatt, reported in *Product Engineering*.

## FACTOR OF SAFETY AND ALLOWABLE STRESS IN DESIGN

Determine the cross section dimension $b$ of a uniform square bar of structural steel having a tensile yield strength $s_y$ = 33,000 lb/in$^2$ (227,535.0 kPa), if the bar carries a center load of 1000 lb (4448.2 N) as a beam with a span of 24 in (61.0 cm) with simply supported ends. Use both the allowable-stress and ultimate-strength methods to determine the required dimension.

## Calculation Procedure:

### 1. *Design first on the basis of allowable stress*

From Table 61, in the section on buildings, the working or allowable stress $s_w = 0.6s_y = 0.6(33{,}000) = 19{,}800$ lb/in$^2$ (136,521.0 kPa).

### 2. *Compute the maximum bending moment*

For a central load and simple supports, the maximum bending moment is, from earlier sections of this handbook, $M_m = (W/2)(L/2)$, where $W$ = load on beam, lb; $L$ = length of beam, in. Thus, $M_m = (1000/2)(24/2) = 6000$ in · lb (677.9 N · m).

### 3. *Compute the required cross section dimension*

For a beam of square cross section, $I/c = [b(b)^3/12]/(b/2) = b^3/6$. Also, $s_w = M_mc/I$. Substituting and solving for $b$ gives $b^3 = (6000)(6)/19{,}800$; $b = 1.22$ in (3.1 cm).

**TABLE 61** Illustrative Allowable Stresses and Factors of Safety*

| Application | Materials | Allowable stress $s_w$ | Approximate factor of safety |
|---|---|---|---|
| Buildings and other structures | Structural steel | Direct tension $s_w = 0.6s_y$; $0.45s_y$ on net section at pin holes; bending $s_w = 0.6s_y$; shear $s_w = 0.45s_y$ on rivets, $0.40s_y$ on girder webs; bearing $s_w = 1.35s_y$ on rivets in double shear; $1s_y$ in single shear | 1.70 for beams; 1.85 for continuous frames |
| | Structural aluminum 6061-T6, 6062-T6 $s_u = 38{,}000$ lb/in$^2$ (262.010 kPa), $s_y = 35{,}000$ lb/in$^2$ (241,325 kPa) | Direct tension $s_w = 17{,}000$ lb/in$^2$ (117,215 kPa); bending structural shapes, $s_w = 17{,}000$ lb/in$^2$ (117,215 kPa); rectangular sections $s_w = 23{,}000$ lb/in$^2$ (158,585 kPa); shear $s_w = 10{,}000$ lb/in$^2$ (68,950 kPa) on cold-driven rivets, 11,000 lb/in$^2$ (75,845 kPa) on girder webs; bearing $s_w = 30{,}000$ lb/in$^2$ (206,850 kPa) on rivets | 1.8 for beams; 2 for columns |
| | Reinforced concrete | Bending, compression in concrete, $s_w = 0.45s_c'$; bending, tension in steel, $s_w = 0.40s_y$; bending, tension in plain concrete footings, $s_w = 0.03s_c'$; shear, on concrete in unreinforced web, $s_w = 0.03s_c'$; compression in concrete column, $s_w = 0.225s_c'$ | |
| | Wood | Bending, $s_w = \frac{1}{6}s'$; long compression, $s_w = \frac{1}{6}s_c'$; transverse-compression, $s_w = \frac{1}{2}$ elastic limit (400 for Douglas fir); shear parallel to grain $s_w = 120$ for Douglas fir | 6 in general |
| Bridges | Structural metals and reinforced concrete | $s_w$ about $0.9s_w$ for buildings | 2.05 |
| | Wood | $s_w$ same as for buildings | 6 |

| | | | |
|---|---|---|---|
| Machinery | Steel (shafts, etc.) | Steady tension, compression or bending, $s_w = s_y/n$; pure shear, $s_w = s_y/2n$; tension $s_t$ plus shear $s_s$; $s_t^2 + 4s_s^2 \leqq s_y/n$; alternating stress $s_a$ plus mean stress $s_m$: point representing an alternating stress $nk_f s_a$ and a mean stress $ns_m$ must lie below the Goodman diagram, Fig. 1 | $n$ usually between 1.5 and 2 |
| | Steel (SAE 1095), leaf springs, thickness = $t$ in $t > 0 < 0.10$ | Static loading $s_w = 230{,}000 - 1{,}000{,}000t$ lb/in² (1,585,850 − 6,895,000$t$ kPa); variable loading, $10^7$ cycles, $s_w = 200{,}000 - 800{,}000t$ lb/in² (1,379,000 − 5,516,000$t$ kPa); dynamic loading, $10^7$ cycles, $s_w = 155{,}000 - 600{,}000t$ lb/in² (1,068,725 − 4,137,000$t$ kPa) | |
| | Steel (wire, ASTM-A228, helical springs) | $s_w = 100{,}000$ lb/in² (689,500 kPa) [for $d = 0.2$ in (0.5 cm); $10^7$ cycles repeated stress] | |
| Pressure vessels (unfired) | Carbon steel | Membrane stress, $s_w = 0.211s_u$; membrane plus discontinuity stresses, $s_w = 0.9s_y$ or $0.6s_u$ | 5 |
| | Alloy steels | Membrane stress, $s_w = 0.25s_u$; membrane plus discontinuity stresses, $s_w = 0.95s_y$ or $0.6s_u$ | 4 |
| | Cast iron | Membrane stress, $s_w = 0.1s_u$; bending stress $s_w = 0.15s_u$ | 10, 6.67 |
| | Nonferrous metals | Same rule as alloy steels | 4 |
| Airplanes | Aluminum alloy and steel | Ultimate strength design | 1.5 against ultimate<br>1 against yield |
| | Wood | Ultimate strength design | |

*From Roark—*Formulas for Stress and Strain*, 4th ed., McGraw-Hill.

*Note:* $s_w$ = allowable or working stress; $s_u$ = ultimate tensile strength; $s_y$ = tensile yield strength; $s'$ = modulus of rupture in cross-bending of rectangular bar; $s_s'$ = ultimate shear strength; $s_c'$ = ultimate compressive strength; $s_c$ = endurance limit or endurance strength for specified life; $n$ = dividing factor applied to $s_u$, $s_y$, or $s_c$ to obtain $s_w$.

### 4. *Design the beam on the basis of ultimate strength and a factor of safety*

The safety factor from Table 61 is 1.70 for beams. This safety factor is applied by designing the beam to fail just under a load of 1.7(1000 lb) = 1700 lb (7562.0 N). Thus, to generalize, the design failure load = (factor of safety)(load on part, lb) = $W_u$.

For structural steel and other materials capable of fully plastic behavior, a simple beam or other member will collapse when the maximum moment equals the *plastic moment* $M_p$, which is developed when the stress throughout the section becomes equal to $s_y$. Hence, $M_p = s_y z$, where $Z$ = plastic modulus = arithmetical sum of the static moments of the upper and lower parts of the cross section about the horizontal axis that divides the area in half. For a square with its edges horizontal and vertical, as assumed here, $Z = (\frac{1}{2}b^2)(\frac{1}{4}b) + (\frac{1}{2}b^2)(\frac{1}{4}b) = \frac{1}{4}b^3$. Hence, $M_p = 33{,}000(\frac{1}{4}b^3)$.

Set $M_p$ = the ultimate bending moment, or $(W_u/2)(L/2)$, where $W_u$ = design failure load = (factor of safety)(load on part, lb) = (1.70)(1000) = 1700 lb (7562.0 N) for this beam. Thus, $M_p = 33{,}000(\frac{1}{4}b^3) = (1700/2)(24/2)$; $b$ = 1.07 in (2.7 cm).

By using the allowable stress, $b$ = 1.22 in (3.1 cm), as compared with 1.07 in (2.7 cm). The difference in results for the two methods (about 12 percent) can be traced to the ratio of $Z$ to $I/c$, called the *shape factor*. For the case under consideration, this ratio is $(\frac{1}{4}b^3)/(\frac{1}{6}d^3) = 1.5$.

### 5. *Determine the beam size for a vertical diagonal*

Here $Z = 0.2357b^3$, $I/c = 0.1178b^3$, and the shape factor $= Z/(I/c) = 0.2357/0.1178 = 2.0$.

Designing by allowable stress, using steps 1, 2, and 3, gives $(0.1178b^3)(19{,}800) = 6000$; $b$ = 1.37 in (3.5 cm). Designing by ultimate strength gives $(0.2357b^3)(33{,}000) = 10{,}200$; $b$ = 1.095 in (2.8 cm).

This computation shows that a more economical design is generally obtained by ultimate-strength design, with the advantage becoming greater as the shape factor becomes larger.

**Related Calculations:** For conventional structural sections, such as I beams, the shape factor is not much greater than 1, and the two methods yield about the same result for statically determinate problems, such as this one. If the problem involves a statically indeterminate beam or a rigid frame, the advantage of the ultimate-strength method becomes more apparent because it takes account of the fact that collapse cannot occur, as a rule, until the plastic moment is developed at each of two or more sections.

The method given here is the work of Professor Raymond J. Roark, reported in *Product Engineering*.

## RUPTURE FACTOR AND ALLOWABLE STRESS IN DESIGN

Determine the proper thickness $t$ of a circular plate 40 in (101.6 cm) in diameter if the edge of the plate is simply supported and the plate carries a uniformly distributed pressure of 200 lb/in² (1379.0 kPa). The plate is made of cast iron having an ultimate strength of 50,000 lb/in² (344,750.0 kPa).

### Calculation Procedure:

### 1. *Design on the basis of allowable stress*

From Table 61 of the previous calculation procedure, note in the section for pressure vessels that the value of the working or allowable stress $s_w$ lb/in² for tension due to bending is 0.15 $s_u$, where $s_u$ = ultimate tensile strength of the material, lb/in². Thus, $s_w$ = 0.15(50,000) = 7500 lb/in² (51,712.5 kPa).

### 2. *Compute the required plate thickness*

The maximum stress for a simply supported plate is, from Roark—*Formulas for Stress and Strain*, $s_{max} = (3W/8\pi t^2)(3 + v)$, where $W$ = total load on plate, lb; $t$ = plate thickness, in; $v$ = Poisson's ratio. For this plate, $W = (200 \text{ lb/in}^2)(\pi)(20)^2$ = 251,330 lb (1,117,971.6 N). Assuming $v = 0.3$ and solving for $t$, we find $7500 = [(3)(251{,}330)/8\pi t^2](3 + 0.3)$; $t$ = 3.63 in (9.2 cm).

### 3. *Design on the basis of ultimate strength*

For ultimate-strength design, from Table 61 of the previous calculation procedure, use the value of 6.67 for the factor of safety. Design the plate to break, theoretically, under a load. Hence, the breaking load would be (factor of safety) $W$, or 6.67(251,330) = 1,667,000 lb (7,415,186.1 N).

### 4. *Apply the rupture factor to the design*

With a brittle material like cast iron, the concept of the plastic moment does not apply. Use instead the *rupture factor,* which is the ratio of the calculated maximum tensile stress at rupture to the ultimate tensile strength, both expressed in lb/in². Whereas the rupture factor must be determined experimentally, a number of typical values are given in Table 62 for a variety of cases.

For case 2, which corresponds to this problem, the rupture factor for cast iron is $R_i = 1.9$. Using this in the same equation for $s_w$ as in step 1, $s_w = 1.9(50{,}000) = 95{,}000$ lb/in² (655,025.0 kPa). Then, by using the procedure and equation in step 2 but substituting the breaking load for the plate, $95{,}000 = [(3)(1{,}677{,}000)/8\pi t^2](3 + 0.3)$; $t = 2.63$ in (6.7 cm).

**Related Calculations:** The reliability of the solution using the ultimate-strength design technique depends on the accuracy of the rupture factor. When experimental or tabulated values of $R_i$ are not available, the modulus of rupture may be used in place of $R_i s_u$. Typical values of Poisson's ratio $v$ for various materials are given in Table 63.

The method presented here is the work of Professor Raymond J. Roark, reported in *Product Engineering.*

## FORCE AND SHRINK FIT STRESS, INTERFERENCE, AND TORQUE

A 0.5-in (1.3-cm) thick steel band having a modulus of elasticity of $E = 30 \times 10^6$ lb/in² (206.8 $\times 10^9$ Pa) is to be forced on a 4-in (10.2-cm) diameter steel shaft. The maximum allowable stress in the band is 24,000 lb/in² (165,480.0 kPa). What interference should be used between the band and the shaft? How much torque can the fit develop if the band is 3 in (7.6 cm) long and the coefficient of friction is 0.20?

### Calculation Procedure:

### 1. *Compute the required interference*

Use the relation $i = sd/E$, where $i$ = the required interference to produce the maximum allowable stress in the band, in; $s$ = stress in band or hub, lb/in²; $d$ = shaft diameter, in; $E$ = modulus of elasticity of band or hub, lb/in². For this fit, $i = (24{,}000)(4.0)/(30 \times 10^6) = 0.0032$ in (0.081 mm).

### 2. *Compute the torque the fit will develop*

Use the relation $T = Eitl\pi f$, where $T$ = fit torque, lb·in; $t$ = band or hub thickness, in; $l$ = band or hub length, in; $f$ = coefficient of friction between the materials. For this joint, $T = 30 \times 10^6 \times 0.0032 \times 0.5 \times 3.0 \times \pi \times 0.20 = 90{,}432$ lb·in (10,217.4 N·m).

**Related Calculations:** Use this general procedure for either shrink or press fits. The axial force required for a press fit of two members made of the same material is $F_a$ = axial force for the press fit, lb; $p_c$ = radial pressure between the two members, lb/in² $= iE(d_c^2 - d_i^2)(d_o^2 - d_c^2)/zd_c^3(d_o^2 - d_i^2)$, where $d_o$ = outside diameter of the external member, in; $d_c$ = nominal diameter of the contact surfaces, in; $d_i$ = inside diameter of the inner member, in.

## HYDRAULIC SYSTEM PUMP AND DRIVER SELECTION

Choose the pump and the driver horsepower for a rubber-tired tractor bulldozer having four-wheel drive. The hydraulic system must propel the vehicle, operate the dozer, and drive the winch. Each main wheel will be driven by a hydraulic motor at a maximum wheel speed of 59.2 r/min and a maximum torque of 30,000 lb·in (3389.5 N·m). The wheel speed at maximum torque will be 29.6 r/min; maximum torque at low speed will be 74,500 lb·in (8417.4 N·m). The tractor speed must be adjustable in two ways: for overall forward and reverse motion and for turning, where the outside wheels turn at a faster rate than do the inside ones. Other operating details are given in the appropriate design steps below.

### Calculation Procedure:

### 1. *Determine the propulsion requirements of the system*

Usual output requirements include speed, torque, force, and power for each function of the system, through the full capacity range.

**TABLE 62** Values of the Rupture Factor for Brittle Materials[*]

| Form of member and manner of loading | Rupture factor; ratio of computed maximum stress at rupture to ultimate tensile strength | | Ratio of computed maximum stress at rupture to modulus of rupture in bending or torsion | |
|---|---|---|---|---|
| | Cast iron | Plaster | Cast iron | Plaster |
| 1. Rectangular beam, end support, center loading, $l/d$ = 8 or more 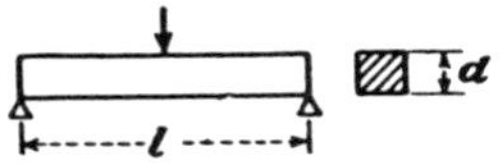 | 1.70 | 1.60 | 1 | 1 |
| 2. Solid circular plate, edge support, uniform loading, $a/t$ = 10 or more 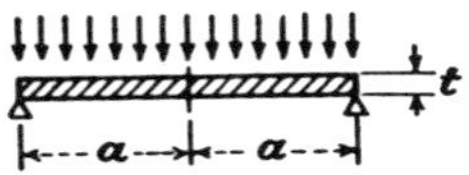 | 1.9 | 1.71 | . . . | 1.07 |
| 3. Solid circular plate edge support, uniform loading on concentric circular area 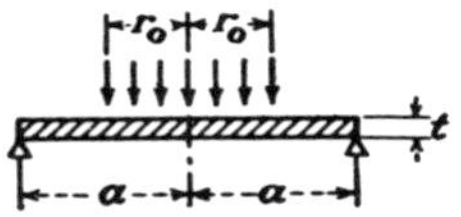 | $2.4–0.5(r_o/a)^{1/6}$ | $2.2–0.5(r_o/a)^{1/6}$ | $1.40–0.3(r_o/a)^{1/6}$ | $1.4–0.3(r_o/a)^{1/6}$ |

[*]From Roark—*Formulas for Stress and Strain*, 4th ed., McGraw-Hill.

**TABLE 63** Poisson's Ratio for Various Materials

| Material | Poisson's ratio |
|---|---|
| Aluminum: | |
| Cast | 0.330 |
| Wrought | 0.330 |
| Brass, cast, 66% Cu, 34% Zn | 0.350 |
| Bronze, cast, 85% Cu, 7.2% Zn, 6.4% Sn | 0.358 |
| Cast iron | 0.260 |
| Copper, pure | 0.337 |
| Phosphor bronze, cast, 92.5% Cu, 7.0% Sn, 0.5% Ph | 0.380 |
| Steel: | |
| Soft | 0.300 |
| 1% C | 0.287 |
| Cast | 0.280 |
| Tin, cast, pure | 0.330 |
| Wrought iron | 0.280 |
| Nickel | 0.310 |
| Zinc | 0.210 |

First analyze the *propel* power requirements. For any propel condition, $hp = Tn/63{,}000$, where $hp$ = horsepower required; $T$ = torque, lb·in, at $n$ r/min. Thus, at maximum speed, $hp$ = (30,000)(59.2)/63,000 = 28.2 hp (21.0 kW). At maximum torque, $hp$ = 74,500 × 29.6/63,000 = 35.0 (26.1 kW); at maximum speed and maximum torque, $hp$ = (74,500)(59.2)/63,000 = 70.0 (52.2 kW).

The drive arrangement for a bulldozer generally uses hydraulic motors geared down to wheel speed. Choose a 3000-r/min step-variable type of motor for each wheel of the vehicle. Then each motor will operate at either of two displacements. At maximum vehicle loads, the higher displacement is used to provide maximum torque at low speed; at light loads, where a higher speed is desired, the lower displacement, producing reduced torque, is used.

Determine from a manufacturer's engineering data the motor specifications. For each of these motors the specifications might be: maximum displacement, 2.1 in$^3$/r (34.4 cm$^3$/r); rated pressure, 6000 lb/in$^2$ (41,370.0 kPa); rated speed, 3000 r/min; power output at rated speed and pressure, 90.5 hp (67.5 kW); torque at rated pressure, 1900 lb·in (214.7 N·m).

The gear reduction ratio GR between each motor and wheel = (output torque required, lb·in)/(input torque, lb·in, × gear reduction efficiency). Assuming a 92 percent gear reduction efficiency, a typical value, we find GR = 74,500/(1900 × 0.92) = 42.6:1. Hence, the maximum motor speed = wheel speed × GR = 59.2 × 42.6 + 2520 r/min. At full torque the motor speed is, by the same relation, 29.6 × 42.6 = 1260 r/min.

The required oil flow for the four motors is, at 1260 r/min, in$^3$/r × 4 motors × (r/min)/(231 in$^3$/gal) = 2.1 × 4 × 1260/231 = 45.8 gal/min (2.9 L/s). With a 10 percent leakage allowance, the required flow = 50 gal/min (3.2 L/s), closely, or 50/4 = 12.5 gal/min (0.8 L/s) per motor.

As computed above, the power output per motor is 35 hp (26.1 kW). Thus, the four motors will have a total output of 4(35) = 140 hp (104.4 kW).

## 2. *Determine the linear auxiliary power requirements*

The dozer uses a linear power output. Two hydraulic cylinders each furnish a maximum force of 10,000 lb (44,482.2 N) to the dozer at a maximum speed of 10 in/s (25.4 cm/s). Assuming that the maximum operating pressure of the system is 3500 lb/in$^2$ (24,132.5 kPa), we see that the piston area required per cylinder is: force developed, lb/operating pressure, lb/in$^2$ = 10,000/3500 = 2.86 in$^2$ (18.5 cm$^2$), or about a 2-in (5.1-cm) cylinder bore. With a 2-in (5.1-cm) bore, the operating pressure could be reduced in the inverse ratio of the piston areas. Or, $2.86/(2^2\pi/4) = p/3500$, where $p$ = cylinder operating pressure, lb/in$^2$. Hence, $p$ = 3180 lb/in$^2$, say 3200 lb/in$^2$ (22,064.0 kPa).

By using a 2-in (5.1-cm) bore cylinder, the required oil flow, gal, to each cylinder = (cylinder volume, in$^3$)(stroke length, in)/(231 in$^3$/gal) = $(2^2\pi/4)(10)/231$ = 0.1355 gal/s, or 0.1355 × (60 s/min) = 8.15 gal/min (0.5 L/s), or 16.3 gal/min (1.0 L/s) for two cylinders. The power input to the two cylinders is $hp$ = 16.3(3200)/1714 = 30.4 hp (22.7 kW).

### 3. *Determine rotary auxiliary power requirements*

The winch will be turned by one hydraulic motor. This winch must exert a maximum line pull of 20,000 lb (88,964.4 N) at a maximum linear speed of 280 ft/min (1.4 m/s) with a maximum drum torque of 200,000 lb·in (22,597.0 N·m) at a drum speed of 53.5 r/min.

Compute the drum hp from $hp = Tn/63{,}000$, where the symbols are the same as in step 1. Or, $hp$ = (200,000)(53.5)/63,000 = 170 hp (126.8 kW).

Choose a hydraulic motor having these specifications: displacement = 6 $in^3$/r (98.3 $cm^3$/r); rated pressure = 6000 lb/$in^2$ (41,370.0 kPa); rated speed = 2500 r/min; output torque at rated pressure = 5500 lb·in (621.4 N·m); power output at rated speed and pressure = 218 hp (162.6 kW). This power output rating is somewhat greater than the computed rating, but it allows some overloading.

The gear reduction ratio GR between the hydraulic motor and winch drum, based on the maximum motor torque, is GR = (output torque required, lb·in)/(torque at rated pressure, lb·in, × reduction gear efficiency) = 20,000/(5500 × 0.92) = 39.5:1. Hence, by using this ratio, the maximum motor speed = 53.5 × 39.5 = 2110 r/min. Oil flow rate to the motor = $in^3$/r × (r/min)/231 = 6 × 2110/231 = 54.8 gal/min (3.5 L/s), without leakage. With 5 percent leakage, flow rate = 1.05(54.8) = 57.2 gal/min (3.6 L/s).

### 4. *Categorize the required power outputs*

List the required outputs and the type of motion required—rotary or linear. Thus: propel = rotary; dozer = linear; winch = rotary.

### 5. *Determine the total number of simultaneous functions*

There are two simultaneous functions: (*a*) propel motors and dozer cylinders; (*b*) propel motors at slow speed and drive winch.

For function *a*, maximum oil flow = 50 + 16.3 = 66.3 gal/min (4.2 L/s); maximum propel motor pressure = 6000 lb/$in^2$ (41,370.0 kPa); maximum dozer cylinder pressure = 3200 lb/$in^2$ (22,064.0 kPa). Data for function *a* came from previous steps in this calculation procedure.

For function *b*, the maximum oil flow need not be computed because it will be less than for function *a*.

### 6. *Determine the number of series nonsimultaneous functions*

These are the dozer, propel, and winch functions.

### 7. *Determine the number of parallel simultaneous functions*

These are the propel and dozer functions.

### 8. *Establish function priority*

The propel and dozer functions have priority over the winch function.

### 9. *Size the piping and values*

Table 64 lists the normal functions required in this machine and the type of valve that would be chosen for each function. Each valve incorporates additional functions: The step variable selector valve has a built-in check valve; the propel directional valve and winch directional valve have built-in relief valves and motor overload valves; the dozer directional valve has a built-in relief valve and a fourth position called *float*. In the float position, all ports are interconnected, allowing the dozer blade to move up or down as the ground contour varies.

### 10. *Determine the simultaneous power requirements*

These are: Horsepower for propel and dozer = (gpm)(pressure, lb/$in^2$)/1714 for the propel and dozer functions, or (50)(6000)/1714 + (16.3)(3200)/1714 = 205.4 hp (153.2 kW). Winch horsepower, by the same relation, is (57.2)(6000)/1714 = 200 hp (149.1 kW). Since the propel-dozer functions do not operate at the same time as the winch, the prime mover power need be only 205.4 hp (153.2 kW).

### 11. *Plan the specific circuit layouts*

To provide independent simultaneous flow to each of the four propel motors, plus the dozer cylinders, choose two split-flow piston-type pumps having independent outlet ports. Split the dis-

**TABLE 64** Hydraulic-System Valving and Piping

| Valving | |
|---|---|
| Function | Type of valve |
| Step variable selector | Three-way, two-position |
| Propel directional | Four-way three-position, tandem-center |
| Winch directional | Four-way, three-position, tandem-center |
| Dozer directional | Four-way, four-position |

| Piping | | | |
|---|---|---|---|
| Branch of circuit | Propel motor | Dozer cylinder | Winch motor |
| Maximum flow, gal/min (L/s) | 12.5 (0.8) | 16.3 (1.0) | 57.2 (3.6) |
| Maximum pressure, lb/in$^2$ (kPa) | 6000 (41,370) | 3200 (22,064) | 6000 (41,370) |
| Tube size, in (cm) | ¾ (1.9) | ¾ (1.9) | 1½ (3.8) |
| Tube material, ASTM | 4130 | 4130 | 4130 |
| Tube wall, in (mm) | 0.120 (3.05) | 0.109 (2.77) | 0.250 (6.35) |

charge of each pump into three independent flows. Two pumps rated at 66.3/2 = 33.15 gal/min (2.1 L/s) each at 6000 lb/in$^2$ (41,370.0 kPa) will provide the needed oil.

When the vehicle is steered, additional flow is required by the outside wheels. Design the circuit so oil will flow from three pump pistons to each wheel motor. Four pistons of one split-flow pump are connected through check valves to all four motors. With this arrangement, oil will flow to the motors with the least resistance.

To make use of all or part of the oil from the propel-dozer circuits for the winch circuit, the outlet series ports of the propel and dozer valves are connected into the winch circuit, since the winch circuit is inoperative only when both the propel *and* the dozer are operating. When only the propel function is in operation, the winch is able to operate slowly but at full torque.

**12. *Investigate adjustment of the winch gear ratio***

As computed in step 3, the winch gear ratio is based on torque. Now, because a known gpm (gallons per minute) is available for the winch motor from the propel and dozer circuits when these are not in use, the gear ratio can be based on the motor speed resulting from the available gpm.

Flow from the propel and dozer circuit = 66.3 gal/min (4.2 L/s); winch motor speed = 2450 r/min; required winch drum speed = 53.5 r/min. Thus, GR = 2450/53.5 = 45.8:1.

With the proposed circuit, the winch gear reduction should be increased from 39.5:1 to 45.8:1. The winch circuit pressure can be reduced to (39.5/45.8)(6000) = 5180 lb/in$^2$ (35,716.1 kPa). The required size of the winch oil tubing can be reduced to 0.219 in (5.6 mm).

**13. *Select the prime mover hp***

Using a mechanical efficiency of 89 percent, we see that the prime mover for the pumps should be rated at 205.4/0.89 = 230 hp (171.5 kW). The prime mover chosen for vehicles of this type is usually a gasoline or diesel engine.

**Related Calculations:** The method presented here is also valid for fixed equipment using a hydraulic system, such as presses, punches, and balers. Other applications for which the method can be used include aircraft, marine, and on-highway vehicles. Use the method presented in an earlier section of this handbook to determine the required size of the connecting tubing.

The procedure presented above is the work of Wes Master, reported in *Product Engineering*.

## SELECTING BOLT DIAMETER FOR BOLTED PRESSURIZED JOINT

Select a suitable bolt diameter for the typical bolted joint in Fig. 45 when the joint is used on a pressurized cylinder having a flanged head clamped to the body of the cylinder by eight equally spaced bolts. The vessel internal pressure, which may be produced by hydraulic fluid or steam,

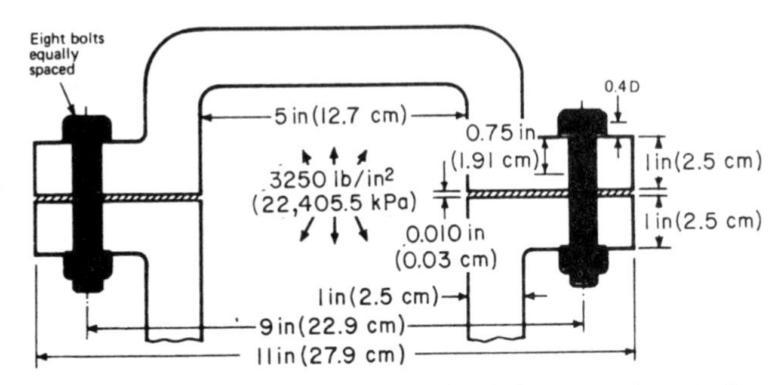


**FIG. 45** Typical bolted joint analyzed in the calculation procedure. *(Product Engineering.)*

varies from 0 to 3250 lb/in$^2$ (0 to 22,405.5 kPa). What clamping force must be applied by each bolt to ensure that no leakage will occur? Check the selected bolt size to ensure long fatigue life under static and fluctuating loading.

## Calculation Procedure:

### 1. *Determine the axially pressurized area of the cylinder*

The internal diameter of the cylinder is, as shown, 5 in (12.7 cm). The axially pressurized area of the cylinder head is $A = \pi D^2/4 = 5^2/4 = 19.63$ in$^2$ (126.62 cm$^2$).

### 2. *Find the applied working load on each bolt*

Since eight bolts are specified for this flanged joint, each bolt will carry one-eighth of the total load found from $F_A = PA/8$, where $F_A$ = axial applied working load, lb; $P$ = maximum pressure in vessel, lb/in$^2$; $A$ = axially pressurized area, in$^2$. Or, $F_A = 3250(19.63)/8 = 7947.7$ lb, say 8000 lb (3633.4 kg) for calculation purposes.

### 3. *Compute the bolt load produced by torquing*

An air-stall power wrench will be used to tighten (torque) the nuts on these bolts. Such a wrench has a torque tightening factor $C_T$ of 2.5 maximum, as shown in Table 65. Using $C_T = 2.5$, we find the load on each bolt $P_Y$, lb, causing a yield stress is $P_Y = C_T F_A = 2.5(8000) = 20{,}000$ lb (9090.9 kg).

### 4. *Find the ultimate tensile strength $P_U$ for the bolt*

With the yield strength typically equal to 80 percent of the ultimate tensile strength of the bolt, $P_U = P_Y/0.80 = 20{,}000/0.80 = 25{,}000$ lb (11,363.6 kg).

**TABLE 65** Torque Tightening Factor*

| Method | $C_T$ |
|---|---|
| Electronic bolt-torquing systems | 1.0 to 1.5 |
| Torque or power wrench with direct torque control | 1.6 to 1.8 |
| Power wrench by elongation measurement of calibrated bolts with the original clamped part | 1.4 to 1.6 |
| Power wrench using air-stall principle | 1.7 to 2.5 |

**Product Engineering.*

### 5. *Determine the required bolt area and diameter*

Select a grade 8 bolt having an ultimate strength of 150,000 lb/in$^2$ (1034.1 MPa). The nominal bolt area must then be $A_b = P_U/U_s = 25{,}000/150{,}000 = 0.1667$ in$^2$ (1.08 cm$^2$). Then the bolt diameter is $D_b = (4A_b/\pi)^{0.5} = [4(0.1667)/\pi]^{0.5} = 0.4607$ in (1.17 cm).

The closest standard bolt size is 0.5 in (1.27 cm). Choose 0.5-13NC bolts, keeping in mind that coarse-thread bolts generally have stronger threads.

### 6. *Find the spring rate of the bolt chosen*

The general equation for the spring rate $K$, lb/in, for a part under tension loading is $K = AE/$

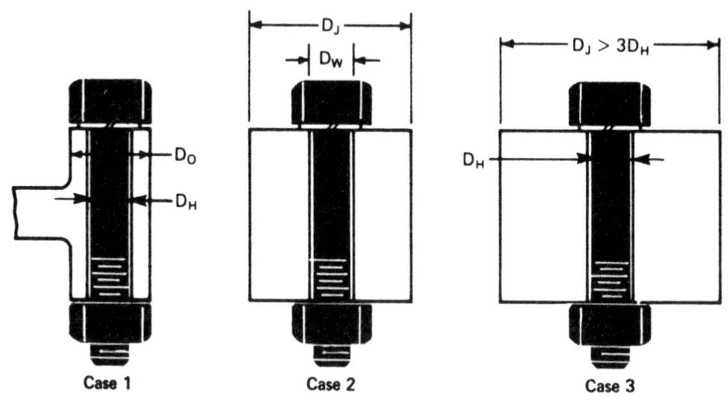


**FIG. 46** Joint size influences the general equations for spring rate of the assembly. *(Product Engineering.)*

$L$, where $A$ = part cross-sectional area, in$^2$; $E$ = Young's modulus, lb/in$^2$; $L$ = length of section, in. To find the spring rate of a part with different cross-sectional areas, as a bolt has, add the reciprocal of the spring rate of each section, or $1/K_{total} = 1/K_1 + 1/K_2 + \cdots$.

For a bolt, which consists of three parts (head, unthreaded portion, and nut), the spring rate determined by G. H. Junker, a consultant with Unbrako-SPS European Division, is given by $1/K_B = (1/E)[(0.4D/A_1) + (L_1/A_1) + (L_T/A_M) + (0.4D_M/A_M)]$, where $D$ = nominal bolt diameter, in; $A_1$ = cross-sectional area of unthreaded portion of bolt, in$^2$; $A_m$ = cross-sectional area, in$^2$, of minor threaded diameter $D_M$; $L_1$ = length of unthreaded portion, 0.75 in (1.91 cm) for this bolt; $L$ = length of threaded portion being clamped, in. Values of $0.4D$ and $0.4D_M$ pertain to the elastic deformation in the head and nut areas, respectively, and were derived in tests in Germany.

For a 0.5-in (1.27-cm) diameter bolt, the cross-sectional area $A_1$ of the unthreaded portion of the body is $A_1 = \pi(0.5)^2/4 = 0.1964$ in$^2$ (1.27 cm$^2$).

The minor thread diameter of the 0.5-13NC thread bolt is, from a table of thread dimensions, 0.4056 in (1.03 cm). Thus, $A_M = \pi(0.4056)^2/4 = 0.1292$ in$^2$ (0.83 cm$^2$).

Substituting the appropriate values in the spring-rate equation for the bolt gives $1/K_B = (1/30 \times 10^6)[0.4 \times 0.5/0.1964 + 0.75/0.1964 + (2.010 - 0.75)/0.1292 + 0.4 \times 0.4056/0.1292]$; $K_B = 1.893 \times 10^6$ lb/in (338,752 kg/cm).

### 7. *Compute the spring rate of the joint*

Calculations of the spring rate of the joint can be simplified by assuming that the bolt head and nut, when compressing the joint as the bolt is tightened, will cause a stress distribution in the shape of a hollow cylinder, with most of the joint compression occurring in the vicinity under the bolt head and nut.

To calculate an equivalent area for the joint $A_J$ in$^2$ for use in the spring-rate equation for the joint, the designer has a choice of one of three equations:

*Case 1*—When most of the outside diameter of the joint is equal to or smaller than the bolt-head diameter, as when parts of a bushing are clamped, Fig. 46, then $A_J = (\pi/4)(D_0^2 - D_H^2)$, where $D_0$ = outside diameter of the joint or bushing, in; $D_H$ = diameter of bolt hole, in.

*Case 2*—When the outer diameter of the joint $D_J$ in is greater than the effective bolt-head diameter or washer $D_W$ in, but less than $3D_H$, then $A_J = (\pi/4)(D_W^2 - D_N^2) + (\pi/8)(D_J/D_W - 1)(D_W L_J/5 + L_J^2/100)$.

*Case 3*—When the joint diameter $D_J$ in is equal to or greater than $3D_H$, then $A_J = (\pi/4)(D_W + L_J/10 - D_H^2)$.

For this bolted joint, case 2 applies. Assuming that the bearing diameter of the head or nut is 0.75 in (1.9 cm), then $A_J = (\pi/4)(0.75^2 - 0.5^2) + (\pi/8)(2/0.75 - 1)[0.75(2.010)/5 + (2.010)^2/$

100] = 0.4692 in$^2$(3.03 cm$^2$). Inserting this value of $A_J$ in the spring-rate equation of step 6 gives $K_J$ = 0.4692(30 × 10$^6$)/2.010 = 7.003 × 10$^6$ lb/in (1.253 × 10$^6$ kg/cm).

**8. *Find the portion of the working load that unloads the clamped joint***

The loading constant $C_L$ considers the bolt and joint elasticity and is given by $C_L = K_B/(K_B + K_J)$. For this joint, $C_L$ = 1.893/(1.893 + 7.003) = 0.2128. Then $F_p$, the portion of the working load $F_A$ that unloads the clamped joint, is $F_P = F_A(1 - C_L)$ = 8000(1 − 0.2128) = 6298 lb (2862.7 kg).

**9. *Determine the loss of clamping force due to embedding***

Some embedding occurs after a bolt is tightened at its assembly. A recommended value is 10 percent. Or specific values can be obtained from tests. Thus, the loss of clamping force $F_Z$, lb = $0.10F_A$ = 0.10(8000) = 800 lb (363.6 kg).

**10. *Find the clamping force required for the joint***

Since the working load in the vessel fluctuates from 0 to 8000 lb (0 to 3636.4 kg), the minimum required clamping force $F_K$ = 0. Now the maximum required clamping force $F_M$, lb, can be found from $F_M = C_T(F_P + F_K + F_Z)$, since all the variables are known. Or, $F_M$ = 2.5(6298 + 0 + 800) = 17,745 lb (8065.9 kg).

Since the bolt has a yield strength $P_Y$ of 20,000 lb (9090.9 kg), it has sufficient strength for static loading. For dynamic loading, however, the additional loading is $F_S = F_A C_L$, where $F_S$ = portion of working load that additionally loads the bolt, lb. Or, $F_S$ = 8000(0.2128) = 1702 lb (773.6 kg).

**11. *Check the endurance limit of the selected bolt***

The endurance limit $S_E$, lb, should not be exceeded for long-life operation of the joint. Exact values of $S_E$ can be obtained from the bolt manufacturer or computed from standard endurance-limit equations. For grade 8 bolts, $S_E = 4600D^{1.59}$, or $S_E$ = 4600(0.5)$^{1.59}$ = 1527 lb (694.1 kg).

The value for $F_S$ that was computed for the 0.5-in (1.27-cm) bolt, 1702 lb (773.6 kg), should not have exceeded 1527 lb (694.1 kg). Since it did, a bolt with a larger diameter must now be selected. Checking a 0.625-in (1.59-cm) bolt (0.625-11), we find the endurance limit becomes $S_E$ = 4600(0.625)$^{1.59}$ = 2179 lb (990.5 kg).

Thus, although the 0.5-in (1.27-cm) bolt would have sufficed for static loading, it would not have provided the necessary endurance strength for long-life operation. So the 0.625-in (1.59-cm) bolt is selected, and a new clamping force $F_M$ = 16,970 lb (7713.6 kg) is calculated, by using the same series of steps detailed above.

**Related Calculations:** The procedure given here can be used to analyze bolted joints used in pressure vessels in many different applications—power plants, hydraulic systems, aircraft, marine equipment, structures, and piping systems. New thread forms permit longer fatigue life for bolted joints. For this reason, the bolted joint is becoming more popular than ever. Further, electronic bolt-tightening equipment allows a bolt to be tightened precisely up to its yield point safely. This gets the most out of a particular bolt, resulting in product and assembly cost savings.

The key to economical bolted joints is finding the minimum bolt size and clamping force to provide the needed seal. Further, the bolt size chosen and clamping force used must be correct. But some designers avoid the computations for these factors because they involve bolt loading, elasticity of the bolt and joint, reduction in preload resulting from embedding of the bolt, and the method used to tighten the bolt. The procedure given here, developed by G. H. Junker, a consultant with Unbrako-SPS European Division, simplifies the computations. Data given here were presented in *Product Engineering* magazine, edited by Frank Yeaple, M.E.

## DETERMINING REQUIRED TIGHTENING TORQUE FOR A BOLTED JOINT

Determine the bolt-tightening torque required for the 0.625-in 11NC (1.59-cm) bolt analyzed in the previous calculation procedure. The bolt must provide a 17,000-lb (7727.3-kg) preload. The coefficient of friction between threads $f_T$ is assumed to be 0.12, while the coefficient of friction between the nut and the washer $f_N$ is assumed to be 0.14.

## Calculation Procedure:

### 1. *Determine the dimensions of the bolt*

From tables of thread dimensions, the minor thread diameter $D_M$ = 0.5135 in (1.3 cm), and the pitch diameter of the bolt threads $D_P$ = 0.5660 in (1.44 cm). The mean bearing diameter of the nut is $D_N = 1.25D$, where $D$ = nominal bolt diameter, in. Or, $D_N = 1.25(0.625) = 0.781$ in (1.98 cm).

For the thread coefficient of friction, $\tan \phi = 0.12$; $\phi = 6.84°$. The helix angle of the thread, $\beta = 2.36°$, is found from $\sin \beta = 1/\pi\,(11)(0.5660) = 0.05112$.

### 2. *Find the dimensionless thread angle factor $C_A$*

The thread angle factor $C_A = \tan(\beta + \phi)/\cos\alpha = \tan(2.36 + 6.84)/\cos 30° = 0.2$.

### 3. *Compute the tightening torque applied*

The tightening torque applied to the nut or bolt head, in·lb, is $T = F_M(D_N f_N + C_A D_p)/2$. Substituting gives $T = 17{,}000[0.781(0.14) + 0.566(0.2)]/2 = 1886$ in·lb = 157 ft·lb (212.7 J).

### 4. *Find the combined stress induced in the bolt*

The combined stress $S_C$ in a bolt in a bolted joint is $S_C = T\{0.89 + 1.66[1 + (5.2C_A D_P/D_M)^2]^{0.5}\}/D_M^2(f_n D_N + C_A D_P)$. Substituting, we find $S_C = 1886\{0.89 + 1.66[1 + (5.2 \times 0.2 \times 0.57/0.51)^2]^{0.5}\}/0.51^2[0.14(0.781) + 0.2(0.57)] = 98{,}381$ lb/in² (678,238.6 kPa).

Generally, $S_C$ should be kept within 68 percent of the ultimate strength $S_U$. Thus, $S_U = 1.47S_C$. Substituting for the above bolt gives $S_U = 1.47(998{,}381) = 144{,}620$ lb/in² (1.0 MPa).

**Related Calculations:** This procedure can be used for determining the bolt-tightening torque for any application in which a bolted joint is used. Numerous applications are listed in the previous calculation procedure. The equations and approach given here are those of Bernie J. Cobb, Mechanical Engineer, Missile Research and Development Command, Redstone Arsenal, as reported in *Product Engineering*, edited by Frank Yeaple.

## SELECTING SAFE STRESS AND MATERIALS FOR PLASTIC GEARS

Determine the safe stress, velocity, and material for a plastic spur gear to transmit 0.125 hp (0.09 W) at 350 r/min 8 h/day under a steady load. Number of teeth in the gear = 75; diametral pitch = 32; pressure angle = 20°; pitch diameter = 2.34375 in (5.95 cm); face width = 1.75 in (4.45 cm).

## Calculation Procedure:

### 1. *Compute the velocity at the gear pitch circle*

Use the relation $V = rpm(D_p)\pi/12$, where $V$ = velocity at pitch-circle diameter, ft/min; $rpm$ = gear speed, r/min; $D_P$ = pitch diameter, in. Solving yields $V = 350(2.34375)\pi/12 = 215$ ft/min (2.15 m/s).

### 2. *Find the safe stress for the gear*

Use the relation $S_S = 55(600 + V)PC_S H_P/(FYV)$, where $S_S$ = safe stress on the gear, lb/in²; $P$ = diametral pitch, in; $C_S$ = service factor from Table 66; $H_P$ = horsepower transmitted by the

**TABLE 66** Service Factor $C_S$ for Horsepower Equations*

| Type of load | 8–10 h/day | 24 h/day | Intermittent, 3 h/day | Occasional, 0.5 h/day |
|---|---|---|---|---|
| Steady | 1.00 | 1.25 | 0.80 | 0.50 |
| Light shock | 1.25 | 1.50 | 1.00 | 0.80 |
| Medium shock | 1.50 | 1.75 | 1.25 | 1.00 |
| Heavy shock | 1.75 | 2.00 | 1.50 | 1.25 |

* *Product Engineering.*

**TABLE 67** Tooth-Form Factor Y for Horsepower Equations*

| Number of teeth | 14½° Involute or cycloidal | 20° Full-depth involute | 20° Stub-tooth involute | 20° Internal full depth | |
|---|---|---|---|---|---|
| | | | | Pinion | Gear |
| 50 | 0.352 | 0.408 | 0.474 | 0.437 | 0.613 |
| 75 | 0.364 | 0.434 | 0.496 | 0.452 | 0.581 |
| 100 | 0.371 | 0.446 | 0.506 | 0.462 | 0.581 |
| 150 | 0.377 | 0.459 | 0.518 | 0.468 | 0.565 |
| 300 | 0.383 | 0.471 | 0.534 | 0.478 | 0.534 |
| Rack | 0.390 | 0.484 | 0.550 | .... | .... |

*Product Engineering.*

gear; $F$ = face width of the gear, in; $Y$ = tooth form factor, Table 67. Substituting, we find $S_S$ = 55(600 + 215)(32)(1.0)(0.125)/[0.375(0.434)(215)] = 5124 lb/in² (35,561 kPa).

**3. *Select the gear material***

Enter Table 68 at the safe stress, 5124 lb/in² (35,561 kPa), and choose either nylon or polycarbonate gear material because the safe stress falls within the allowed range, 600 lb/in² (41,364 kPa), for these two materials. If a glass-reinforced gear is to be used, any of a number of materials listed in Table 68 would be suitable.

**Related Calculations:** Two other v equations are used in the analysis of plastic gears. For helical gears: $H_P - S_S FYV/[423(78 + V^{0.5})P_N C_S]$. For straight bevel gears, $H_P = S_S FYV(C - F)^4/[55(600 + V)PCC_S]$, where all the symbols are as given earlier; $C$ = pitch-cone diameter, in; $P_N$ = normal diametral pitch.

In growing numbers of fractional-horsepower applications up to 1.5 hp (1.12 kW), gears molded of plastics are being chosen. Typical products are portable power tools, home appliances, instrumentation, and various automotive components.

The reasons for this trend are many. Besides offering the lowest initial cost, plastic gears can be molded as one piece to include other functional parts such as cams, ratchets, lugs, and other gears without need for additional assembly or finishing operations. Moreover, plastic gears are lighter and quieter than metal gears, and are self-lubricating, corrosion-resistant, and relatively free from maintenance. Also, they can be molded inexpensively in colors for coding during assembly or just for looks.

Improved molding techniques achieve high accuracy. Also new special gear-tooth forms improve strength and wear. "We now can hold tooth-to-tooth composite error to within 0.0005 in (0.00127 cm)," says Samuel Pierson, president of ABA Tool & Die Co. It is made possible by electric-discharge machining of the metal molds and computerized analysis of the effects of moisture absorption and other factors on size change of the plastic gears.

**TABLE 68** Safe Stress Values for Horsepower Equations*

| Plastic | Unfilled | | Glass-reinforced | |
|---|---|---|---|---|
| | lb/in² | kPa | lb/in² | kPa |
| ABS | 3,000 | 20,682 | 6,000 | 41,364 |
| Acetal | 5,000 | 34,470 | 7,000 | 48,258 |
| Nylon | 6,000 | 41,364 | 12,000 | 82,728 |
| Polycarbonate | 6,000 | 41,364 | 9,000 | 62,046 |
| Polyester | 3,500 | 24,129 | 8,000 | 55,152 |
| Polyurethane | 2,500 | 17,235 | ..... | ..... |

*Product Engineering.*

"Furthermore, we now are recommending special tooth forms we developed to utilize full-fillet root radii for increased fatigue strength and tip relief for more uniform motion when teeth flex under load. Usually, it is too expensive for designers of machined metal gears to deviate from standard AGMA tooth forms. But with molded plastic gears, deviations add little to the cost," reports Mr. Pierson in *Product Engineering* magazine.

Plastic gears, however, are weaker than metal gears, have a relatively high rate of thermal expansion, and are temperature-limited.

If performance cannot be achieved solely with change in tooth form, try fillers that stabilize the molded part, boost load-carrying abilities, and improve self-lubricating and wear characteristics. Popular fillers include glass, polytetrafluoroethylene (PTFE), molybdenum disulfide, and silicones.

In short hairlike fibers, miniscule beads, or fine-milled powder, glass can markedly increase the tensile strength of a gear and reduce thermal expansion to as little as one-third the original value. Molybdenum disulfide, PTFE, and silicones, as built-in lubricants, reduce wear. Plastic formulations containing both glass and lubricant are becoming increasingly popular to combine strength and lubricity.

Six common plastics for molded gears are nylon, acetal, ABS, polycarbonate, polyester, and polyurethane.

The most popular gear plastic is still nylon. This workhorse has good strength, high abrasion resistance, and a low coefficient of friction. Furthermore, it is self-lubricating and unaffected by most industrial chemicals. Numerous manufacturers make nylon resins.

Nylon's main drawback is its tendency to absorb moisture. This is accompanied by an increase in the gear dimensions and toughness. The effects must be predicted accurately. Also, nylon is harder to mold than acetal—one of its main competitors.

All plastics have higher coefficients of thermal expansion than metals do. The coefficient for steel between 0 and 30°C is $1.23 \times 10^5$ per degree, whereas for nylon it lies between 7 and 10 $\times 10^{-5}$.

The grade usually preferred is nylon 6/6. Frequently it is filled with about 25 percent short glass fibers and a small amount of lubricant fillers. Do not hesitate to ask the injection molder to custom-blend resins and fillers.

The combination of a nylon gear running against an acetal gear results in a lower coefficient of friction than either a nylon against nylon or an acetal against acetal can. A good idea is to intersperse nylon with acetal in gear trains. Acetal is less expensive than nylon and is generally easier to mold. Strength is lower, however.

Two types of acetals are popular for gears: acetal homopolymer, which was the original acetal developed by DuPont, available under the tradename of Delrin; and acetal copolymer, manufactured by Celanese under the tradename Celcon. Other examples are glass-filled acetals (Fulton 404, from LNP Corp.).

All acetals are easily processed and have good natural lubricity, creep resistance, chemical resistance, and dimensional stability. But they have a high rate of mold shrinkage.

The main attraction of ABS plastic is its low cost, probably the lowest of the six classes of plastics used for gears. Some ABS is translucent and has a high gloss surface. It also offers ease of processibility, toughness, and rigidity. Two typical tradenames of ABS are Cycolac (Borg Warner), and Kralastic (Uniroyal).

The polycarbonates have high impact strength, high resistance to creep, a useful temperature range of −60 to 240°F (−51 to 115.4°C), low water absorption and thermal expansion rates, and ease and accuracy in molding.

Because of a rather high coefficient of friction, polycarbonate formulations are available to boost flexible strength of the gear teeth.

Recently, thermoplastic polyesters have been available for high-performance injection-molded parts. Some polyesters are filled with reinforcing glass fibers for gear applications. The adhesion between the polyester matrix and the glass fibers results in a substantial increase in strength and produces a rigid material that is creep-resistant at elevated temperature [330°F(148.7°C)].

The glossy surface of some polyesters seems to improve lubricity against other thermoplastics and metals. It withstands most organic solvents and chemicals at room temperature and has long-term resistance to gasoline, motor oil, and transmission fluids up to 140°F (60°C).

The polyurethanes are elastomeric resins sought for noise dampening or shock absorption, as in gears for bedroom clocks or sprockets for snowmobiles. Many proprietary polyurethane versions are available, including Cyanaprene (American Cyanimide), Estane (B. F. Goodrich), Texin (Mobay Chemical), and Voranol (Dow Chemical).

## TOTAL DRIVING AND SLIP TORQUE FOR EXTERNAL-SPRING CLUTCHES

Determine the capacity of a light, steel, external-spring clutch in which the spring is made of 0.05-in (0.13-cm) diameter round wire. The spring rides on 1-in (2.54-cm) diameter shafts. The inner diameter of the spring at rest is 0.98 in (2.49 cm). There are 16 coils, 8 per shaft, in the spring. The coefficient of friction of the steel $\mu = 0.10$, and the modulus of elasticity $E = 30 \times 10^6$ lb/in$^2$ ($208.2 \times 10^6$ kPa). What are the transmitted and slip torques for this overrunning clutch?

### Calculation Procedure:

#### 1. *Find the transmitted torque for this clutch*

Use the relation $T_t = [\pi(Ed^4i)/32D^4](e^{2\pi\mu N} - 1)$, where $T_t$ = transmitted torque, lb·in; $d$ = diameter of spring wire, in; $i$ = diametral interference (the difference between the spring helix diameter and shaft diameter), in; $D$ = drive shaft diameter, in; $\mu$ = coefficient of friction; $N$ = number of spring coils per shaft; other symbols as before. Substituting gives $T_t = [\pi(30 \times 10^6)(0.05)^4(0.02)/32(1)^4](e^{2\pi(0.1)(8)} - 1) = (0.368)(152.4 - 1) = 55.7$ lb·in (6.29 N·m).

#### 2. *Compute the maximum slip torque for the clutch*

Use the relation $T_s = [\pi(Ed^4i)/32D^4]\,[1/(e^{2\pi\mu N} - 1)]$, where the symbols are as defined earlier and $T_s$ = slip torque, lb·in. Substituting, we find $T_s = 0.368(1/52.4 - 1) = -0.366$ lb·in (−0.04 N·m). This shows that $T_s \simeq -M$, where $M$ = spring moment, lb·in. It also demonstrates that although the capacity of a spring clutch depends on the spring moment, such a clutch can be made relatively insensitive to changes in the coefficient of friction or number of coils. These are the factors appearing in the exponent of the above equations.

For many years, spring clutches have been almost ignored in the clutch clan, even in the relatively small and specialized family of overrunning clutches. A number of popular design texts omit the spring clutch entirely, so the key design formulas are hard to come by. Yet most spring or overrunning clutches function in two modes: either they are locked and driving the output (the usual mode), or they are overrunning, with the input stopped and the output continuing to rotate undriven.

The spring clutch may well be the simplest overrunning clutch. It has a helical spring, wound from rectangular steel stock and coiled inside the shaft bores, bridging the gap between them (Fig. 47*a*). The helix diameter at either end of the spring is slightly larger than that of the shaft in which it nestles, so the steel coils remain in continuous contact with the shaft. (This internal design differs from most other versions of the spring clutch, in which the spring is wrapped around the outsides of two butting, coaxial shafts.)

When the drive shaft rotates in the same sense as the twist of the helix, that is, when the rotation tends to unwind the internal spring, then the coils try to expand, gripping both shaft bores tightly and pushing the output.

When the relative rotation of the input shaft reverses, however, when the motion tends to wind the internal spring tighter, as it would if the input stopped or slowed, leaving the output to spin under its own momentum, then the helix contracts, loosening its grip and slipping freely inside the bores—*freewheeling*.

The spring inevitably drags a bit in the overrunning mode; it could not tighten its grip again if it did not. But the residual friction is relatively small and well controlled. The consequent low wear rate and high reliability are the internal spring's principal advantage. And the internal-spring design is virtually immune to centrifugal effects, another critical consideration for high-speed use, unlike the external-spring version, which can loosen its grip at high speeds.

The wrapped-spring clutch, Fig. 47*b*, does lend itself to a number of sophisticated applications, despite this limitation. It can serve as an overrunning clutch, although its operation is the mirror image of that of the internal spring. The wrapped spring loosens and freewheels when driven by an unwinding torque; and when subjected to a winding torque, it squeezes down on the shafts like a child's "Chinese Handcuffs" toy, pulling the output along after the input.

By bending the spring's input end outward (Fig. 47*b*), clutch makers can fashion an on/off

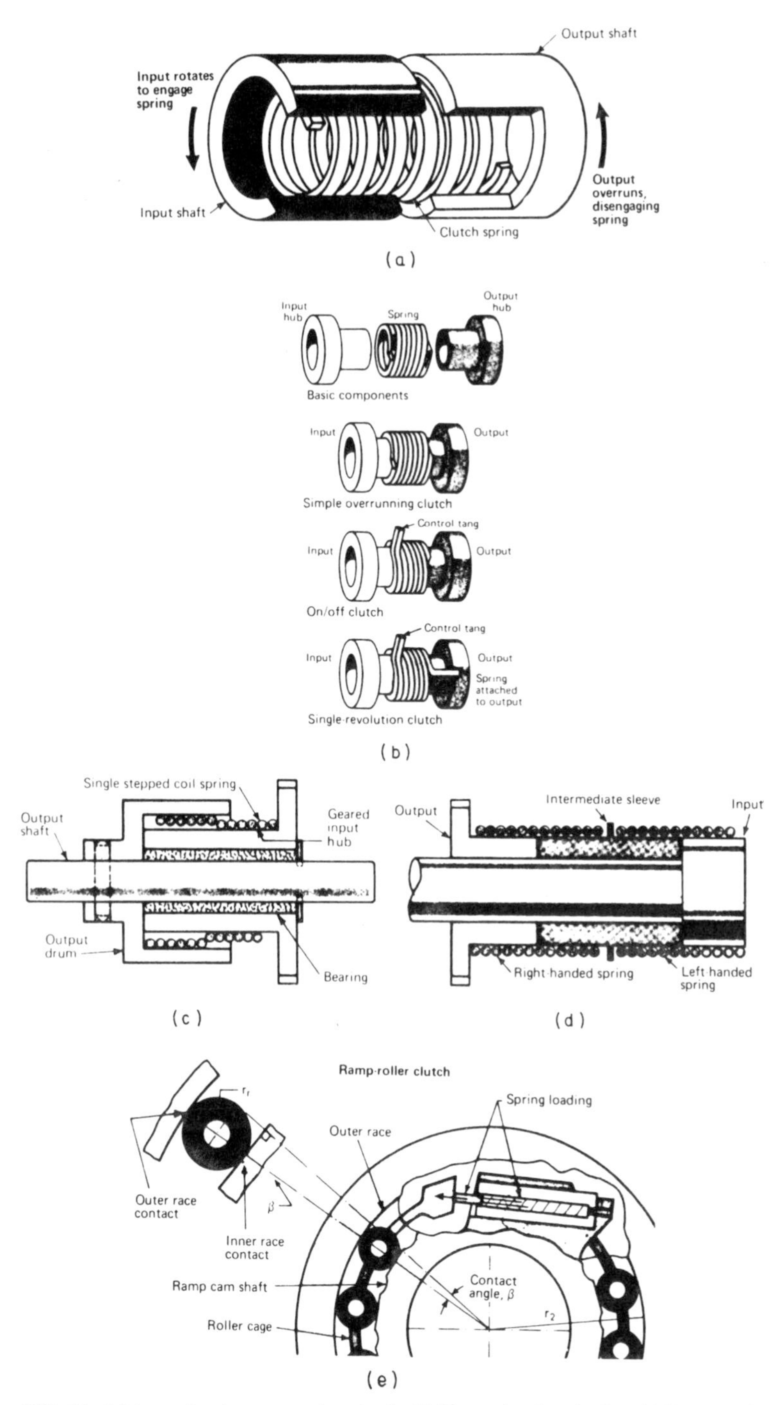


**FIG. 47** (*a*) Internal-spring overrunning clutch. (*b*) Wrapped-spring clutches. (*c*) Two-way slip clutch. (*d*) Mated right- and left-handed spring slip clutch. (*e*) Ramp-roller clutch. *(Design Engineering.)*

clutch. When this tang is released, the spring wraps down on the shafts for a positive drive; when the tang is engaged again, the spring unwinds and releases the output.

In a variation, the other end of the spring is permanently attached to the output shaft, giving single-revolution control. When the tang is engaged, the input clutches out and the spring stops the output (Fig. 47*b*).

The spring clutch also can be modified to work as a slip clutch or drag brake. Such a device is, in a sense, the opposite of an overrunning clutch: It works by beefing up residual drag so that the external spring can continue to transmit torque, even when it is being "unwound" in the overrunning direction.

By varying the spring dimensions and material, this drag can be precisely controlled to produce an extremely useful slip clutch. Indeed, spring clutches have been modified to give a predetermined slip for either direction of rotation. In one version, Fig. 47*c*, a stepped helix creates a dual external/internal spring. Another version, Fig. 47*d*, attaches a right-handed spring to a left-handed spring through an intermediate sleeve. In each case, one spring component slips with a controlled drag on one shaft while its partner clamps down, Chinese-handcuff-like, on the other shaft. If the input rotation is reversed, the two springs shift roles. Spring clutches are available from many sources.

Two factors control any application of wrapped spring clutches, whether overrunning or slip: the torque capacity in the driving or locked mode (where the spring grips the shaft) and the frictional drag torque capacity in the overrunning or slip direction. For overrunning applications, one would obviously want high driving torque and low drag torque. In a slip clutch, one would want to adjust the drag torque to the specific job. Happily, each factor can be calculated, by using equations derived by Joseph Kaplan of Machine Components.

The total driving torque $T_t$ is the product of two factors: an exponential function of the coefficient of friction and number of coils wrapping each shaft; and the spring's moment $M$, a function of shaft diameter and the spring's dimensions and modulus of elasticity.

The drag or slip torque capacity $T_s$ is the product of the same spring moment $M$ and an inverse exponential function of the number of coils and the coefficient of friction. Since this exponential factor generally approaches $-1$, the approximation of $T_s = -M$ is good enough for most applications.

These clutch-analysis equations, using the symbols given earlier, are useful: *For spring clutches:* The maximum torque in the locked or drive direction is $T_t = M(e^{2\mu\pi N} - 1)$ lb·in. The overrunning slip or friction torque is $T_s = M(e^{-2\mu\pi N} - 1)$. Where $M$ is the spring moment for circular cross-sectional wire, $M = \pi E d^4 \delta/(32D^2)$. For rectangular cross-sectional wire, $M = Ebt^3\delta/(6D^2)$. Large drums or high speeds may add a centrifugal correction $\Delta\delta$ to the diametral interference $\delta$; $\Delta\delta = 2\rho D^5\omega^2/ed^2$; $\Delta\delta = 1.5\rho D^5\omega^2/et^2$. *For sprag clutches:* $T_t = (r_1 nLK \tan\alpha)/(1/r_1) + 1/r_s$. *For ramp-roller clutches:* $T_t = nr_r r_2 LK \tan(\beta/2)$. Table 69 shows the results of a study by Sikorsky Aircraft of 10 types of overrunning clutches. Figures of merit were developed for each.

Additional symbols in these equations are $e$ = Euler's number $\simeq$ 2.7183; $K$ = hertzian stress constant factor from Table 70; $L$ = length of roller or sprag, in; $M$ = spring moment, lb·in; $N$ = number of spring coils per shaft; $n$ = number of sprags or rollers; $r_1$ = radius of inner race, in; $r_2$ = radius of outer race, in $r_r$ = radius of roller, in; $r_s$ = radius of sprag at contact of inner race, in; $T_t$ = transmitted torque, lb·in; $t$ = thinness of rectangular wire, in; $\alpha$ = sprag gripping angle, degrees; $\beta$ = ramp-race contact angle; $\delta$ = diametral interference (difference between spring helix diameter and shaft diameter), in; $\rho$ = coil material density, lb/in$^3$; $\omega$ = shaft angular velocity, rad/s.

This procedure, containing information from a study conducted by Sikorsky Aircraft, was reported in *Design Engineering* magazine in an article by Doug McCormick, Associate Editor, and uses equations derived by Joseph Kaplan of Machine Components.

## DESIGN METHODS FOR NONCIRCULAR SHAFTS

Find the maximum shear stress and the angular twist per unit length produced by a torque of 20,000 lb·in (2258.0 N·m) imposed on a double-milled steel shaft with a four-splined hollow core, Fig. 48. The shaft outer diameter is 2 in (5.1 cm); the inner diameter is 1 in (2.5 cm). Modulus of rigidity of the shaft is $G = 12 \times 10^6$ lb/in$^2$ ($83.3 \times 10^6$ kPa). The flat surfaces of the outer, milled

**TABLE 69** Sikorsky's Ratings of 10 Clutches*

| | Weight | Cost | Reliability | Maintainability | Centrifugal effects | Vibration | Transient torque | Thermal effects | Startup control | Failure modes | Operation limits | Unusual testing | Multi-engine | Bearing brinelling | Figure of merit |
|---|---|---|---|---|---|---|---|---|---|---|---|---|---|---|---|
| Spring clutch | 11.1 | 9.4 | 22.7 | 8.2 | 7.0 | 3.5 | 3.0 | 2.0 | 2.3 | 2.6 | 3.9 | 3.0 | 1.6 | 1.0 | 81.3 |
| Sprag, type A | 12.3 | 10.9 | 18.4 | 7.4 | 6.8 | 4.7 | 2.8 | 1.3 | 3.2 | 1.5 | 3.3 | 3.0 | 1.5 | 0.3 | 77.4 |
| Sprag, type B | 12.2 | 11.0 | 20.5 | 7.7 | 2.5 | 4.5 | 3.0 | 1.3 | 3.2 | 2.5 | 3.3 | 3.0 | 1.5 | 0.3 | 76.5 |
| Ramp roller (RR) | 10.8 | 8.9 | 21.3 | 7.6 | 6.5 | 4.5 | 2.8 | 1.3 | 3.2 | 2.4 | 3.3 | 3.0 | 1.4 | 0.7 | 77.7 |
| Actuated RR | 9.2 | 7.7 | 17.1 | 8.1 | 7.0 | 4.5 | 2.8 | 0.7 | 3.1 | 3.0 | 3.7 | 3.0 | 2.0 | 0.7 | 72.6 |
| Ball bearing RR | 11.9 | 10.9 | 13.9 | 5.3 | 1.0 | 5.2 | 2.8 | 1.1 | 1.6 | 2.4 | 3.3 | 3.0 | 1.4 | 0.3 | 64.1 |
| Roller gear RR | 10.4 | 9.9 | 14.5 | 6.7 | 1.0 | 5.0 | 2.8 | 1.1 | 3.2 | 2.3 | 3.3 | 3.0 | 1.3 | 0.7 | 65.2 |
| Positive ratchet | 10.4 | 8.2 | 18.1 | 8.7 | 6.8 | 5.2 | 2.8 | 2.0 | 3.6 | 1.8 | 3.9 | 2.0 | 1.8 | 0.7 | 76.0 |
| Face ratchet | 10.8 | 8.8 | 16.4 | 8.4 | 1.8 | 4.9 | 2.8 | 0.7 | 3.8 | 1.8 | 3.7 | 2.0 | 1.8 | 0.7 | 68.4 |
| Link ratchet | 11.3 | 12.3 | 19.0 | 8.8 | 4.5 | 5.9 | 2.8 | 2.0 | 3.8 | 2.8 | 3.3 | 2.0 | 1.8 | 0.3 | 80.6 |
| Of a possible | 13 | 13 | 30 | 9 | 7 | 6 | 3 | 2 | 4 | 3 | 4 | 3 | 2 | 1 | 100 |

* *Design Engineering.*

**TABLE 70** *K* Factors for Various Materials

| Material° | *K* at number of cycles | | |
|---|---|---|---|
| | $10^6$ | $10^7$ | $10^8$ |
| Gray cast-iron rolls (130–180 BHN) | 4,000 | 2,000 | 1,300† |
| GM Meehanite cast-iron rolls (190–240 BHN) | 4,000 | 2,500 | 1,950† |
| Gray cast-iron rolls with hardened-steel rolls: | | | |
| Cast iron, 140–160 BHN | 2,600 | 1,400 | 1,000† |
| Cast iron, 160–190 BHN | 3,200 | 1,900 | 1,300† |
| Cast iron, 270–290 BHN | 5,500 | 4,000 | 3,100† |
| Medium-carbon-steel rolls with hardened-steel rolls (medium-carbon steel, 270–300 BHN) | 14,000 | 11,000 | 9,000 |
| Carburized-carbon-steel rolls with hardened-steel rolls (carburized-carbon steel, 55–58 $R_c$) | 24,000 | 18,000 | 15,000‡ |

°Lubricated with straight mineral oil.
†At $4 \times 10^7$ cycles.
‡At $5 \times 10^8$ cycles.

shaft are cut down 0.1 in (0.25 cm), while the inner contour has a radius $r_i$ = 0.5 in (1.27 cm) with four splines of half-width $A$ = 0.1 in (0.25 cm) and height $B$ = 0.1 in (0.25 cm).

## Calculation Procedure:

### 1. *Find the torsional-stiffness and shear-stress factors for the outer shaft*

The proportionate mill height $H/R$ for the outer shaft is, from Fig. 49 and the given data, $H/R = 0.1/1 = 0.1$. Entering Fig. 49 at $H/R = 0.1$, project to the torsional-stiffness factor ($V$) curve for shaft type 1, and read $V = 0.71$. Likewise, from Fig. 49, $f$ = shear-stress factor = 0.82.

### 2. *Find the torsional-stiffness factor for the inner contour*

The spline dimensions, from Fig. 50, are $A/B = 0.1/0.1 = 1.0$, and $B/R = 0.1/0.5 = 0.2$. Entering Fig. 50 at $B/R = 0.2$, project upward to the $A/B = 1.0$ curve, and read the torsional-stiffness factor $V = 0.85$.

### 3. *Compute the angular twist of the shaft*

The composite torsional stiffness is defined as $V_tR^4$, where $V_t$ = composite-shaft stiffness factor and $R$ = the internal diameter. For a composite shaft, $V_tR^4 = \Sigma V_t r_i^4 = (0.71)(1^4) - (0.85)(0.5)^4$

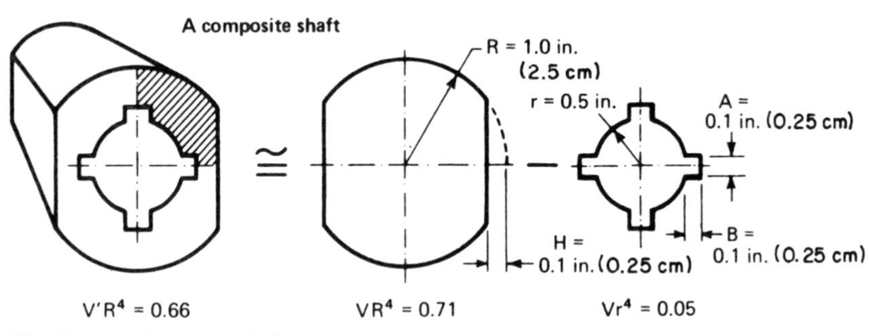


**FIG. 48** Typical composite shaft. *(Design Engineering.)*

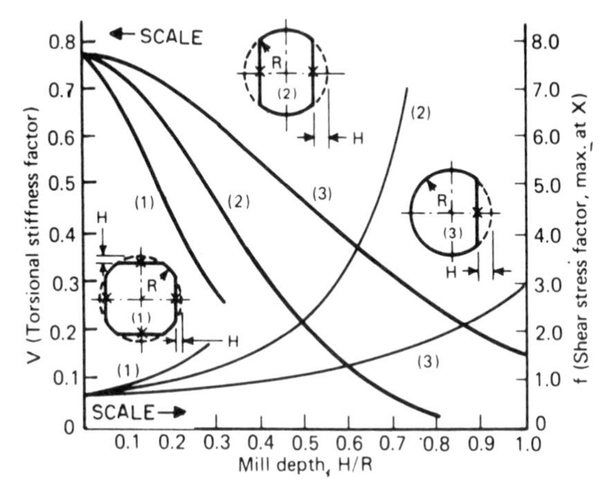


**FIG. 49** Milled shafts. *(Design Engineering.)*

= 0.66 for this shaft. (The first term in this summation is the outer shaft; the second is the inner cross section.)

To find the angular twist $\theta$ of this composite shaft, substitute in the relation $T = 2G\theta V_t R^4$, where $T$ = torque acting on the shaft, lb·in; other symbols as defined earlier. Substituting, we find $20{,}000 = 2(12 \times 10^6)\theta(0.66)$; $\theta = 0.0012$ rad/in (0.000472 rad/cm).

### 4. *Find the maximum shear stress in the shaft*

For a *solid* double-milled shaft, the maximum shear stress, $S_s$, lb/in$^2$ $= TfR^3$, where the symbols are as given earlier. Substituting for this shaft, assuming for now that the shaft is solid, we get $S_s = TfR^3 = (20{,}000)(0.82)(1^3) = 16{,}400$ lb/in$^2$ (113,816 kPa), where the value of $f$ is from step 1.

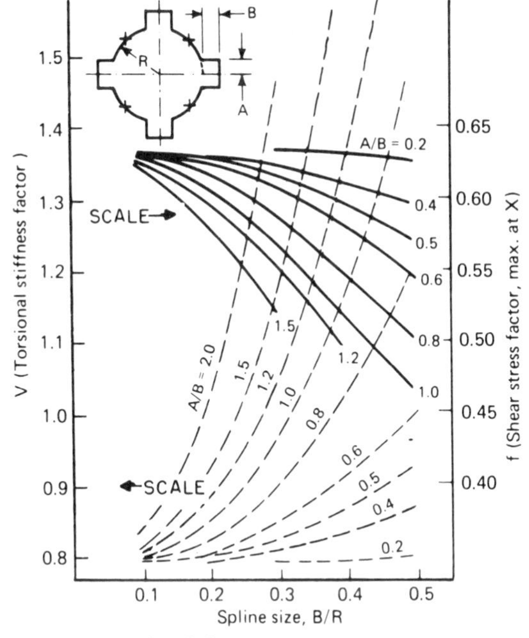


**FIG. 50** Four-spline shaft. *(Design Engineering.)*

Because the shaft is hollow, however, the solid-shaft maximum shear stress must be multiplied by the ratio $V/V_t = 0.71/0.66 = 1.076$, where the value of $V$ is from step 1 and the value of $V_t$ is from step 3.

The hollow-shaft maximum shear stress is $S_s' = S_s(V/V_t) = 16{,}400(1.076) = 17{,}646$ lb/in$^2$ (122,993 kPa).

**Related Calculations:** By using the charts (Figs. 49 through 58) and equations presented here, quantitative and performance factors can be calculated to well within 5 percent for a variety of widely used shafts. Thus, designers and engineers now have a solid analytical basis for choosing shafts, instead of having to rely on rules of thumb, which can lead to application problems.

Although design engineers are familiar with torsion and shear stress analyses of uniform circular shafts, usable solutions for even the most common noncircular shafts are often not only unfamiliar, but also unavailable. As a circular bar is twisted, each infinitesimal cross section rotates about the bar's longitudinal axis: plane cross sections remain plane, and the radii within each cross section remain straight. If the shaft cross section deviates even slightly from a circle, however, the situation changes radically and calculations bog down in complicated mathematics.

The solution for the circular cross section is straightforward: The shear stress at any point is proportional to the point's distance from the bar's axis; at each point, there are two equal stress vectors perpendicular to the radius through the point, one stress vector lying in the plane of the cross section and the other parallel to the bar's axis. The maximum stress is tangent to the shaft's outer surface. At the same time, the shaft's torsional stiffness is a function of its material, angle of twist, and the polar moment of inertia of the cross section.

The stress and torque relations can be summarized as $\theta = T/(JG)$, or $T = G\theta J$, and $S_s = TR/J$ or $S_s = G\theta R$, where $J$ = polar moment of inertia of a circular cross section ($= \pi R^4/2$); other symbols are as defined earlier.

If the shaft is splined, keyed, milled, or pinned, then its cross sections do not remain plane in torsion, but warp into three-dimensional surfaces. Radii do not remain straight, the distribution of shear stress is no longer linear, and the directions of shear stress are no longer perpendicular to the radius.

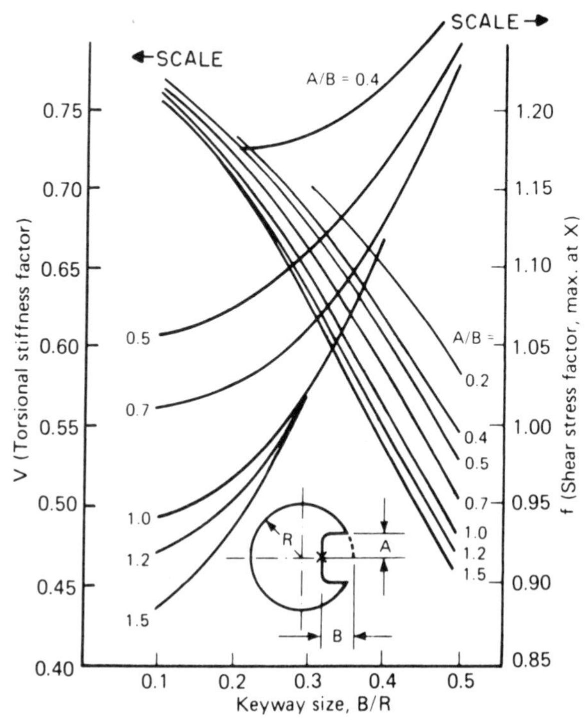


**FIG. 51** Single-keyway shaft. *(Design Engineering.)*

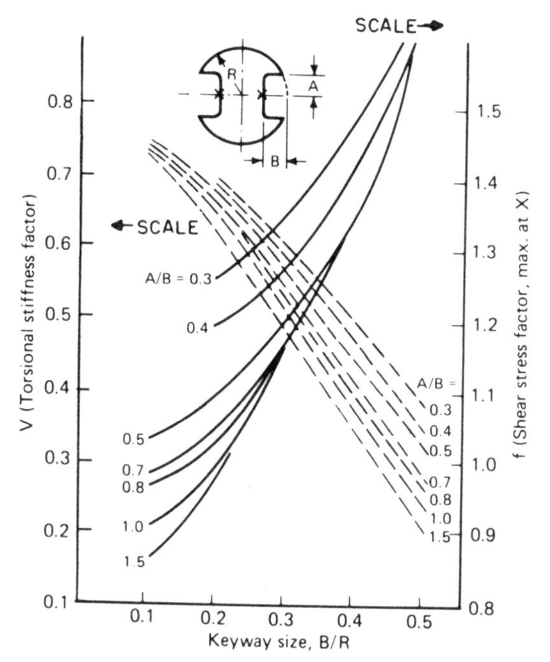


**FIG. 52** Two-keyway shaft. *(Design Engineering.)*

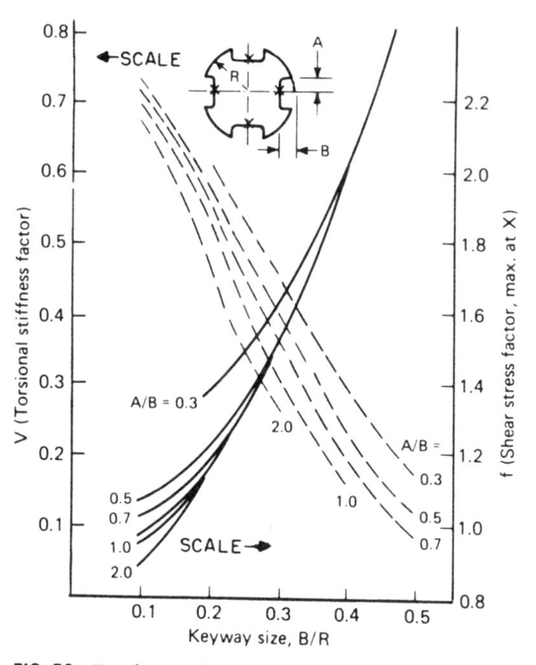


**FIG. 53** Four-keyway shaft. *(Design Engineering.)*

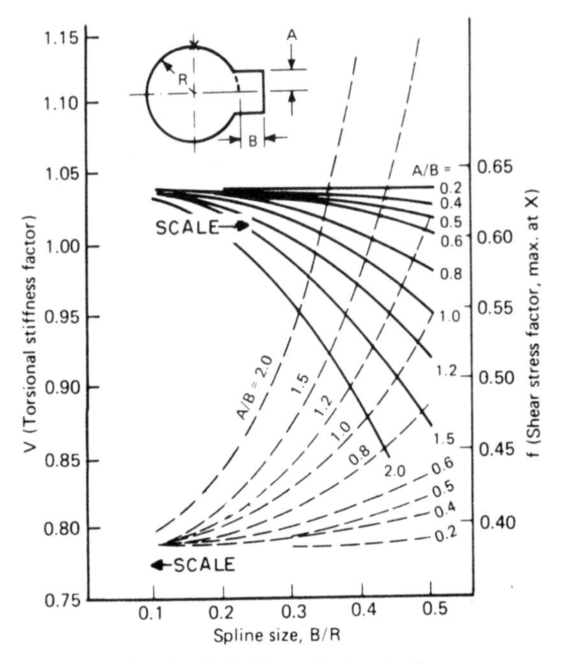


**FIG. 54** Single-spline shaft. (*Design Engineering.*)

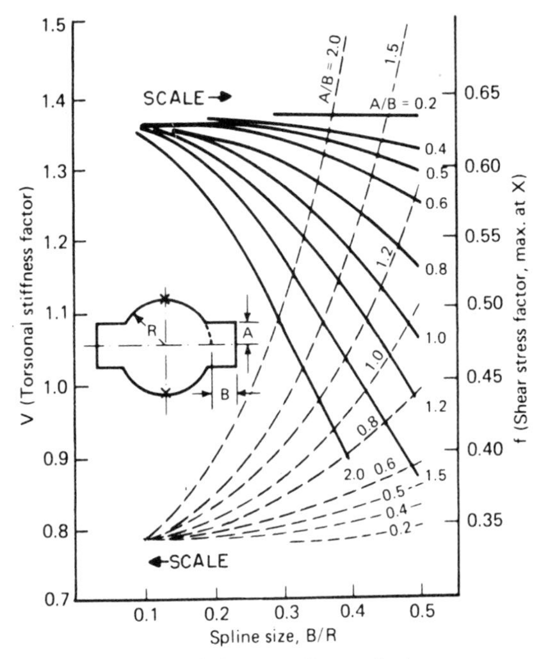


**FIG. 55** Two-spline shaft. (*Design Engineering.*)

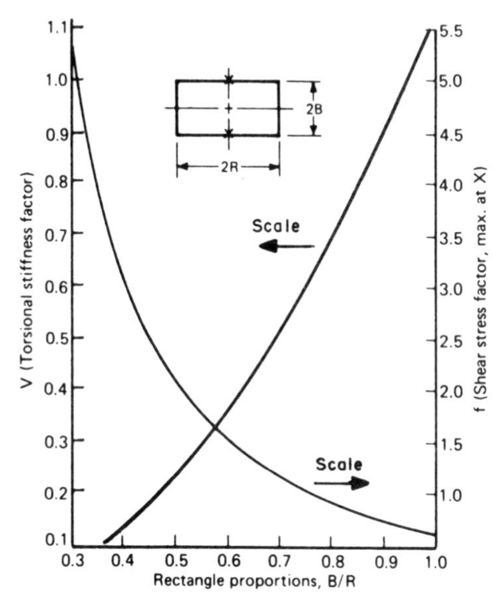


**FIG. 56** Rectangular shafts. *(Design Engineering.)*

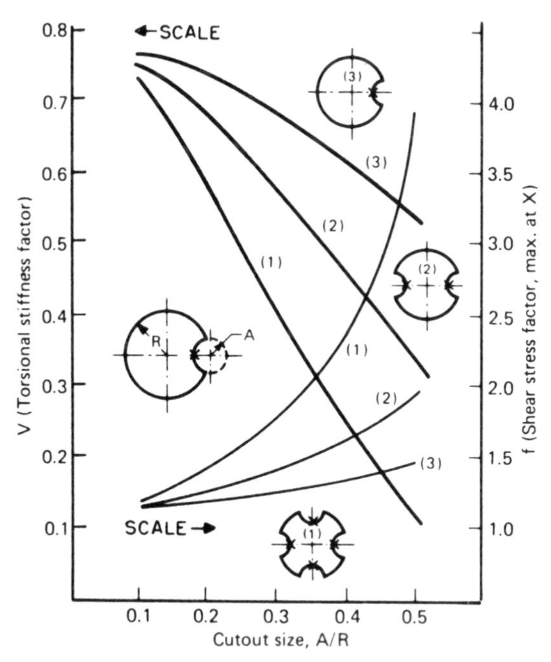


**FIG. 57** Pinned shaft. *(Design Engineering.)*

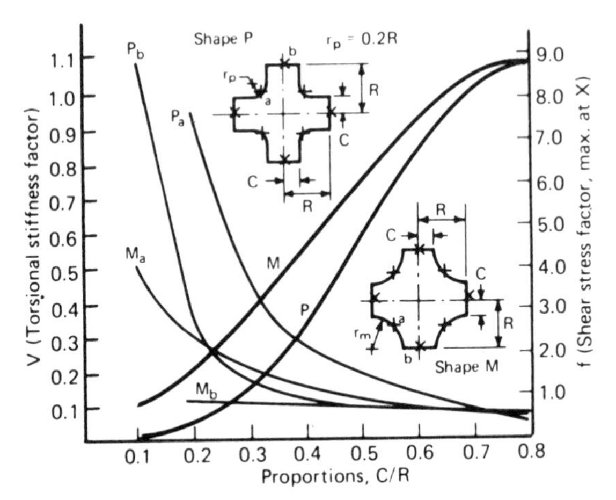


**FIG. 58** Cross-shaped shaft. *(Design Engineering.)*

These changes are described by partial differential equations drawn from Saint-Venant's theory. The equations are unwieldy, so unwieldy that most common shaft problems cannot be solved in closed form, but demand numerical approximations and educated intuition.

Of the methods of solving for Saint-Venant's torsion stress functions $\Phi$, one of the most effective is the technique of finite differences. A finite-difference computer program (called SHAFT) was developed for this purpose by the Scientific and Engineering Computer Applications Division of U.S. Army ARRADCOM (Dover, New Jersey).

SHAFT analyzed 10 fairly common transmission-shaft cross sections and (in the course of some 50 computer runs for each cross section) generated dimensionless torsional-stiffness and shear-stress factors for shafts with a wide range of proportions. Since the factors were calculated for unit-radius and unit-side cross sections, they may be applied to cross sections of any dimensions. These computer-generated factors, labeled $V$, $f$, and $d\phi/ds$, are derived from Prandtl's "membrane analogy" of the $\Phi$ function.

Because the torsional-stiffness factor $V$ may be summed for parallel shafts, $V$ values for various shaft cross sections may be adjusted for differing radii and then added or subtracted to give valid results for composite shaft shapes. Thus, the torsional stiffness of a 2-in (5.1-cm) diameter eight-splined shaft may be calculated (to within 1 percent accuracy) by adding the $V$ factors of two four-splined shafts and then subtracting the value for one 2-in (5.1-cm) diameter circle (to compensate for the overlapping of the central portion of the two splined shafts).

Or, a hollow shaft (like that analyzed above) can be approximated by taking the value of $VR^4$ for the cross section of the hollow and subtracting it from the $VR^4$ value of the outer contour.

In general, any composite shaft will have its own characteristic torsional-stiffness factor $V$, such that $V_t R^4 = \Sigma V_i r_i^4 t$, or $V_t = \Sigma V_i (r_i/R)^4$, where $R$ is the radius of the outermost cross section and $V_i$ and $r_i$ are the torsional-stiffness factors and radii for each of the cross sections combined to form the composite shaft.

By the method given here, a total of 10 different shaft configurations can be analyzed: single, two, and four keyways; single, two, and four splines; milled; rectangular; pinned; and cross-shaped. It is probably the most versatile method of shaft analysis to be developed in recent years. It was published by Robert I. Isakower, Chief, Scientific and Engineering Computer Applications Division, U.S. Army ARRADCOM, in *Design Engineering*.

## HYDRAULIC PISTON ACCELERATION, DECELERATION, FORCE, FLOW, AND SIZE DETERMINATION

What net acceleration force is needed by a horizontal cylinder having a 10,000-lb (4500-kg) load and 500-lb (2.2-kN) friction force, if 1500 lb/in$^2$ (gage) (10,341 kPa) is available at the cylinder port, there is zero initial piston velocity, and a 100-ft/min (30.5-m/min) terminal velocity is reached after 3-in (76.2-mm) travel at constant acceleration with the rod extending? Determine the required piston diameter and maximum fluid flow needed.

What pressure will stop a piston and load within 2 in (50.8 mm) at constant deceleration if the cylinder is horizontal, the rod is extending, the load is 5000 lb (2250 kg), there is a 500-lb (2224-N) friction force, the driving pressure at the head end is 800 lb/in$^2$ (gage) (5515.2 kPa), and the initial velocity is 80 ft/min (24.4 m/min)? The rod diameter is 1 in (25.4 mm), and the piston diameter is 1.5 in (38.1 mm).

## Calculation Procedure:

### 1. *Find the needed accelerating force*

Use the relation $F_A = Ma = M\,\Delta V/\Delta t$, where $F_A$ = net accelerating force, lb (N); $M$ = mass, slugs or lb·s$^2$/ft (N·s$^2$/m); $a$ = linear acceleration, ft/s$^2$(m/s$^2$), assumed constant; $\Delta V$ = velocity change during acceleration, ft/s (m/s); $\Delta t$ = time to reach terminal velocity, s. Substituting for this cylinder, we find $M = 10{,}000/32.17 = 310.85$ slugs.

Next $\Delta S = 3$ in/(12 in/ft) = 0.25 ft (76.2 mm). Also $\Delta V$ = (100 ft/min)/(60 s/min) = 1.667 ft/s (0.51 m/s). Then $F_A = 0.5(310.85)(1.667)^2/0.25 = 1727.6$ lb (7684.4 N).

### 2. *Determine the piston area and diameter*

Add the friction force and compute the piston area and diameter thus: $\Sigma F = F_A + F_F$, where $\Sigma F$ = sum of forces acting on piston, i.e., pressure, friction, inertia, load, lb; $F_F$ = friction force, lb. Substituting gives $\Sigma F = 1727.6 + 500 = 2227.6$ lb (9908.4 N).

Find the piston area from $A = \Sigma F/P$, where $P$ = fluid gage pressure available at the cylinder port, lb/in$^2$. Or, $A = 2227.6/1500 = 1.485$ in$^2$ (9.58 cm$^2$). The piston diameter $D$, then, is $D = (4A/\pi)^{0.5} = 1.375$ in (34.93 mm).

### 3. *Compute the maximum fluid flow required*

The maximum fluid flow $Q$ required is $Q = VA/231$, where $Q$ = maximum flow, gal/min; $V$ = terminal velocity of the piston, in/s; $A$ = piston area, in$^2$. Substituting, we find $Q = (100 \times 12)(1.485)/231 = 7.7$ gal/min (0.49 L/s).

### 4. *Determine the effective driving force for the piston with constant deceleration*

The driving force from pressure at the head end is $F_D$ = [fluid pressure, lb/in$^2$ (gage)](piston area, in$^2$). Or, $F_D = 800(1.5)^2\pi/4 = 1413.6$ lb (6287.7 N). However, there is a friction force of 500 lb (2224 N) resisting this driving force. Therefore, the effective driving force is $F_{ED} = 1413.6 - 500 = 913.6$ lb (4063.7 N).

### 5. *Compute the decelerating forces acting*

The mass, in slugs, is $M = F_A/32.17$, from the equation in step 1. By substituting, $M = 5000/32.17 = 155.4$ slugs.

Next, the linear piston travel during deceleration is $\Delta S = 2$ in/(12 in/ft) = 0.1667 ft (50.8 mm). The velocity change is $\Delta V = 80/60 = 1.333$ ft/s (0.41 m/s) during deceleration.

The decelerating force $F_A = 0.5M(\Delta V^2)/\Delta S$ for the special case when the velocity is zero at the start of acceleration or the end of deceleration. Thus $F_A = 0.5(155.4)(1.333)^2/0.1667 = 828.2$ lb (3684 N).

The total decelerating force is $\Sigma F = F_A + F_{ED} = 827.2 + 913.6 = 1741.8$ lb (7748 N).

### 6. *Find the cushioning pressure in the annulus*

The cushioning pressure is $P_c = \Sigma F/A$, where $A$ = differential area = piston area − rod area, both expressed in in$^2$. For this piston, $A = \pi(1.5)^2/4 - \pi(1.0)^2/4 = 0.982$ in$^2$ (6.34 cm$^2$). Then $P = \Sigma F/A = 1741.8/0.982 - 1773.7$ lb/in$^2$ (gage) (12,227.9 kPa).

**Related Calculations:** Most errors in applying hydraulic cylinders to accelerate or decelerate loads are traceable to poor design or installation. In the design area, miscalculation of acceleration and/or deceleration is a common cause of problems in the field. The above procedure for determining acceleration and deceleration should eliminate one source of design errors.

Rod buckling can also result from poor design. A basic design rule is to allow a compressive stress in the rod of 10,000 to 20,000 lb/in$^2$ (68,940 to 137,880 kPa) as long as the effective rod length-to-diameter ratio does not exceed about 6:1 at full extension. A firmly guided rod can help prevent buckling and allow at least four times as much extension.

With rotating hydraulic actuators, the net accelerating, or decelerating torque in lb·ft (N·m) is given by $T_A = J\alpha = MK^2$ rad/s$^2$ = $0.1047MK\ \Delta N/\Delta T = WK^2\ \Delta N/(307)\ \Delta t$, where $J$ = mass moment of inertia, slugs·ft$^2$, or lb·s$^2$·ft; $\alpha$ = angular acceleration (or deceleration), rad/s$^2$; $K$ = radius of gyration, ft; $\Delta N$ = rpm change during acceleration or deceleration; other symbols as given earlier.

For the special case where the rpm is zero at the start of acceleration or end of deceleration, $T_A = 0.0008725MK^2\ (\Delta N)^2/\Delta revs$; in this case, $\Delta revs$ = total revolutions = average rpm $\times\ \Delta t/60 = 0.5\ \Delta N \Delta T/60$; $\Delta t = 120(\Delta revs/\Delta t)$. For the linear piston and cylinder where the piston velocity at the start of acceleration is zero, or at the end of deceleration is zero, $\Delta t = \Delta S$/average velocity = $\Delta S/(0.5\ \Delta V)$.

High water base fluids (HWBF) are gaining popularity in industrial fluid power cylinder applications because of lower cost, greater safety, and biodegradability. Cylinders function well on HWBF if the cylinder specifications are properly prepared for the specific application. Some builders of cylinders and pumps offer designs that will operate at pressures up to several thousand pounds per square inch, gage. Most builders, however, recommend a 1000-lb/in$^2$ (gage) (6894-kPa) limit for cylinders and pumps today.

Robotics is another relatively recent major application for hydraulic cylinders. There is nothing quite like hydrostatics for delivering high torque or force in cramped spaces.

This procedure is the work of Frank Yeaple, Editor, *Design Engineering*, as reported in that publication.

## COMPUTATION OF REVOLUTE ROBOT PROPORTIONS AND LIMIT STOPS

Determine the equations for a two-link revolute robot's maximum and minimum paths, the shape and area of the robot's workspace, and the maximum necessary reach. Give the design steps to follow for a three-link robot.

### Calculation Procedure:

#### 1. *Give the equations for the four arcs of the robot*

Use the procedure developed by Y. C. Tsai and A. H. Soni of Oklahoma State University which gives a design strategy for setting the proportions and limit stops of revolute robot arms, as reported in the ASME *Journal of Mechanical Design*. Start by sketching the general workspace of the two-link robot arm, Fig. 59*a*.

Seen from the side, a 3*R* mechanism like that in Fig. 59*a* resolves itself into a 2*R* projection. This allows simple calculation of the robot's maximum and minimum paths.

In the vertical plane, the revolute robot's workspace is bounded by a set of four circular arcs. The precise positions and dimensions of the arcs are determined by the lengths of the robot's limbs and by the angular motion permitted in each joint. In the *xy* plot in Fig. 59*a*, the coordinates are determined by these equations:

$$x = \mathrm{l}_1 \sin\theta_1 + \mathrm{l}_2 \sin(\theta_1 + \theta_2) \qquad y = {}_1 \cos\theta_1 + l_2 \cos(\theta_1 + \theta_2)$$

These resolve into:

$$f_1 : (x - l_1 \sin\theta_1)^2 + (y - l_1 \cos\theta_1)^2 = l_2^2$$

$$f_2 : x^2 + y^2 = l_1^2 + l_2^2 + 2l_1 l_2 \cos\theta_2$$

In these relations, $f_2$ is the equation of a circle with a radius equal to the robot's forearm, $l_2$; the center of the circle varies with the inclination of the robot's upper arm from the vertical, $\theta_1$.

The second function, $f_2$, also describes a circle. This circle has a fixed center at (0,0), but the radius varies with the angle between the upper arm and the forearm. In effect, crooking the elbow shortens the robot's reach.

From the above relations, in turn, we get the equations for the four arcs: $DF = f_1(\theta_{1,\min})$; $EB = f_1(\theta_{1,\max})$; $DE = f_2(\theta_{2,\max})$; $FB = f_2(\theta_{2,\min})$.

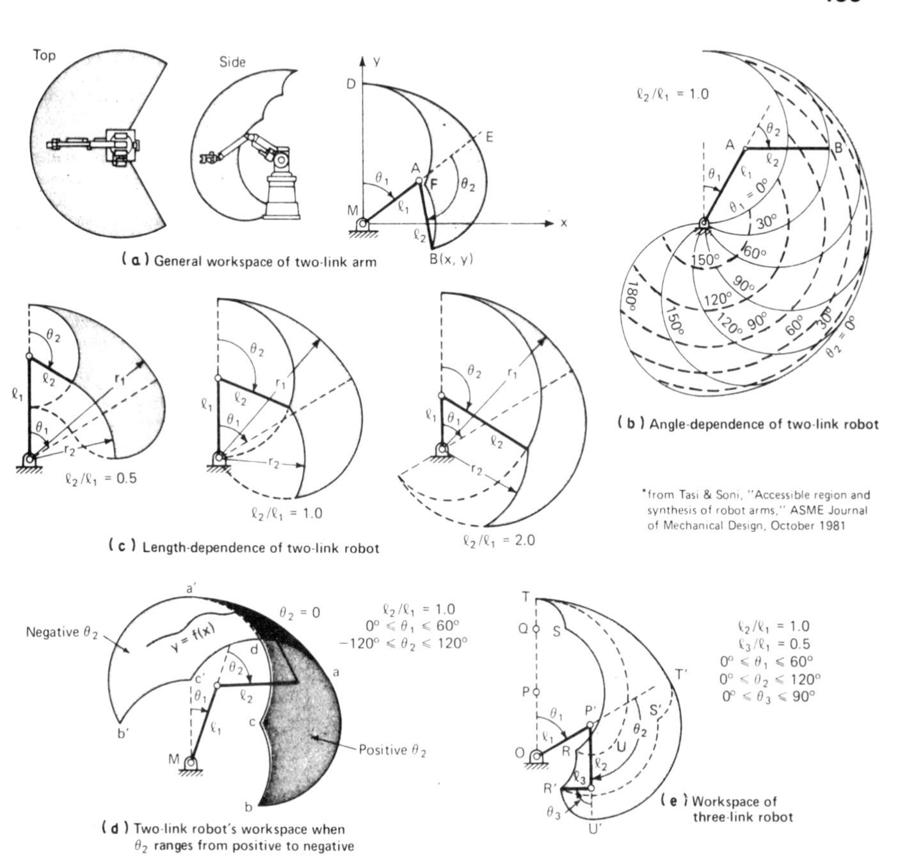


(a) General workspace of two-link arm

(b) Angle-dependence of two-link robot

(c) Length-dependence of two-link robot

(d) Two-link robot's workspace when $\theta_2$ ranges from positive to negative

(e) Workspace of three-link robot

**FIG. 59** Revolute robots are common in industrial applications. The robot's angular limits and the relative length of its limbs determine the size and shape of the workspace of the robot. *(Tasi and Soni, ASME Journal of Mechanical Design and Design Engineering.)*

## 2. *Define the shape and area of the robot's workspace*

Angular travel limitations are particularly important on robots whose major joints are powered by linear actuators, generally hydraulic cylinders. Figure 59*b* shows how maximum and minimum values for $\theta_1$ and $\theta_2$ affect the workspace envelope of planar projection of a common 3*R* robot.

Other robots—notably those powered by rotary actuators or motor-reducer sets—may be double-jointed at the elbow; $\theta$ may be either negative or positive. These robots produce the reflected workspace cross sections shown in Fig. 59*d*.

The relative lengths of the upper arm and forearm also strongly influence the shape of the two-link robot's workspace, Fig. 59*c*. Tsai and Soni's calculations show that, for a given total reach $L = l_1 + l_2$, the area bounded by the four arcs—the workspace—is greatest when $l_1/l_2 = 1.0$.

Last, the shape and area of the workspace depend on the ratio $l_2/l_1$, on $\theta_{2,\max}$, and on the difference $(\theta_{1,\max} - \theta_{1,\min})$. And given a constant rate of change for $\theta_2$, Tsai and Soni found that the arm can cover the most ground when the elbow is bent 90°.

## 3. *Specify parameters for a two-link robot able to reach any collection of points*

To reach any collection of points $(x_i, y_i)$ in the cross-sectional plane, Tsai and Soni transform the equations for $f_1$ and $f_2$ into a convenient procedure by turning $f_1$ and $f_2$ around to give equations for the angles $\theta_{1_i}$ and $\theta 2_i$ needed to reach each of the points $(x_i, y_i)$. These equations are:

$$\theta_{1i} = \cos^{-1}\left[\frac{y_i - y_0}{\sqrt{(x_i - x_0)^2 + (y_i - y_0)^2}}\right] - \cos^{-1}\left[\frac{(x_i - x_0)^2 + (y_i - y_0)^2 + l_1^2 - l_2^2}{2l_1\sqrt{(x_i - x_0)^2 + (y_i - y_0)}}\right]$$

$$\theta_{2i} = \cos^{-1}\left[\frac{(x_i - x_0)^2 + (y_i - y_0)^2 - l_1^2 + l_2^2}{2l_1l_2}\right]$$

Whereas the original equations assumed that the robot's shoulder is located at (0,0), these equations allow for a center of rotation $(x_0, y_0)$ anywhere in the plane.

Using the above equations, the designer then does the following: (*a*) She or he finds $x_{min}$, $y_{min}$, $x_{max}$, and $y_{max}$ among all the values $(x_i,y_i)$. (*b*) If the location of the shoulder of the robot is constrained, the designer assigns the proper values $(x_0,y_0)$ to the center of rotation. If there are no constraints, the designer assumes arbitrary values; the optimum position for the shoulder may be determined later.

(*c*) The designer finds the maximum necessary reach $L$ from among all $L_i = [(x_i - x_0)^2 + (y_i - y_0)^2]^{0.5}$. Then set $l_1 = l_2 = L/2$. (*d*) Compute $\theta_{1i}$ and $\theta_{2i}$ from the equations above for every point $(x_i,y_i)$. Then find the maximum and minimum values for both angles.

(*e*) Compute the area $A_2$ of the accessible region from $A = F(\theta_{1,max} - \theta_{1,min})(l_1 + l_2)^2$, where $F = (l_2/l_1)(\cos\theta_{2,min} - \cos\theta_{2,max})/[1 + (l_2/l_1)^2]$. (*f*) Use a grid search method, repeating steps *b* through *e* to find the optimum values for $(x_0,y_0)$, the point at which $A$ is at a minimum.

As Tsai and Soni note, this procedure can be computerized. The end result by either manual or computer computation is a set of optimum values for $x_0$, $y_0$, $l_1$, $l_2$, $\theta_{1,max}$, $\theta_{1,min}$, $\theta_{2,max}$, and $\theta_{2,min}$.

### 4. *List the steps for three-link robot design*

In practice, the pitch link of a robot's wrist extends the mechanism to produce a three-link 4*R* robot—equivalent to a 3*R* robot in the cross-sectional plane, Fig. 59. This additional link changes the shape and size of the workspace; it is generally short, and the additions are often minor.

Find the shape of the workspace thus: (*a*) Fix the first link at $\theta_{1,min}$ and treat the links $l_2$ and $l_3$ (that is, *PQ* and *QT*) as a two-link robot to determine their accessible region *RSTU*. (*b*) Rotate the workspace *RSTU* through the whole permissible angle $\theta_{1,max} - \theta_{1,min}$. The region swept out is the workspace.

The third link increases the workspace and permits the designer to specify the attitude of the last link and the "precision points" through which the arm's endpoint must pass.

Besides specifying a set of points $(x_i,y_i)$, the designer may specify for each point a unit vector $\mathbf{e}_i$. In operation, the end link *QT* will point along $\mathbf{e}_i$. Thus, the designer specifies the location of two points: the endpoint *T* and the base of the third link *Q*, Fig. 59*e*.

Designing such a three-link device is quite similar to designing a two-link version. The designer must add three steps at the start of the design sequence: (*a*) Select an appropriate length $l_3$ for the third link. (*b*) Specify a unit vector $\mathbf{e}_i = \mathbf{e}_{xi}\mathbf{i} + \mathbf{e}_{yi}\mathbf{j}$, for each prescribed accessible point $(x_i,y_i)$. (*c*) From these, specify a series of precision points $(x'_i,y'_i)$ for the endpoint *Q* of the two-link arm $l_1 + l_2$; $x'_i = x_i - \mathbf{e}_{xi}\, l_3$, $y'_i = y_i - \mathbf{e}_{xi} l_3$.

The designer then creates a linkage that is able to reach all precision points $(x'_i,y'_i)$, using the steps outlined for a two-link robot. Tsai and Soni also synthesize five-bar mechanisms to generate prescribed coupler curves. They also show how to design equivalent single- and dual-cam mechanisms for producing the same motion.

**Related Calculations:** The robot is becoming more popular every year for a variety of industrial activities such as machining, welding, assembly, painting, stamping, soldering, cutting, grinding, etc. Kenichi Ohmae, a director of McKinsey and Company, refers to robots as "steel-collar workers." Outside of the industrial field robots are finding other widespread applications. Thus, on the space shuttle *Columbia* a 45-ft (13.7-m) robot arm hauled a 65,000-lb (29,545-kg) satellite out of earth orbit. Weighing only 905 lb (362 kg), the arm has a payload capacity 70 times its own weight. In the medical field, robots are helping disabled people and others who are incapacitated to lead more normal lives. Newer robots are being fitted with vision devices enabling them to distinguish between large and small parts. Designers look forward to the day when vision can be added to medical robots to further expand the life of people having physical disabilities.

Joseph Engelberger, pioneer roboticist, classifies robots into several different categories. Chief among these are as follows: (1) A cartesian robot must move its entire mass linearly during any *x*

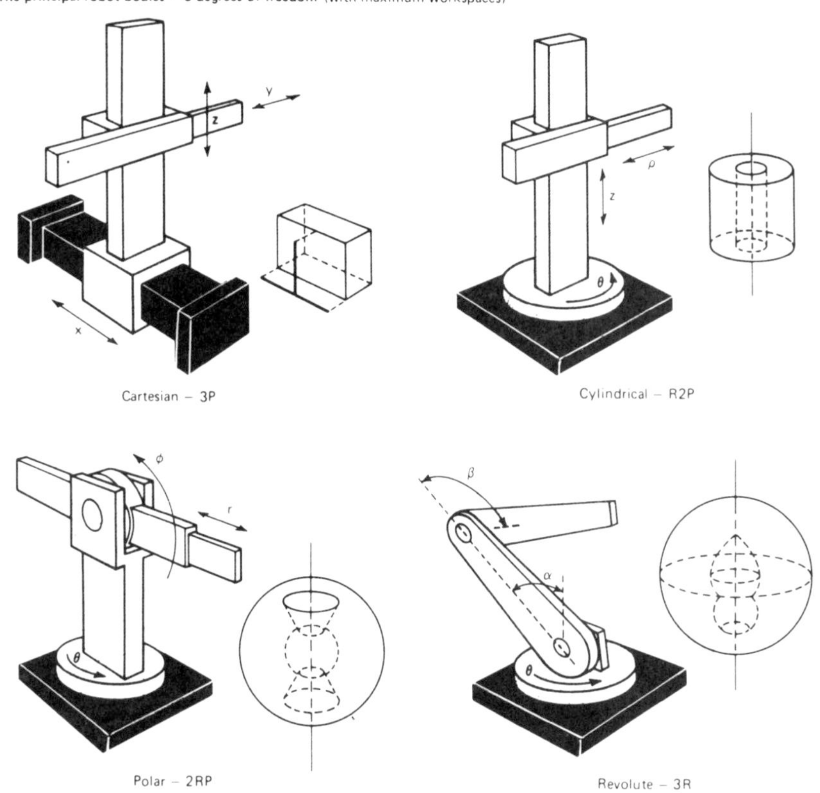


**FIG. 60** Types of robot bodies. *(Design Engineering, after Engelberger.)*

axis translation; this robot is well adapted for dealing with wide flat sheets as in painting and welding. The cartesian robot might be an inefficient choice for jobs needing many fast left-and-right moves. (2) Spherical-body robots might be best suited for loading machine tools. (3) Likewise, cylindrical robots are adapted to loading machine tools. (4) Revolute robots find a wide variety of applications in industry. Figure 60 shows a number of different robot bodies.

In the human body we get 7 degrees of freedom from just three joints. Most robots get only 6 degrees of freedom from six joints. This comparison gives one an appreciation of the construction of the human body compared to that of a robot. Nevertheless, robots are replacing humans in a variety of activities, saving labor and money for the organization using them.

This calculation procedure provides the designer with a number of equations for designing industrial, medical, and other robots. In designing a robot the designer must be careful not to use a robot which is too complex for the activity performed. Where simple operations are performed, such as painting, loading, and unloading, usually a simple one-directional robot will be satisfactory. Using more expensive multidirectional robots will only increase the cost of performing the operation and reduce the savings which might otherwise be possible.

Ohmae cities four ways in which robots are important in industry: (1) They reduce labor costs in industries which have a large labor component as part of their total costs. (2) Robots are easier to schedule in times of recession than are human beings. In many plants robots will reduce the breakeven point and are easier to "lay off" than human beings. (3) Robots make it easier for a small firm to enter precision manufacturing businesses. (4) Robots allow location of a plant to be made independent of the skilled-labor supply. For these reasons, there is a growing interest in the use of robots in a variety of industries.

A valuable reference for designers is Joseph Engelberger's book *Robotics in Practice*, published by Amacon, New York. This pioneer roboticist covers many topics important to the modern designer.

At the time of this writing, the robot population of the United States was increasing at the rate of 150 robots per month. The overhead cost of a robot in the automotive industry is currently under \$5 per hour, compared to about \$14 per hour for hourly employees. Robot maintenance cost is about 50 cents per hour of operation, while the operating labor cost of a robot is about 40 cents per hour. Downtime for robots is less than 2 percent, according to *Mechanical Engineering* magazine of the ASME. Mean time between failures for robots is about 500 h.

The procedure given here is the work of Y. C. Tsai and A. H. Soni of Oklahoma State University, as reported in *Design Engineering* magazine in an article by Doug McCormick, Associate Editor.

## HYDROPNEUMATIC ACCUMULATOR DESIGN FOR HIGH FORCE LEVELS

Design a hydropneumatic spring to absorb the mechanical shock created by a 300-lb (136.4-kg) load traveling at a velocity of 20 ft/s (6.1 m/s). Space available to stop the load is limited to 4 in (10.2 cm).

### Calculation Procedure:

#### 1. *Determine the kinetic energy which the spring must absorb*

Figure 61 shows a typical hydropneumatic accumulator which functions as a spring. The spring is a closed system made up of a single-acting cylinder (or sometimes a rotary actuator) and a gas-filled accumulator. As the load drives the piston, fluid (usually oil) compresses the gas in the flexible rubber bladder. Once the load is removed, either partially or completely, the gas pressure drives the piston back for the return cycle.

The flow-control valve limits the speed of the compression and return strokes. In custom-designed springs, flow-control valves are often combinations of check valves and fixed or variable orifices. Depending on the orientation of the check valve, the compression speed can be high with low return speed, or vice versa. Within the pressure limits of the components, speed and stroke length can be varied by changing the accumulator precharge. Higher precharge pressure gives shorter strokes, slower compression speed, and faster return speed.

The kinetic energy that must be absorbed by the spring is given by $E_k = 12WV^2/2g$, where $E_k$ = kinetic energy that must be absorbed, in·lb; $W$ = weight of load, lb; $V$ = load velocity, ft/s; $g$ = acceleration due to gravity, 32.2 ft/s$^2$. From the given data, $E_k = 12(300)(20)^2/2(32.3) = 22{,}380$ in·lb (2524.5 N·m).

#### 2. *Find the final pressure of the gas in the accumulator*

To find the final pressure of the gas in the accumulator, first we must assume an accumulator size and pressure rating. Then we check the pressure developed and the piston stroke. If they are

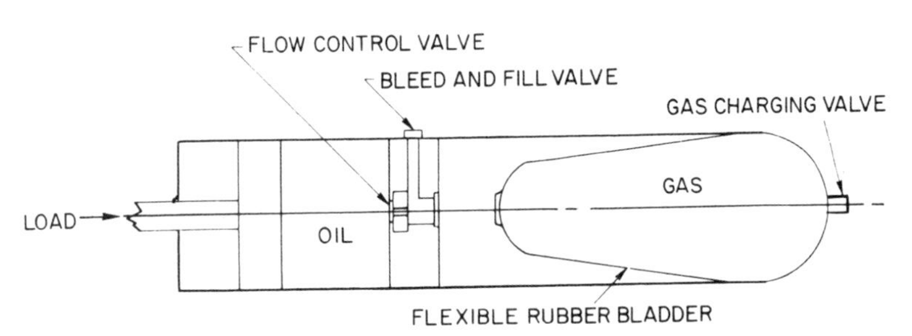


**FIG. 61** Typical hydropneumatic accumulator. *(Machine Design.)*

within the allowable limits for the application, the assumptions were correct. If the limits are exceeded, we must make new assumptions and check the values again until a suitable design is obtained.

For this application, based on the machine layout, assume that a 2.5-in (6.35-cm) cylinder with a 60-$in^3$ (983.2-$cm^3$) accumulator is chosen and that both are rated at 2000 lb/$in^2$ (13,788 kPa) with a 1000-lb/$in^2$ (abs) (6894-kPa) precharge. Check that the final loaded pressure and volume are suitable for the load.

The final load pressure $p_2$ lb/$in^2$ (abs) is found from $p_2^{(n-1)/n} = p_1^{(n-1)/n}[E^{n-1}(p_1 v_1) + 1]$, where $p_1$ = precharge pressure of the accumulator, lb/$in^2$ (abs); $n$ = the polytropic gas constant = 1.4 for nitrogen, a popular changing gas; $v_1$ = accumulator capacity, $in^3$. Substituting gives $p_2^{(1.4-1)/1.4} = 1000^{(1.4-1)/1.4}[22{,}380^{1.4-1}/1000(60) + 1]$; $p_2 = 1616$ lb/$in^2$ (abs) (11,202.8 kPa). Since this is within the 2000-lb/$in^2$ (abs) limit selected, the accumulator is acceptable from a pressure standpoint.

### 3. *Determine the final volume of the accumulator*

Use the relation $v_2 = v_1(p_1/p_2)^{1/n}$, where $v_2$ = final volume of the accumulator, $in^3$; $v_1$ = initial volume of the accumulator, $in^3$; other symbols as before. Substituting, we get $v_2 = 60(1000/1616)^{1/1.4} = 42.67$ $in^3$ (699.3 $cm^3$).

### 4. *Compute the piston stroke under load*

Use the relation $L = 4(v_1 - v_2)/\pi D^2$, where $L$ = length of stroke under load, in; $D$ = piston diameter, in. Substituting yields $L = 4(60 - 42.67)/\pi(2.5)^2 = 3.53$ in (8.97 cm). Since this is within the allowable travel of 4 in (10 cm), the system is acceptable.

**Related Calculations:** Hydropneumatic accumulators have long been used as shock dampers and pulsation attenuators in hydraulic lines. But only recently have they been used as mechanical shock absorbers, or springs.

Current applications include shock absorption and seat-suspension systems for earth-moving and agricultural machinery, resetting mechanisms for plows, mill-roll loading, and rock-crusher loading. Potential applications include hydraulic hammers and shake tables.

In these relatively high-force applications, hydropneumatic springs have several advantages over mechanical springs. First, they are smaller and lighter, which can help reduce system costs. Second, they are not limited by metal fatigue, as mechanical springs are. Of course, their life is not infinite, for it is limited by wear of rod and piston seals.

Finally, hydropneumatic springs offer the inherent ability to control load speeds. With an orifice check valve or flow-control valve between actuator and accumulator, cam speed can be varied as needed.

One reason why these springs are not more widely used is that they are not packaged as off-the-shelf items. In the few cases where packages exist, they are often intended for other uses. Thus, package dimensions may not be those needed for spring applications, and off-the-shelf springs may not have all the special system parameters needed. But it is not hard to select individual off-the-shelf accumulators and actuators for a custom-designed system. The procedure given here is an easy method for calculating needed accumulator pressures and volumes. It is the work of Zeke Zahid, Vice President and General Manager, Greer Olaer Products Division, Greer Hydraulics, Inc., as reported in *Machine Design*.

## POWER SAVINGS ACHIEVABLE IN INDUSTRIAL HYDRAULIC SYSTEMS

An industrial hydraulic system can be designed with three different types of controls. At a flow rate of 100 gal/min (6.31 L/s), the pressure drop across the controls is as follows: Control *A*, 500 lb/$in^2$ (3447 kPa); control *B*, 1000 lb/$in^2$ (6894 kPa); control *C*, 2000 lb/$in^2$ (13,788 kPa). Determine the power loss and the cost of this loss for each control if the cost of electricity is 15 cents per kilowatthour. How much more can be spent on a control if it operates 3000 h/year?

### Calculation Procedure:

### 1. *Compute the horsepower lost in each control*

The horsepower lost during pressure drop through a hydraulic control is given by $hp = 5.82(10^{-4})Q\ \Delta P$, where $Q$ = flow rate through the control, gal/min; $\Delta P$ = pressure loss through the control. Substituting for each control and using the letter subscript to identify it, we find $hp_A = 5.82(10^{-4})(100)(500) = 29.1$ hp (21.7 kW); $hp_B = 5.82(10^{-4})(100)(1000) = 5.82$ hp (43.4 kW); $hp_c = 5.82(10^{-4})(100)(2000) = 116.4$ hp (86.8 kW).

### 2. *Find the cost of the pressure loss in each control*

The cost in dollars per hour wasted $w$ = kW($/kWh) = $hp$(0.746)($/kWh). Substituting and using a subscript to identify each control, we get $w_A$ = 21.7($0.15) = $3.26; $w_B$ = 43.4($0.15) = $6.51; $w_C$ = 86.8($0.15) = $13.02.

The annual loss for each control with 3000-h operation is $w_{A,\text{an}}$ = 3000($3.26) = $9780; $w_{B,\text{an}}$ = 3000($6.51) = $19,530; $w_{C,\text{an}}$ = 3000($13.02) = $39,060.

### 3. *Determine the additional amount that can be spent on a control*

Take one of the controls as the base or governing control, and use it as the guide to the allowable extra cost. Using control $C$ as the base, we can see that it causes an annual loss of $39,060. Hence, we could spend up to $39,060 for a more expensive control which would provide the desired function with a smaller pressure (and hence, money) loss.

The time required to recover the extra money spent for a more efficient control can be computed easily from ($39,060 − loss with new control, $), where the losses are expressed in dollars per year.

Thus, if a new control costs $2500 and control $C$ costs $1000, while the new control reduced the annual loss to $20,060, the time to recover the extra cost of the new control would be ($2500 − $1000)/($39,060 − $20,060) = 0.08 year, or less than 1 month. This simple application shows the importance of careful selection of energy control devices.

And once the new control is installed, it will save $39,060 − $20,060 = $19,000 per year, assuming its maintenance cost equals that of the control it replaces.

**Related Calculations:** This approach to hydraulic system savings can be applied to systems serving industrial plants, aircraft, ships, mobile equipment, power plants, and commercial installations. Further, the approach is valid for any type of hydraulic system using oil, water, air, or synthetic materials as the fluid.

With greater emphasis in all industries on energy conservation, more attention is being paid to reducing unnecessary pressure losses in hydraulic systems. Dual-pressure pumps are finding wider use today because they offer an economical way to provide needed pressures at lower cost. Thus, the alternative control considered above might be a dual-pressure pump, instead of a throttling valve.

Other ways that pressure (and energy) losses are reduced is by using accumulators, shutting off the pump between cycles, modular hydraulic valve assemblies, variable-displacement pumps, electronic controls, and shock absorbers. Data in this procedure are from *Product Engineering* magazine, edited by Frank Yeaple.

## Metalworking

**REFERENCES:** Blazynski—*Metal Forming: Tool Profiles and Flow*, Halsted Press; Lippmann—*Engineering Plasticity: Theory of Metal Forming Processes*, Springer-Verlag; Ross—*Handbook of Metal Treatments and Testing*, Tavistock (England); Le Grand—*American Machinist's Handbook*, McGraw-Hill; Boston—*Metal Processing*, Wiley; Nordhoff—*Machine-Shop Estimating*, McGraw-Hill; *Machinery's Handbook*, Industrial Press; *Welding Handbook*, American Welding Society; ASTME—*Tool Engineer's Handbook*, McGraw-Hill; *Procedure Handbook of Arc Welding Design and Practice*, The Lincoln Electric Company; Black—*Theory of Metal Cutting*, McGraw-Hill; Doyle—*Manufacturing Processes and Materials for Engineers*, Prentice-Hall; Brierly and Siekmann—*Machining Principles and Cost Control*, McGraw-Hill; Reason—*The Measurement of Surface Texture*, Cleaver-Hume; Bolz—*Production Processes: Their Influence on Design*, Penton; Harris—*A Handbook of Woodcutting*, HMSO, London; *Application Data, Cemented Carbides, Cemented Oxides*, Metallurgical Products Department, General Electric Company; Wood—*Final Report on Advanced Theoretical Formability Manufacturing Technology*, LTV, Inc., and USAF; Maynard—*Handbook of Business Administration*,

McGraw-Hill; Niedzwiedzki—*Manual of Machinability and Tool Evaluation*, Huebner Publications, Cleveland; Hendriksen—*Chipbreakers*, The National Machine Tool Builders Association; ASTME—*Fundamentals of Tool Design*, Prentice-Hall; Crane—*Plastic Working of Metals and Power Press Operations*, Wiley; Jones—*Die Design and Die Making Practice*, Industrial Press; DeGarmo—*Materials and Processes in Manufacturing*, Macmillan; Jevons—*The Metallurgy of Deep Drawing and Pressing*, Wiley; Stanley—*Punches and Dies*, McGraw-Hill.

## TOTAL ELEMENT TIME AND TOTAL OPERATION TIME

The observed times for a turret-lathe operation are as follows: (1) material to bar stop, 0.0012 h; (2) index turret, 0.0010 h; (3) point material, 0.0005 h; (4) index turret, 0.0012 h; (5) turn 0.300-in (0.8-cm) diameter part, 0.0075 h; (6) clear hexagonal turret, 0.0009 h; (7) advance cross-slide tool, 0.0008 h; (8) cutoff part, 0.0030 h; (9) aside with part, 0.0005 h. What is the total element time? What is the total operation time if 450 parts are processed? Pointing of the material was later found unnecessary. What effect does this have on the element and operation total time?

### Calculation Procedure:

**1. *Compute the total element time***

Compute the total element time by finding the sum of each of the observed times in the operation, or sum steps 1 through 9: 0.0012 + 0.0010 + 0.0005 + 0.0012 + 0.0075 + 0.0009 + 0.0008 + 0.0030 + 0.0005 = 0.0166 h = 0.0166 (60 min/h) = 0.996 minute per element.

**2. *Compute the total operation time***

The total operation time = (element time, h)(number of parts processed). Or, (0.0166)(450) = 7.47 h.

**3. *Compute the time savings on deletion of one step***

When one step is deleted, two or more times are usually saved. These times are the machine preparation and machine working times. In this process, they are steps 2 and 3. Subtract the sum of these times from the total element time, or 0.0166 − (0.0010 + 0.0005) = 0.0151 h. Thus, the total element time decreases by 0.0015 h. The total operation time will now be (0.0151)(450) = 6.795 h, or a reduction of (0.0015)(450) = 0.6750 h. Checking shows 7.470 − 6.795 = 0.675 h.

**Related Calculations:** Use this procedure for any multiple-step metalworking operation in which one or more parts are processed. These processes may be turning, boring, facing, threading, tapping, drilling, milling, profiling, shaping, grinding, broaching, hobbing, cutting, etc. The time elements used may be from observed or historical data.

## CUTTING SPEEDS FOR VARIOUS MATERIALS

What spindle rpm is needed to produce a cutting speed of 150 ft/min (0.8 m/s) on a 2-in (5.1-cm) diameter bar? What is the cutting speed of a tool passing through 2.5-in (6.4-cm) diameter material at 200 r/min? Compare the required rpm of a turret-lathe cutter with the available spindle speeds.

### Calculation Procedure:

**1. *Compute the required spindle rpm***

In a rotating tool, the spindle rpm $R = 12C/\pi d$, where $C$ = cutting speed, ft/min; $d$ = work diameter, in. For this machine, $R = 12(150)/\pi(2) = 286$ r/min.

**2. *Compute the tool cutting speed***

For a rotating tool, $C = R\pi d/12$. Thus, for this tool, $C = (200)(\pi)(2.5)/12 = 131$ ft/min (0.7 m/s).

The cutting-speed equation is sometimes simplified to $C = Rd/4$. Using this equation for the above machine, we see $C = 200(2.5)/4 = 125$ ft/min (0.6 m/s). In general, it is wiser to use the exact equation.

### 3. *Compare the required rpm with the available rpm*

Consult the machine nameplate, *American Machinist's Handbook*, or a manufacturer's catalog to determine the available spindle rpm for a given machine. Thus, one Warner and Swasey turret lathe has a spindle speed of 282 compared with the 286 r/min required in step 1. The part could be cut at this lower spindle speed, but the time required would be slightly greater because the available spindle speed is $286 - 282 = 4$ r/min less than the computed spindle speed.

When preparing job-time estimates, be certain to use the available spindle speed, because this is frequently less than the computed spindle speed. As a result, the actual cutting time will be longer when the available spindle speed is lower.

**Related Calculations:** Use this procedure for a cutting tool having a rotating cutter, such as a lathe, boring mill, automatic screw machine, etc. Tables of cutting speeds for various materials (metals, plastics, etc.) are available in the *American Machinist's Handbook*, as are tables of spindle rpm and cutting speed.

## DEPTH OF CUT AND CUTTING TIME FOR A KEYWAY

What depth of cut is needed for a ¾-in (1.9-cm) wide keyway in a 3-in (7.6-cm) diameter shaft? The keyway length is 2 in (5.1 cm). How long will it take to mill this keyway with a 24-tooth cutter turning at 130 r/min if the feed is 0.005 per tooth?

### Calculation Procedure:

### 1. *Sketch the shaft and keyway*

Figure 1 shows the shaft and keyway. Note that the depth of cut $D$ in $= W/2 + A$, where $W$ = keyway width, in; $A$ = distance from the key horizontal centerline to the top of the shaft, in.

### 2. *Compute the distance from the centerline to the shaft top*

For a machined keyway, $A = [d - (d^2 - W^2)^{0.5}]/2$, where $d$ = shaft diameter, in. With the given dimensions, $A = [3 - (3^2 - 0.75^2)^{0.5}]/2 = 0.045$ in (1.1 mm).

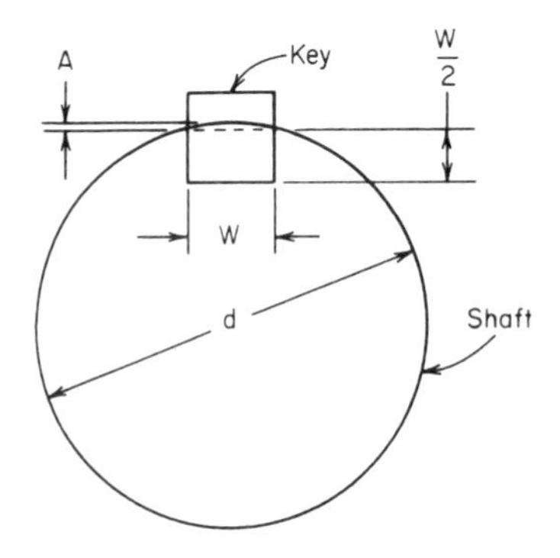


**FIG. 1** Keyway dimensions.

### 3. *Compute the depth of cut for the keyway*

The depth of cut $D = W/2 + A = 0.75/2 + 0.045 = 0.420$ in (1.1 cm).

### 4. *Compute the keyway cutting time*

For a single milling cutter, cutting time, min = length of cut, in/[(feed per tooth) × (number of teeth on cutter)(cutter rpm)]. Thus, for this keyway, cutting time = $2.0/[(0.005)(24)(130)] = 0.128$ min.

**Related Calculations:** Use this procedure for square or rectangular keyways. For Woodruff key-seat milling, use the same cutting-time equation as in step 4. A Woodruff key seat is almost a semicircle, being one-half the width of the key *less than* a semicircle. Thus, a 9/16-in (1.4-cm) deep Woodruff key seat containing a ⅜-in (1.0-cm) wide key will be (⅜)/2 = 3/16 in (0.5 cm) less than a semicircle. The key seat would be cut with a cutter having a radius of 9/16 + 3/16 = 12/16, or ¾ in (1.9 cm).

## MILLING-MACHINE TABLE FEED AND CUTTER APPROACH

A 12-tooth milling cutter turns at 400 r/min and has a feed of 0.006 per tooth per revolution. What table feed is needed? If this cutter is 8 in (20.3 cm) in diameter and is facing a 2-in (5.1-cm) wide part, determine the cutter approach.

**Calculation Procedure:**

1. ***Compute the required table feed***

For a milling machine, the table feed $F_T$ in/min $= f_t nR$, where $f_t$ = feed per tooth per revolution; $n$ = number of teeth in cutter; $R$ = cutter rpm. For this cutter, $F_T = (0.006) \times (12)(400) = 28.8$ in/min (1.2 cm/s).

2. ***Compute the cutter approach***

The approach of a milling cutter $A_c$ in $= 0.5D_c - 0.5(D_c^2 - w^2)^{0.5}$, where $D_c$ = cutter diameter, in; $w$ = width of face of cut, in. For this cutter, $A_c = 0.5(8) - 0.5(8^2 - 2^2)^{0.5} = 0.53$ in (1.3 cm).

**Related Calculations:** Use this procedure for any milling cutter whose dimensions and speed are known. These cutters can be used for metals, plastics, and other nonmetallic materials.

## DIMENSIONS OF TAPERS AND DOVETAILS

What are the taper per foot (TPF) and taper per inch (TPI) of an 18-in (45.7-cm) long part having a large diameter $d_l$ of 3 in (7.6 cm) and a small diameter $d_s$ of 1.5 in (3.8 cm)? What is the length of a part with the same large and small diameters as the above part if the TPF is 3 in/ft (25 cm/m)? Determine the dimensions of the dovetail in Fig. 2 if $B = 2.15$ in (5.15 cm), $C = 0.60$ in (1.5 cm), and $a = 30°$. A ⅜-in (1.0-cm) diameter plug is used to measure the dovetail.

**Calculation Procedure:**

1. ***Compute the taper of the part***

For a round part TPF in/ft $= 12(d_l - d_s)/L$, where $L$ = length of part, in; other symbols as defined above. Thus for this part, TPF $= 12(3.0 - 1.5)/18 = 1$ in/ft (8.3 cm/m). And TPI in/in $= (d_l - d_s)/L_2$, or $(3.0 - 1.5)/18 = 0.0833$ in/in (0.0833 cm/cm).

The taper of round parts may also be expressed as the angle measured from the shaft centerline, that is, one-half the included angle between the tapered surfaces of the shaft.

2. ***Compute the length of the tapered part***

Converting the first equation of step 1 gives $L = 12(d_l - d_s)/\text{TPF}$. Or, $L = 12(3.0 - 1.5)/3.0 = 6$ in (15.2 cm).

3. ***Compute the dimensions of the dovetail***

For external and internal dovetails, Fig. 2, with all dimensions except the angles in inches, $A = B + CF = I + HF$; $B = A - CF = G - HF$; $E = P \cot (90 + a/2) + P$; $D = P \cot (90 - a/2) + P$; $F = 2 \tan a$; $Z = A - D$. Note that $P$ = diameter of plug used to measure the dovetail, in.

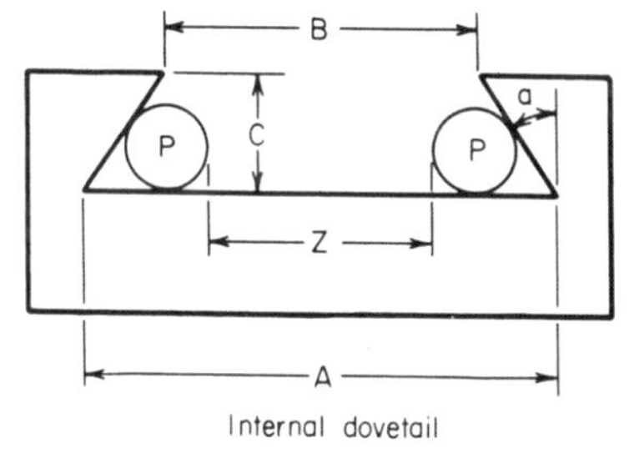


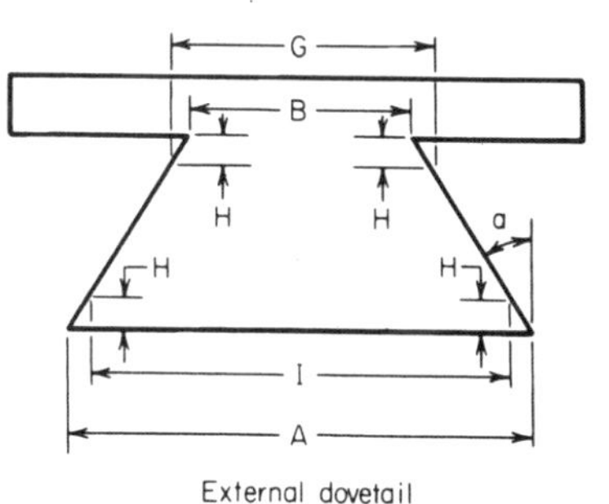


**FIG. 2** Dovetail dimensions.

With the given dimensions, $A = B + CF$, or $A = 2.15 + (0.60)(2 \times 0.577) = 2.84$ in (7.2 cm). Since the plug $P$ is ⅜ in (1.0 cm) in diameter, $D = P \cot (90 - a/2) + P = 0.375 \cot (90 - {}^{30}\!/_{2}) + 0.375 = 1.025$ in (2.6 cm). Then $Z = A - D = 2.840 - 1.025 = 1.815$ in (4.6 cm). Also $E = P \cot (90 + a/2) + P = 0.375 \cot (90 + {}^{30}\!/_{2}) + 0.375 = 0.591$ in (1.5 cm).

With flat-cornered dovetails, as at $I$ and $G$, and $H$ = ⅛ in (0.3 cm), $A = I + HF$. Solving for $I$, we get $I = A - HF = 2.84 - (0.125)(2 \times 0.577) = 2.696$ in (6.8 cm). Then $G = B + HF = 2.15 + (0.125)(2 \times 0.577) = 2.294$ in (5.8 cm).

**Related Calculations:** Use this procedure for tapers and dovetails in any metallic and nonmetallic material. When a large number of tapers and dovetails must be computed, use the appropriate tables in the *American Machinist's Handbook*.

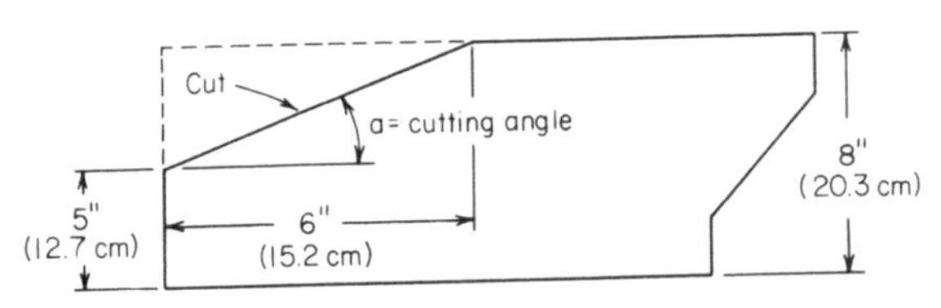


**FIG. 3** Length of cut of a part.

## ANGLE AND LENGTH OF CUT FROM GIVEN DIMENSIONS

At what angle must a cutting tool be set to cut the part in Fig. 3? How long is the cut in this part?

### Calculation Procedure:

**1. *Compute the angle of the cut***

Use trigonometry to compute the angle of the cut. Thus, tan $a$ = opposite side/adjacent side = $(8 - 5)/6 = 0.5$. From a table of trigonometric functions, $a$ = cutting angle = 26° 34′, closely.

**2. *Compute the length of the cut***

Use trigonometry to compute the length of cut. Thus, sin $a$ = opposite side/hypotenuse, or 0.4472 = (8 − 5)/hypotenuse; length of cut = length of hypotenuse = 3/0.4472 = 6.7 in (17.0 cm).

**Related Calculations:** Use this general procedure to compute the angle and length of cut for any metallic or nonmetallic part.

## TOOL FEED RATE AND CUTTING TIME

A part 3.0 in (7.6 cm) long is turned at 100 r/min. What is the feed rate if the cutting time is 1.5 min? How long will it take to cut a 7.0-in (17.8-cm) long part turning at 350 r/min if the feed is 0.020 in/r (0.51 mm/r)? How long will it take to drill a 5-in (12.7-cm) deep hole with a drill speed of 1000 r/min and a feed of 0.0025 in/r (0.06 mm/r)?

### Calculation Procedure:

**1. *Compute the tool feed rate***

For a tool cutting a rotating part, $f = L/(Rt)$, where $t$ = cutting time, min. For this part, $f$ = 3.0/[(100)(1.5)] = 0.02 in/r (0.51 mm/r).

**2. *Compute the cutting time for the part***

Transpose the equation in step 1 to yield $t = L/(Rf)$, or $t$ = 7.0/[(350)(0.020)] = 1.0 min.

**3. *Compute the drilling time for the part***

Drilling time is computed using the equation of step 2, or $t$ = 5.0/[(1000)(0.0025)] = 2.0 min.

**Related Calculations:** Use this procedure to compute the tool feed, cutting time, and drilling time in any metallic or nonmetallic material. Where many computations must be made, use the feed-rate and cutting-time tables in the *American Machinist's Handbook*.

## TRUE UNIT TIME, MINIMUM LOT SIZE, AND TOOL-CHANGE TIME

What is the machine unit time to work 25 parts if the setup time is 75 min and the unit standard time is 5.0 min? If one machine tool has a setup standard time of 9 min and a unit standard time of 5.0 min, how many pieces must be handled if a machine with a setup standard of 60 min and a unit standard time of 2.0 min is to be more economical? Determine the minimum lot size for an operation requiring 3 h to set up if the unit standard time is 2.0 min and the maximum increase in the unit standard may not exceed 15 percent. Find the unit time to change a lathe cutting tool if the operator takes 5 min to change the tool and the tool cuts 1.0 min/cycle and has a life of 3 h.

## Calculation Procedure:

### 1. *Compute the true unit time*

The true unit time for a machine $T_u = S_u/N + U_s$, where $S_u$ = setup time, min; $N$ = number of pieces in lot; $U_s$ = unit standard time, min. For this machine, $T_u = 75/75 + 5.0 = 6.0$ min.

### 2. *Determine the most economical machine*

Call one machine $X$, the other $Y$. Then (unit standard time of $X$, min)(number of pieces) + (setup time of $X$, min) = (unit standard time of $Y$, min)(number of pieces) + (setup time of $Y$, min). For these two machines, since the number of pieces Z is unknown, $5.0Z + 9 = 2.0Z + 60$. So Z = 17 pieces. Thus, machine $Y$ will be more economical when 17 or more pieces are made.

### 3. *Compute the minimum lot size*

The minimum lot size $M = S_u/(U_sK)$, where $K$ = allowable increase in unit-standard time, percent. For this run, $M = (3 \times 60)/[(2.0)(0.15)] = 600$ pieces.

### 4. *Compute the unit tool-changing time*

The unit tool-changing time $U_t$ to change from dull to sharp tools is $U_t = T_cC_t/l$, where $T_c$ = total time to change tool, min; $C_t$ = time tool is in use during cutting cycle, min; $l$ = life of tool, min. For this lathe, $U_t = (5)(1)/[(3)(60)] = 0.0278$ min.

**Related Calculations:** Use these general procedures to find true unit time, the most economical machine, minimum lot size, and unit tool-changing time for any type of machine tool—drill, lathe, milling machine, hobs, shapers, thread chasers, etc.

# TIME REQUIRED FOR TURNING OPERATIONS

Determine the time to turn a 3-in (7.6-cm) diameter brass bar down to a 2½-in (6.4-cm) diameter with a spindle speed of 200 r/min and a feed of 0.020 in (0.51 mm) per revolution if the length of cut is 4 in (10.2 cm). Show how the turning-time relation can be used for relief turning, pointing of bars, internal and external chamfering, hollow mill work, knurling, and forming operations.

## Calculation Procedure:

### 1. *Compute the turning time*

For a turning operation, the time to turn $T_t$ min $= L/(fR)$, where $L$ = length of cut, in; $f$ = feed, in/r; $R$ = work rpm. For this part, $T_t = 4/[(0.02)(200)] = 1.00$ min.

### 2. *Develop the turning relation for other operations*

For *relief turning* use the same relation as in step 1. Length of cut is the length of the relief, Fig. 4. A small amount of time is also required to hand-feed the tool to the minor diameter of the relief. This time is best obtained by observation of the operation.

The time required to *point a bar*, called *pointing*, is computed by using the relation in step 1. The length of cut is the distance from the end of the bar to the end of the tapered point, measured parallel to the axis of the bar, Fig. 4.

Use the relation in step 1 to compute the time to cut an internal or external chamfer. The length of cut of a chamfer is the horizontal distance $L$, Fig. 4.

A hollow mill reduces the external diameter of a part. The cutting time is computed by using the relation in step 1. The length of cut is shown in Fig. 4.

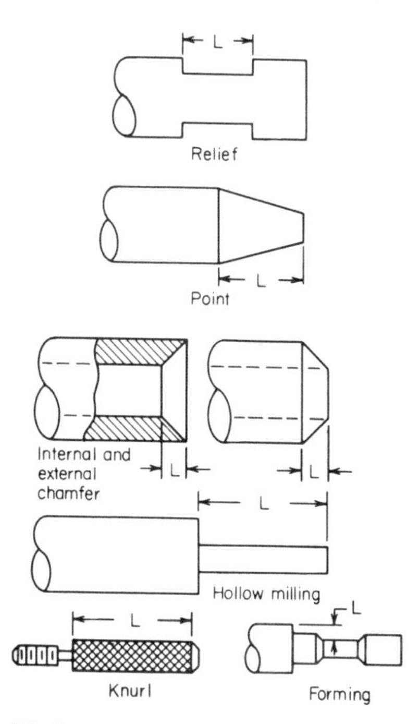


**FIG. 4** Turning operations.

Compute the time to knurl, using the relation in step 1. The length of cut is shown in Fig. 4. Compute the time for forming, using the relation in step 1. Length of cut is shown in Fig. 4.

## TIME AND POWER TO DRILL, BORE, COUNTERSINK, AND REAM

Determine the time and power required to drill a 3-in (7.6-cm) deep hole in an aluminum casting if a ¾-in (1.9-cm) diameter drill turning at 1000 r/min is used and the feed is 0.030 in (0.8 mm) per revolution. Show how the drilling relations can be used for boring, countersinking, and reaming. How long will it take to drill a hole through a 6-in (15.2-cm) thick piece of steel if the cone height of the drill is 0.5 in (1.3 cm), the feed is 0.002 in/r (0.05 mm/r), and the drill speed is 100 r/min?

### Calculation Procedure:

**1. *Compute the time required for drilling***

The time required to drill $T_d$ min $= L/fR$, where $L$ = depth of hole = length of cut, in. In most drilling calculations, the height of the drill cone (point) is ignored. (Where the cone height is used, follow the procedure in step 4.) For this hole, $T_d = 3/[(0.030) \times (1000)] = 0.10$ min.

**2. *Compute the power required to drill the hole***

The power required to drill, in hp, is $hp = 1.3LfCK$, where $C$ = cutting speed, ft/min, sometimes termed surface feet per minute $sfpm = \pi DR/12$; $K$ = power constant from Table 1. For an aluminum casting, $K = 3$. Then $hp = (1.3)(3)(0.030)(\pi \times 0.75 \times 1000/12)(3) = 66.0$ hp (49.2 kW). The factor 1.3 is used to account for dull tools and for overcoming friction in the machine.

**3. *Adapt the drill relations to other operations***

The time and power required for boring are found from the two relations given above. The length of the cut = length of the bore. Also use these relations for undercutting, sometimes called *internal relieving* and for counterboring. These same relations are also valid for countersinking, center

**TABLE 1** Power Constants for Machining

| Material | Power constant |
|---|---|
| Carbon steel C1010 to C1025 | 6 |
| Manganese steel T1330 to T1350 | 9 |
| Nickel steel 2015 to 2320 | 7 |
| Molybdenum | 9 |
| Chromium | 10 |
| Stainless steels | 11 |
| Cast iron: | |
| Soft | 3 |
| Medium | 3 |
| Hard | 4 |
| Aluminum alloys: | |
| Castings | 3 |
| Bar | 4 |
| Copper | 4 |
| Brass (except manganese) | 4 |
| Monel metal | 10 |
| Magnesium alloys | 3 |
| Malleable iron: | |
| Soft | 3 |
| Medium | 4 |
| Hard | 5 |

drilling, start or spot drilling, and reaming. In reaming, the length of cut is the total depth of the hole reamed.

**4. *Compute the time for drilling a deep hole***

With parts having a depth of 6 in (15.2 cm) or more, compute the drilling time from $T_d = (L + h)/(fR)$, where $h$ = cone height, in. For this hole, $T_d = (6 + 0.5)/[(0.002)(100)] = 32.25$ min. This compares with $T_d = L/fR = 6/[(0.002)(100)] = 30$ min when the height of the drill cone is ignored.

## TIME REQUIRED FOR FACING OPERATIONS

How long will it take to face a part on a lathe if the length of cut is 4 in (10.2 cm), the feed is 0.020 in/r (0.51 mm/r) and the spindle speed is 50 r/min? Determine the facing time if the same part is faced by an eight-tooth milling cutter turning at 1000 r/min and having a feed of 0.005 in (0.13 mm) per tooth per revolution. What table feed is required if the cutter is turning at 50 r/min? What is the feed per tooth with a table feed of 4.0 in/min (1.7 mm/s)? What added table travel is needed when a 4-in (10.2-cm) diameter cutter is cutting a 4-in (10.2-cm) wide piece of work?

### Calculation Procedure:

**1. *Compute the lathe facing time***

For lathe facing, the time to face $T_f$ min $= L/(fR)$, where the symbols are the same as given for previous calculation procedures in this section. For this part, $T_f = 4/[(0.02)(50)] = 4.0$ min.

**2. *Compute the facing time using a milling cutter***

With a milling cutter, $T_f = L/(f_t nR)$, where $f_t$ = feed per tooth, in/r; $n$ = number of teeth on cutter; other symbols as before. For this part, $T_f = 4/[(0.005)(8) \times (1000)] = 0.10$ min.

**3. *Compute the required table feed***

In a milling machine, the table feed $F_t$ in/min $= f_t nR$. For this machine, $F_t = (0.005) \times (8)(50) = 2.0$ in/min (0.85 mm/s).

**4. *Compute the feed per tooth***

For a milling machine, the feed per tooth, in/r, $f_t = F_t/Rn$. In this machine, $f_t = 4.0/[(50)(8)] = 0.01$ in/r (0.25 mm/r).

**5. *Compute the added table travel***

In face milling, the added table travel $A_t$ in $= 0.5[D_c - (D_c^2 - W^2)^{0.5}]$, where the symbols are the same as given earlier. For this cutter and work, $A_t = 0.5[4 - (4^2 - 4^2)^{0.5}] = 2.0$ in (5.1 cm).

## THREADING AND TAPPING TIME

How long will it take to cut a 4-in (10.2-cm) long thread at 100 r/min if the rod will have 12 threads per inch and a button die is used? The die is backed off at 200 r/min. What would the threading time be if a self-opening die were used instead of a button die? What will the threading time be for a single-pointed threading tool if the part being threaded is aluminum and the back-off speed is twice the threading speed? The rod is 1 in (2.5 cm) in diameter. How long will it take to tap a 2-in (5.1-cm) deep hole with a 1-14 solid tap turning at 100 r/min? How long will it take to mill-thread a 1-in (2.5-cm) diameter bolt having 15 threads per inch 3 in (7.6 cm) long if a 4-in (10.2-cm) diameter 20-flute thread-milling hob turning at 80 r/min with a 0.003 in (0.08-mm) feed is used?

### Calculation Procedure:

**1. *Compute the button-die threading time***

For a multiple-pointed tool, the time to thread $T_t = Ln_t/R$, where $L$ = length of cut = length of thread measured parallel to thread longitudinal axis, in; $n_t$ = number of threads per inch. For this button die, $T_t = (4)(12)/100 = 0.48$ min. This is the time required to cut the thread.

**TABLE 2** Number of Cuts and Cutting Speed for Dies and Taps

| | No. of cuts* | Cutting speed† ft/min | Cutting speed† m/s |
|---|---|---|---|
| Aluminum | 4 | 30 | 0.15 |
| Brass (commercial) | 3 | 30 | 0.15 |
| Brass (naval) | 4 | 30 | 0.15 |
| Bronze (ordinary) | 5 | 30 | 0.15 |
| Bronze (hard) | 7 | 20 | 0.10 |
| Copper | 5 | 20 | 0.10 |
| Drill rod | 8 | 10 | 0.05 |
| Magnesium | 4 | 30 | 0.15 |
| Monel (bar) | 8 | 10 | 0.05 |
| Steel (mild) | 5 | 20 | 0.10 |
| Steel (medium) | 7 | 10 | 0.05 |
| Steel (hard) | 8 | 10 | 0.05 |
| Steel (stainless) | 8 | 10 | 0.05 |

*Single-pointed threading tool; maximum spindle speed, 250 r/min.
†Maximum recommended speed for single- and multiple-pointed tools; maximum spindle speed for multiple-pointed tools = 150 r/min for dies and taps.

Compute the back-off time $B$ min from $B = Ln_t/R_B$, where $R_B$ = back-off rpm, or $B = (4)(12)/200 = 0.24$ min. Hence, the total time to cut and back off $= T_t + B = 0.48 + 0.24 = 0.72$ min.

**2. *Compute the self-opening die threading time***

With a self-opening die, the die opens automatically when it reaches the end of the cut thread and is withdrawn instantly. Therefore, the back-off time is negligible. Hence the time to thread $= T_t = Ln_t/R = (4)(12)/100 = 0.48$ min. One cut is usually sufficient to make a suitable thread.

**3. *Compute the single-pointed tool cutting time***

With a single-pointed tool, more than one cut is usually necessary. Table 2 lists the number of cuts needed with a single-pointed tool working on various materials. The maximum cutting speed for threading and tapping is also listed.

Table 2 shows that four cuts are needed for an aluminum rod when a single-pointed tool is used. Before computing the cutting time, compute the cutting speed to determine whether it is within the recommended range given in Table 2. From a previous calculation procedure, $C = R\pi d/12$, or $C = (100)(\pi)(1.0)/12 = 26.2$ ft/min (13.3 cm/s). Since this is less than the maximum recommended speed of 30 r/min, Table 2, the work speed is acceptable.

Compute the time to thread from $T_t = Ln_tc/R$, where $c$ = number of cuts to thread, from Table 2. For this part, $T_t = (4)(12)(4)/100 = 1.92$ min.

If the tool is backed off at twice the threading speed, and the back-off time $B = Ln_tc/R_B$, $B = (4)(12)(4)/200 = 0.96$ min. Hence, the total time to thread and back off $= T_t + B = 1.92 + 0.96 = 2.88$ min. In some shapes, a single-pointed tool may not be backed off; the tool may instead be repositioned. The time required for this approximates the back-off time.

**4. *Compute the tapping time***

The time to tap $T_t$ min $= Ln_t/R$. With a solid tap, the tool is backed out at twice the tapping speed. With a collapsing tap, the tap is withdrawn almost instantly without reversing the machine or tap.

For this hole, $T_t = (2)(14)/100 = 0.28$ min. The back-off time $B = Ln_t/R_B = (2)(14)/200 = 0.14$ min. Hence, the total time to tap and back off $= T_t + B = 0.28 + 0.14 = 0.42$ min.

The maximum spindle speed for tapping should not exceed 250 r/min. Use the cutting-speed values given in Table 2 in computing the desirable speed for various materials.

### 5. *Compute the thread-milling time*

The time for thread milling is $T_t = L/(fnR)$, where $L$ = length of cut, in = circumference of work, in; $f$ = feed per flute, in; $n$ = number of flutes on hob; $R$ = hob rpm. For this bolt, $T_t = 3.1416/[(0.003)(20)(80)] = 0.655$ min.

Note that neither the length of the threaded portion nor the number of threads per inch enters into the calculation. The thread hob covers the entire length of the threaded portion and completes the threading in one revolution of the work head.

## TURRET-LATHE POWER INPUT

How much power is required to drive a turret lathe making a ½-in (1.3-cm) deep cut in cast iron if the feed is 0.015 in/r (0.38 mm/r), the part is 2.0 in (5.1 cm) in diameter, and its speed is 382 r/min? How many 1.5-in (3.8-cm) long parts can be cut from a 10-ft (3.0-m) long bar if a ¼-in (6.4-mm) cutoff tool is used? Allow for end squaring.

### Calculation Procedure:

### 1. *Compute the surface speed of the part*

The cutting, or surface, speed, as given in a previous calculation procedure, is $C = R\pi d/12$, or $C = (382)(\pi)(2.0)/12 = 200$ ft/min (1.0 m/s).

### 2. *Compute the power input required*

For a turret lathe, the hp input $hp = 1.33DfCK$, where $D$ = cut depth, in; $f$ = feed, in/r; $K$ = material constant from Table 3. For cast iron, $K = 3.0$. Then $hp = (1.33)(0.5)(0.015)(200)(3.0) = 5.98$, say 6.0 hp (4.5 kW).

**TABLE 3** Turret-Lathe Power Constant

| Material | Constant $K$ |
|---|---|
| Bronze | 3 |
| Cast iron | 3 |
| SAE steels: | |
| 1020 | 6 |
| 1045 | 8 |
| 3250 | 9 |
| 4150 | 9 |
| 4615 | 6 |
| X1315 | 6 |
| Straight tubing | 6 |
| Steel castings and forgings | 9 |
| Heat-treated steels: | |
| 4150 | 10 |
| 52100 | 10 |

### 3. *Compute the number of parts that can be cut*

Allow 2 in (5.1 cm) on the bar end for checking and ½ in (1.3 cm) on the opposite end for squaring. With an original length of 10 ft = 120 in (304.8 cm), this leaves 120 − 2.5 = 117.5 in (298.5 cm) for cutting.

Each part cut will be 1.5 in (3.8 cm) long + 0.25 in (6.4 mm) for the cutoff, or 1.75 in (4.4 cm) of stock. Hence, the number of pieces which can be cut = 117.5/1.75 = 67.1, or 67 pieces.

**Related Calculations:** Use this procedure to find the turret-lathe power input for any of the materials, and similar materials, listed in Table 3. The parts cutoff computation can be used for any material—metallic or nonmetallic. Be sure to allow for the width of the cutoff tool.

## TIME TO CUT A THREAD ON AN ENGINE LATHE

How long will it take an engine lathe to cut an acme thread having a length of 5 in (12.7 cm), a major diameter of 2 in (5.1 cm), four threads per inch (1.575 threads per centimeter), a depth of 0.1350 in (3.4 mm), a cutting speed of 70 ft/min (0.4 m/s), and a depth of cut of 0.005 in (0.1 mm) per pass if the material cut is medium steel? How many passes of the tool are required?

### Calculation Procedure:

### 1. *Compute the cutting time*

For an acme, square, or worm thread cut on an engine lathe, the total cutting time $T_t$ min, excluding the tool positioning time, is found from $T_t = Ld_tDn_t/(4Cd_c)$, where $L$ = thread length, in, measured parallel to the thread longitudinal axis; $d_t$ = thread major diameter, in; $D$ = depth of thread, in; $n_t$ = number of threads per inch; $C$ = cutting speed, ft/min; $d_c$ = depth of cut per pass, in.

For this acme thread, $T_t = (5)(2)(0.1350)(4)/[4(70)(0.005)] = 3.85$ min. To this must be added the time required to position the tool for each pass. This equation is also valid for SI units.

**2. *Compute the number of tool passes required***

The depth of cut per pass is 0.005 in (0.1 mm). A total depth of 0.1350 in (3.4 mm) must be cut. Therefore, the number of passes required = total depth cut, in/depth of cut per pass, in = 0.1350/0.005 = 27 passes.

**Related Calculations:** Use this procedure for threads cut in ferrous and nonferrous metals. Table 4 shows typical cutting speeds.

## TIME TO TAP WITH A DRILLING MACHINE

How long will it take to tap a 4-in (10.2-cm) deep hole with a 1½-in (3.8-cm) diameter tap having six threads per inch (2.36 threads per centimeter) if the tap turns at 75 r/min?

**TABLE 4** Thread Cutting Speeds

| Material | Cutting speed | |
|---|---|---|
| | ft/min | m/s |
| Soft nonferrous metals | 250 | 1.25 |
| Mild steel | 100 | 0.50 |
| Medium steel | 75 | 0.38 |
| Hard steel | 50 | 0.25 |

### Calculation Procedure:

**1. *Compute the tap surface speed***

By the method of a previous calculation procedure, $C = R\pi d/12 = (75)(\pi)(1.5)/12 = 29.5$ ft/min (0.15 m/s).

**2. *Compute the time to tap the hole and withdraw the tool***

For tapping with a drilling machine, $T_t = Dn_tD_c\pi/(8\ C)$, where $D$ = depth of cut = depth of hole tapped, in; $n_t$ = number of threads per inch; $D_c$ = cutter diameter, in = tap diameter, in. For this hole, $T_t = (4)(6)(1.5)\pi/[8(29.5)] = 0.48$ min, which is the time required to tap and withdraw the tool.

**Related Calculations:** Use this procedure for tapping ferrous and nonferrous metals on a drill press. The recommended tap surface speed for various metals is: aluminum, soft brass, ordinary bronze, soft cast iron, and magnesium: 30 ft/min (0.15 m/s); naval brass, hard bronze, medium cast iron, copper and mild steel: 20 ft/min (0.10 m/s); hard cast iron, medium steels, and hard stainless steel: 10 ft/min (0.05 m/s).

## MILLING CUTTING SPEED, TIME, FEED, TEETH NUMBER, AND HORSEPOWER

What is the cutting speed of a 12-in (30.5-cm) diameter milling cutter turning at 190 r/min? How many teeth are needed in the cutter at this speed if the feed is 0.010 in (0.3 mm) per tooth, the depth of cut is 0.075 in (1.9 mm), the length of cut is 5 in (12.7 cm), the power available at the cutter is 12 hp (8.9 kW), and the mill is cutting hard malleable iron? How long will it take the mill to make this cut? What is the maximum feed rate that can be used? What is the power input to the cutter if a 20-hp (14.9-kW) machine is used?

### Calculation Procedure:

**1. *Compute the cutter cutting speed***

For a milling cutter, use the simplified relation $C = Rd/4$, where the symbols are as given earlier in this section. Or, $C = (190)(12)/4 = 570$ ft/min (2.9 m/s).

**2. *Compute the number of cutter teeth required***

For a carbide cutter, $n = K_m hp_c/(Df_tLR)$, where $n$ = number of teeth on cutter; $K_m$ = machinability constant or $K$ factor from Table 5; $hp_c$ = horsepower available at the milling cutter; $D$ = depth of cut, in; $f_t$ = cutter feed, inches per tooth; $L$ = length of cut, in; $R$ = cutter rpm.

Table 5 shows that $K_m = 0.90$ for malleable iron. Then $n = (0.90)(12)/[(0.075)(0.01) \times$

**TABLE 5** Machinability Constant $K_m$

| | |
|---|---|
| Aluminum | 2.28 |
| Brass, soft | 2.00 |
| Bronze, hard | 1.40 |
| Bronze, very hard | 0.65 |
| Cast iron, soft | 1.35 |
| Cast iron, hard | 0.85 |
| Cast iron, chilled | 0.65 |
| Cast magnesium | 2.50 |
| Malleable iron | 0.90 |
| Steel, soft | 0.85 |
| Steel, medium | 0.65 |
| Steel, hard | 0.48 |
| Steel: | |
| 100 Brinell | 0.80 |
| 150 Brinell | 0.70 |
| 200 Brinell | 0.65 |
| 250 Brinell | 0.60 |
| 300 Brinell | 0.55 |
| 400 Brinell | 0.50 |

**TABLE 6** Typical Milling-Machine Efficiencies

| Rated power of machine | | Overall efficiency, percent |
|---|---|---|
| hp | kW | |
| 3 | 2.2 | 40 |
| 5 | 3.7 | 48 |
| 7.5 | 5.6 | 52 |
| 10 | 7.5 | 52 |
| 15 | 11.2 | 52 |
| 20 | 14.9 | 60 |
| 25 | 18.6 | 65 |
| 30 | 22.4 | 70 |
| 40 | 29.8 | 75 |
| 50 | 27.3 | 80 |

(5)(190)] = 15.15, say 16 teeth. For general-purpose use, the Metal Cutting Institute recommends that $n$ = 1.5(cutter diameter, in) for cutters having a diameter of more than 3 in (7.6 cm). For this cutter, $n$ = 1.5(12) = 16 teeth. This agrees with the number of teeth computed with the cutter equation.

### 3. *Compute the milling time*

For a milling machine, the time to cut $T_t$ min = $L/(f_t nR)$, where $L$ = length of cut, in; $f_t$ = feed per tooth, inches per tooth per revolution; $n$ = number of teeth on the cutter; $R$ = cutter rpm. Thus, the time to cut is $T_t$ = 5/[(0.01)(16)(190)] = 0.164 min.

### 4. *Compute the maximum feed rate*

For a milling machine, the maximum feed rate $f_m$ in/min = $K_m hp_c/(DI)$, where $L$ = length of cut; other symbols are the same as in step 2. Thus, $f_m$ = (0.90)(12)/[(0.075)(5)] = 28.8 in/min (1.2 cm/s).

### 5. *Compute the power input to the machine*

The power available at the cutter is 12 hp (8.9 kW). The power required $hp_c$ = $DLnRf_t/K_m$, where all symbols are as given above. Thus, $hp_c$ = (0.075)(5)(16)(190)(0.01)/0.90 = 12.68 hp (9.5 kW). This is slightly more than the available horsepower.

Milling machines have overall efficiencies ranging from a low of 40 percent to a high of 80 percent, Table 6. Assume a machine efficiency of 65 percent. Then the required power input is 12.68/0.65 = 19.5hp (14.5 kW). Therefore, a 20- or 25-hp (14.9- or 18.6-kW) machine will be satisfactory, depending on its actual operating efficiency.

**Related Calculations:** After selecting a feed rate, check it against the suggested feed per tooth for milling various materials given in the *American Machinist's Handbook*. Use the method of a previous calculation procedure in this section to determine the cutter approach. With the approach known, the maximum chip thickness, in = (cutter approach, in)(table advance per tooth, in)/(cutter radius, in). Also, the feed per tooth, in = (feed rate, in/min)/[(cutter rpm)(number of teeth on cutter)].

## GANG-, MULTIPLE-, AND FORM-MILLING CUTTING TIME

How long will it take to gang mill a part if three cutters are used with a spindle speed of 70 r/min and there are 12 teeth on the smallest cutter, a feed of 0.015 in/r (0.4 mm/r) and a length of cut of 8 in (20.3 cm)? What will be the unit time to multiple mill four keyways if each of the four cutters has 20 teeth, the feed is 0.008 in (0.2 mm) per tooth, spindle speed is 150 r/min, and the keyway length is 3 in (7.6 cm)? Show how the cutting time for form milling is computed, and how the cutter diameter for straddle milling is computed.

## Calculation Procedure

### 1. *Compute the gang-milling cutting time*

For any gang-milling operation, from the dimensions of the smallest cutter, the time to cut $T_t = L/f_t nR$, where $L$ = length of cut, in; $f_t$ = feed per tooth, in/r; $n$ = number of teeth on cutter; $R$ = spindle rpm. For this part, $T_t = 8/[(0.015)(12)(70)] = 0.635$ min.

Note that in all gang-milling cutting-time calculations, the number of teeth and feed of the *smallest* cutter are used.

### 2. *Compute the multiple-milling cutting time*

In multiple milling, the cutting time $T_t = L/(f_t nR_m)$, where $n$ = number of milling cutter used. In multiple milling, the cutting time is termed the unit time. For this machine, $T_t = 3/[(0.008)(20)(150)(4)] = 0.0303$ min.

### 3. *Show how form milling time is computed*

Form-milling cutters are used on surfaces that are neither flat nor square. The cutters used for form milling resemble other milling cutters. The cutting time is therefore computed from $T_t = L/(f_t nR)$, where all symbols are the same in step 1.

### 4. *Show how the cutter diameter is computed for straddle milling*

In straddle milling, the cutter diameter must be large enough to permit the work to pass under the cutter arbor. The minimum-diameter cutter to straddle mill a part = (diameter of arbor, in) + 2 (face of cut, in + 0.25). The 0.25 in (6.4 mm) is the allowance for clearance of the arbor.

**Related Calculations:** Use the equation of step 1 to compute the cutting time for metal slitting, screw slotting, angle milling, T-slot milling, Woodruff key-seat milling, and profiling and routing of parts. In T-slot milling, two steps are required—milling of the vertical member and milling of the horizontal member. Compute the milling time of each; the sum of the two is the total milling time.

# SHAPER AND PLANER CUTTING SPEED, STROKES, CYCLE TIME, POWER

What is the cutting speed of a shaper making 54-strokes/min if the stroke length is 6 in (15.2 cm)? How many strokes per minute should the ram of a shaper make if it is shaping a 12-in (30.5-cm) long aluminum bar at a cutting speed of 200 ft/min? How long will it take to make a cut across a 12-in (30.5-cm) face of a cast-iron plate if the feed is 0.050 in (1.3 mm) per stroke and the ram makes 50 strokes/min? What is the cycle time of a planer if its return speed is 200 ft/min (1.0 m/s), the acceleration-deceleration constant is 0.05, and the cutting speed is 100 ft/min (0.5 m/s)? What is the planer power input if the depth of cut is ⅛ in (3.2 mm) and the feed is 1/16 in (1.6 mm) per stroke?

## Calculation Procedure:

### 1. *Compute the shaper cutting speed*

For a shaper, the cutting speed, ft/min, is $C = SL/6$, where $S$ = strokes/min; $L$ = length of stroke, in; where the cutting-stroke time = return-stroke time. Thus, for this shaper, $C = (54)(6)/6 = 54$ ft/min (0.3 m/s).

### 2. *Compute the shaper stroke rate*

Transpose the equation of step 1 to $S = 6C/L$. Then $S = 6(200)/12 = 100$ strokes/min.

### 3. *Compute the shaper cutting time*

For a shaper the cutting time, min, is $T_t = L/(fS)$, where $L$ = length of cut, in; $f$ = feed, in/stroke; $S$ = strokes/min. Thus, for this shaper, $T_t = 12/[(0.05)(50)] = 4.8$ min. Multiply $T_t$ by the number of strokes needed; the result is the total cutting time, min.

**TABLE 7** Power Factors for Planers*

| Depth of cut | | Feed | | |
|---|---|---|---|---|
| in | cm | 1/32 in (0.8 mm) per stroke | 1/16 in (1.6 mm) per stroke | 1/8 in (3.2 mm) per stroke |
| 1/8 | 0.3 | 0.0115 | 0.0235 | 0.047 |
| 1/4 | 0.6 | 0.023 | 0.047 | 0.094 |
| 3/8 | 1.0 | 0.035 | 0.070 | 0.141 |
| 1/2 | 1.3 | 0.047 | 0.094 | 0.189 |
| 5/8 | 1.6 | 0.063 | 0.118 | 0.236 |
| 3/4 | 1.9 | 0.080 | 0.142 | 0.284 |
| 7/8 | 2.2 | 0.087 | 0.165 | 0.331 |
| 1 | 2.5 | 0.094 | 0.189 | 0.378 |

*Excerpted from the Cincinnati Planer Company and *American Machinist's Handbook*.

**4. *Compute the planer cycle time***

The cycle time for a planer, min, $= (L/C) + (L/R_c) + k$, where $R_c$ = cutter return speed, ft/min; $k$ = acceleration-deceleration constant. Since the cutting speed is 100 ft/min (0.5 m/s) and the return speed is 200 ft/min (1.0 m/s), the cycle time = (12/100) + (12/200) + 0.05 = 0.23 min.

**5. *Compute the power input to the planer***

Table 7 lists typical power factors for planers planing cast iron and steel. To find the power required, multiply the power factor by the cutter speed, ft/min. For the planer in step 3 with a cutting speed of 100 ft/min (0.5 m/s) and a power factor of 0.0235 for a 1/8-in (3.2-mm) deep cut and a 1/16-in (1.6-mm) feed, $hp_{input}$ = (0.0235)(100) = 2.35 hp (1.8 kW).

For steel up to 40 points carbon, multiply the above result by 2; for steel above 40 points carbon, multiply by 2.25.

**Related Calculations:** Where a shaper has a cutting stroke time that does not equal the return-stroke time, compute its cutting speed from $C = SL/(12)$(cutting-stroke time, min/sum of cutting- and return-stroke time, min). Thus, if the shaper in step 1 has a cutting-stroke time of 0.8 min and a return-stroke time of 0.4 min, $C = (54)(6)/[(12) \times (0.8/1.2)]$ = 40.5 ft/min (0.2 m/s).

## GRINDING FEED AND WORK TIME

What is the feed of a centerless grinding operation if the regulating wheel is 8 in (20.3 cm) in diameter and turns at 100 r/min at an angle of inclination of 5°? How long will it take to rough grind on an external cylindrical grinder a brass shaft that is 3.0 in (7.6 cm) in diameter and 12 in (30.5 cm) long, if the feed is 0.003 in (0.076 mm), the spindle speed is 20 r/min, the grinding-wheel width is 3 in (7.6 cm) and the diameter is 8 in (20.3 cm), and the total stock on the part is 0.015 in (0.38 mm)? How long would it take to make a finishing cut on this grinder with a feed of 0.001 in (0.025 mm), stock of 0.010 in (0.25 mm), and a cutting speed of 100 ft/min (0.5 m/s)?

### Calculation Procedure:

**1. *Compute the feed rate for centerless grinding***

In centerless grinding, the feed, in/min, $f = \pi dR \sin \infty$, where $\pi$ = 3.1416; $d$ = diameter of the regulating wheel, in; $R$ = regulating wheel rpm; $\infty$ = angle of inclination of the regulating wheel. For this grinder, $f = \pi(8)(100)(\sin 5°)$ = 219 in/min (9.3 cm/s). Centerless grinders will grind as many as 50,000 1-in (2.5-cm) parts per hour.

### 2. *Compute the rough-grinding time*

The rough-grinding time $T_t$ min = $Lt_sd/(2WfC)$, where $L$ = length of ground part, in; $t_s$ = total stock on part, in; $W$ = width of grinding-wheel face, in; $C$ = cutting speed, ft/min.

Compute the cutting speed first because it is not known. By the method of previous calculation procedures, $C = \pi dR/12 = \pi(8)(20)/12 = 42$ ft/min (0.2 m/s). Then $T_t = (12)(0.015)(3)/[2(3)(0.003)(42)] = 0.714$ min.

### 3. *Compute the finish-grinding time*

For finish grinding, use the same equation as in step 2, except that the factor 2 is omitted from the denominator. Thus, $T_t = Lt_sd/(WfC)$, or $T_t = (12)(0.010)(3)/[(3)(0.001)(100)] = 1.2$ min.

**Related Calculations:** Use the same equations as in steps 1 and 4 for internal cylindrical grinding. In surface grinding, about 250 $\text{in}^2$/min (26.9 $\text{cm}^2$/s) can be ground 0.001 in (0.03 mm) deep if the material is hard. For soft materials, about 1000 $\text{in}^2$ (107.5 $\text{cm}^2$) and 0.001 in (0.03 mm) deep can be ground per minute.

In honing cast iron, the average stock removal is 0.006 to 0.008 in/(ft·min) [0.008 to 0.011 mm/(m·s)]. With hard steel or chrome plate, the rate of honing averages 0.003 to 0.004 in/ft·min [0.004 to 0.006 mm/(m·s)].

## BROACHING TIME AND PRODUCTION RATE

How long will it take to broach a medium-steel part if the cutting speed is 20 ft/min (0.1 m/s), the return speed is 100 ft/min (0.5 m/s), and the stroke length is 36 in (91.4 cm)? What will the production rate be if starting and stopping occupy 2 s and loading 5 s with an efficiency of 85 percent?

### Calculation Procedure:

### 1. *Compute the broaching time*

The broaching time $T_t$ min = $(L/C) + (L/R_c)$, where $L$ = length of stroke, ft; C = cutting speed, ft/min; $R_c$ = return speed, ft/min; for this work, $T_t = (3/20) + (3/100) = 0.18$ min.

### 2. *Compute the production rate*

In a complete cycle of the broaching machine there are three steps: broaching; starting and stopping; and loading. The cycle time, at 100 percent efficiency, is the sum of these three steps, or $0.18 \times 60 + 2 + 5 = 17.8$ s, where the factor 60 converts 0.18 min to seconds. At 85 percent efficiency, the cycle time is greater, or $17.8/0.85 = 20.9$ s. Since there are 3600 s in 1 h, production rate = $3600/20.9 = 172$ pieces per hour.

## HOBBING, SPLINING, AND SERRATING TIME

How long will it take to hob a 36-tooth 12-pitch brass spur gear having a tooth length of 1.5 in (3.8 cm) by using a 2.75-in (7.0-cm) hob? The whole depth of the gear tooth is 0.1789 in (4.5 mm). How many teeth should the hob have? Hob feed is 0.084 in/r (2.1 mm/r). What would be the cutting time for a 47° helical gear? How long will it take to spline-hob a brass shaft which is 2.0 in (5.1 cm) in diameter, has 12 splines, each 10 in (25.4 cm) long, if the hob diameter is 3.0 in (7.6 cm), cutter feed is 0.050 in (1.3 mm), cutter speed is 120 r/min, and spline depth is 0.15 in (3.8 mm)? How long will it take to hob 48 serrations on a 2-in (5.1-cm) diameter brass shaft if each serration is 2 in (5.1 cm) long, the 18-flute hob is 2.5 in (6.4 cm) in diameter, the approach is 0.3 in (7.6 mm), the feed per flute is 0.008 in (0.2 mm), and the hob speed is 250 r/min?

### Calculation Procedure:

### 1. *Compute the hob approach*

The hob approach $A_c = \sqrt{d_g(D_c - d_g)}$, where $d_g$ = whole depth of gear tooth, in; $D_c$ = hob diameter, in. For this hob, $A_c = \sqrt{0.1798(2.7500 - 0.798)} = 0.68$ in (1.7 cm).

**TABLE 8** Gear-Hobbing Cutting Speeds

| Gear material | Spur gears: Cutting speed, ft/min | Spur gears: Cutting speed, m/s | Helical gears*: Angle, ° | Helical gears*: Percentage of feed to use |
|---|---|---|---|---|
| Brass | 150 | 0.8 | 0–36 | 100 |
| Fiber | 150 | 0.8 | 36–48 | 80 |
| Cast iron (soft) | 100 | 0.5 | 48–60 | 67 |
| Steel (mild) | 100 | 0.5 | 60–70 | 50 |
| Steel (medium) | 75 | 0.4 | 70–90 | 33 |
| Steel (hard) | 50 | 0.3 | | |

*Reduce feed by percentage shown when helical gears are cut.

### 2. *Determine the cutting speed of the hob*

Table 8 shows that a cutting speed of $C$ = 150 ft/min (0.8 m/s) is generally used for brass gears. With a 2.75-in (7.0-cm) diameter hob, this corresponds to a hob rpm of $R = 12C/(\pi D_c) = (12)(150)/[\pi(2.75)] = 208$ r/min.

### 3. *Compute the hobbing time*

The time to hob a spur gear $T_t$ min $= N(L + A_c)/fR$, where $N$ = number of teeth in gear to be cut; $L$ = length of a tooth in the gear, in; $A_c$ = hob approach, in; $f$ = hob feed, in/r; $R$ = hob rpm. For this spur gear, $T_t = (36)(1.5 + 0.68)/[(0.084)(208)] = 4.49$ min.

### 4. *Compute the cutting time for a helical gear*

Table 8 shows that the feed for a 47° helical gear should be 80 percent of that for a spur gear. By the relation in step 3, $T_t = (36)(1.5 + 0.68)/[(0.80)(0.084)(208)] = 5.61$ min.

### 5. *Compute the time to spline hob*

Use the same procedure as for hobbing. Thus, $A_c = \sqrt{d_g(D_c - d_g)} = \sqrt{0.15(3.0 - 0.15)} = 0.654$ in (1.7 cm). Then $T_t = N(L + A)/fR$, where $N$ = number of splines; $L$ = length of spline, in; other symbols as before. For this shaft, $T_t = (12)(10 + 0.654)/[(0.05)(120)] = 21.3$ min.

### 6. *Compute the time to serrate*

The time to hob serrations $T_t$ min $= N(L + A)/(fnR)$, where $N$ = number of serrations; $L$ = length of serration, in; $n$ = number of flutes on hob; other symbols as before. For this shaft, $T_t = (48)(2 + 0.30)/[(0.008)(18)(250)] = 3.07$ min.

## TIME TO SAW METAL WITH POWER AND BAND SAWS

How long will it take to saw a rectangular piece of alloy-plate aluminum 6 in (15.2 cm) wide and 2 in (5.1 cm) thick if the length of cut is 6 in (15.2 cm), the power hacksaw makes 120 strokes/min, and the average feed per stroke is 0.0040 in (0.1 mm)? What would the sawing time be if a band saw with a 200-ft/min (1.0-m/s) cutting speed, 16 teeth per inch (6.3 teeth per centimeter), and a 0.0003-in (0.008-mm) feed per tooth is used?

## Calculation Procedure:

### 1. *Compute the sawing time for a power saw*

For a power saw with positive feed, the time to saw $T_t$ min $= L/(Sf)$, where $L$ = length of cut, in; $S$ = strokes/min of saw blade; $f$ = feed per stroke, in. In this saw, $T_t = (6)/[(120)(0.0040)] = 12.5$ min.

**TABLE 9** Oxyacetylene Cutting Speed and Gas Consumption

| Metal thickness | | Speed | | | | Gas consumption | | | |
|---|---|---|---|---|---|---|---|---|---|
| | | Manual | | Machine | | Oxygen | | Acetylene | |
| in | cm | in/min | mm/s | in/min | mm/s | $ft^3/h$ | $cm^3/s$ | $ft^3/h$ | $cm^3/s$ |
| 0.25 | 0.6 | 16–18 | 6.8–7.6 | 20–26 | 8.5–11.0 | 50–90 | 393.3–707.9 | 8–11 | 62.9–86.5 |
| 0.50 | 1.3 | 12–15 | 5.1–6.4 | 17–22 | 7.2–9.3 | 90–125 | 707.9–983.2 | 10–13 | 78.7–102.3 |
| 1 | 2.5 | 8–12 | 3.4–5.1 | 14–18 | 5.9–7.6 | 130–200 | 1023–1573 | 13–16 | 102.3–125.9 |
| 2 | 5.1 | 5–7 | 2.1–3.0 | 10–13 | 4.2–5.5 | 200–300 | 1573–2360 | 16–20 | 125.9–157.3 |
| 4 | 10.2 | 4–5 | 1.7–2.1 | 7–9 | 3.0–3.8 | 300–400 | 2360–3146 | 21–26 | 165.2–204.5 |
| 6 | 15.2 | 3–4 | 1.3–1.7 | 5–7 | 2.1–3.0 | 400–500 | 3146–3933 | 26–32 | 204.5–251.7 |
| 8 | 20.3 | 3–6 | 1.3–2.5 | 4–6 | 1.7–2.5 | 500–650 | 3933–5113 | 28–35 | 220.2–275.3 |
| 10 | 25.4 | 2–3 | 0.8–1.3 | 3–4 | 1.3–1.7 | 700–1000 | 5506–7860 | 30–38 | 236.0–298.9 |
| 12 | 30.5 | 2.5–3.5 | 1.1–1.5 | 3–4 | 1.3–1.7 | 720–880 | 5663–6922 | 42–52 | 330.4–409.0 |

**2. *Compute the band-saw cutting time***

For a band saw, the sawing time $T_t$ min = $L/(12Cnf)$, where $L$ = length of cut, in; $C$ = cutting speed, ft/min; $n$ = number of saw teeth per inch; $f$ = feed, inches per tooth. With this band saw, $T_t = (6)/[(12)(200)(16)(0.0003)] = 0.521$ min.

**Related Calculations:** When nested round, square, or rectangular bars are to be cut, use the greatest *width* of the nested bars as the length of cut in either of the above equations.

## OXYACETYLENE CUTTING TIME AND GAS CONSUMPTION

How long will it take to make a 96-in (243.8-cm) long cut in a 1-in (2.5-cm) thick steel plate by hand and by machine? What will the oxygen and acetylene consumption be for each cutting method?

### Calculation Procedure:

**1. *Compute the cutting time***

For any flame cutting, the cutting time $T_t$ min = $L/C$, where $L$ = length of cut, in; $C$ = cutting speed, in/min, from Table 9. With manual cutting, $T_t = 96/8 = 12$ min, using the lower manual cutting speed given in Table 9. At the higher manual cutting speed, $T_t = 96/12 = 8$ min. With machine cutting, $T_t = 96/14 = 6.86$ min, by using the lower machine cutting speed in Table 9. At the higher machine cutting speed, $T_t = 96/18 = 5.34$ min.

**2. *Compute the gas consumption***

From Table 9 the oxygen consumption is 130 to 200 $ft^3/h$ (1023 to 1573 $cm^3/s$). Thus, actual consumption, $ft^3$ = (cutting time, min/60) (consumption, $ft^3/h$) = (12/60)(130) = 26 $ft^3$ (0.7 $m^3$) at the minimum cutting speed and minimum oxygen consumption. For this same speed with maximum oxygen consumption, actual $ft^3$ used = (12/60)(200) = 40 $ft^3$ (1.1 $m^3$).

Compute the acetylene consumption in the same manner, or (12/60)(13) = 2.6 $ft^3$ (0.07 $m^3$), and (12/60)(16) = 3.2 $ft^3$ (0.09 $m^3$). Use the same procedure to compute the acetylene and oxygen consumption at the higher cutting speeds.

**Related Calculations:** Use the procedure given here for computing the cutting time and gas consumption when steel, wrought iron, or cast iron is cut. Thicknesses ranging up to 5 ft (1.5 m) are economically cut by an oxyacetylene torch. Alloying elements in steel may require preheating of the metal to permit cutting. To compute the gas required per lineal foot, divide the actual consumption for the length cut, in inches, by 12.

## COMPARISON OF OXYACETYLENE AND ELECTRIC-ARC WELDING

Determine the time required to weld a 4-ft (1.2-m) long seam in a ⅜-in (9.5-mm) plate by the oxyacetylene and electric-arc methods. How much oxygen and acetylene are required? What weight of electrode will be used? What is the electric-power consumption? Assume that one weld bead is run in the joint.

### Calculation Procedure:

**1. *Compute the welding time***

For any welding operation, the time required to weld $T_t$ min = $L/C$, where $L$ = length of weld, in; $C$ = welding speed, in/min. When oxyacetylene welding is used, $T_t = 48/1.0 = 48$ min, when a welding speed of 1.0 in/min (0.4 mm/s) is used. With electric-arc welding, $T_t = 48/18 = 2.66$ min when the welding speed = 18 in/min (7.6 mm/s) per bead. For plate thicknesses under 1 in (2.5 cm), typical welding speeds are in the range of 1 to 2 in/min (0.4 to 0.8 mm/s) for oxyacetylene and 18 in/min (7.6 mm/s) for electric-arc welding. For thicker plates, consult *The Welding Handbook*, American Welding Society.

**2. *Compute the gas consumption***

Gas consumption for oxyacetylene welding is given in cubic feet per foot of weld. Using values from *The Welding Handbook*, or a similar reference, we see that oxygen consumption = ($ft^3$ $O_2$ per ft of weld) (length of weld, ft); acetylene consumption = ($ft^3$ acetylene per ft of weld) (length of weld, ft). For this weld, with only one bead, oxygen consumption = (10.0)(4) = 40 $ft^3$ (1.1 $m^3$); acetylene consumption = (9.0)(4) = 36 $ft^3$ (1.0 $m^3$).

**3. *Compute the weight of electrode required***

*The Welding Handbook* tabulates the weight of electrode for various types of welds—square grooves, 90° grooves, etc., per foot of weld. Then the electrode weight required, lb = (rod consumption, lb/ft) (weld length, ft).

For oxyacetylene welding, the electrode weight required, from data in *The Welding Handbook*, is (0.597)(4) = 2.388 lb (1.1 kg). For electric-arc welding, weight = (0.18)(4) = 0.72 lb (0.3 kg).

**4. *Compute the electric-power consumption***

In electric-arc welding the power consumption is $kW = (V)(A)/(1000)$ (efficiency). *The Welding Handbook* shows that for a ⅜-in (9.5-mm) thick plate, $V$ = 40, $A$ = 450 A, efficiency = 60 percent. Then power consumption = (40)(450)/[(1000)(0.60)] = 30 kW. For this press, $F$ = (8)(0.5)(16.0) = 64 tons (58.1 t).

**Related Calculations:** Where more than one pass or bead is required, multiply the time for one bead by the number of beads deposited. If only 50 percent penetration is required for the bead, the welding speed will be twice that where full penetration is required.

## PRESSWORK FORCE FOR SHEARING AND BENDING

What is the press force to shear an 8-in (20.3-cm) long 0.5-in (1.3-cm) thick piece of annealed bronze having a shear strength of 16.0 tons/in$^2$ (2.24 t/cm$^2$)? What is the stripping load? Determine the force required to produce a U bend in this piece of bronze if the unsupported length is 4 in (10.2 cm), the bend length is 6 in (15.2 cm), and the ultimate tensile strength is 32.0 tons/in$^2$ (4.50 t/cm$^2$).

### Calculation Procedure:

**1. *Compute the required shearing force***

For any metal in which a straight cut is made, the required shearing force, tons = $F = Lts$, where $L$ = length of cut, in; $t$ = metal thickness, in; $s$ = shear strength of metal being cut, tons/in$^2$. Where round, elliptical, or other shaped holes are being cut, substitute the sum of the circumferences of all the holes for $L$ in this equation.

**2. *Compute the stripping load***

For the typical press, the stripping load is 3.5 percent of the required shearing force, or (0.035)(64) = 2.24 tons (2.0 t).

**3. *Compute the required bending force***

When U bends or channels are pressed in a metal, $F = 2Lt^2s_t/W$, where $s_t$ = ultimate tensile strength of the metal, tons/in$^2$; $W$ = width of unsupported metal, in = distance between the vertical members of a channel or U bend, measured to the *outside* surfaces, in. For this U bend, $F = 2(6)(0.5)^2(32)/4$ = 24 tons (21.8 t).

**Related Calculations:** Right-angle edge bends require a bending force of $F = Lt^2s_t/(2W)$, while free V bends with a centrally located load require a bending force of $F = Lt^2s_t/W$. All symbols are as given in steps 1 and 2.

## MECHANICAL-PRESS MIDSTROKE CAPACITY

Determine the maximum permissible midstroke capacity of single- and twin-driven 2-in (5.1-cm) diameter crankshaft presses if the stroke of the slide is 12 in (30.5 cm) for each.

### Calculation Procedure:

**1. *Compute the single-driven press capacity***

For a single-driven crankshaft press with a heat-treated 0.35 to 0.45 percent carbon-steel crankshaft having a shear strength of 6 tons/in$^2$ (0.84 t/cm$^2$), the maximum permissible midstroke capacity $F$ tons $= 2.4d^3/S$, where $d$ = shaft diameter at main bearing, in; $S$ = stroke length, in; or $F = (2.4)(2)^3/12 = 1.6$ tons (1.5 t).

### 2. *Compute the twin-driven press capacity*

Twin-driven presses with main (bull) gears on each end of the crankshaft have a maximum permissible midstroke capacity of $F = 3.6d^3/S$, when the shaft shearing strength is 9 tons/in². For this press, $F = 3.6(2)^3/12 = 2.4$ tons (2.2 t).

**Related Calculations:** Use the equation in step 2 to compute the maximum permissible midstroke capacity of all wide (right-to-left) double-crank presses. Since gear eccentric presses are built in competition with crankshaft presses, their midstroke pressure capacity is within the same limits as in crankshaft presses. The diameters of the fixed pins on which the gear eccentrics revolve are usually made the same as the crankshaft in crankshaft presses of the same rated capacity.

## STRIPPING SPRINGS FOR PRESSWORKING METALS

Determine the force required to strip the work from a punch if the length of cut is 5.85 in (14.9 cm) and the stock is 0.25 in (0.6 cm) thick. How many springs are needed for the punch if the force per inch deflection of the spring is 100 lb (175.1 N/cm)?

### Calculation Procedure:

### 1. *Compute the required stripping force*

The required stripping force $F_p$ lb needed to strip the work from a punch is $F_p = Lt/0.00117$, where $L$ = length of cut, in; $t$ = thickness of stock cut, in. For this punch, $F_p = (5.85)(0.25)/0.00117 = 1250$ lb (5560.3 N).

### 2. *Compute the number of springs required*

Only the first ⅛-in (0.3-cm) deflection of the spring can be used in the computation of the stripping force produced by the spring. Thus, for this punch, number of springs required = stripping force, lb/force, lb, to produce ⅛-in (0.3-cm) deflection of the spring, or $1250/100 = 12.5$ springs. Since a fractional number of springs cannot be used, 13 springs would be selected.

**Related Calculations:** In high-speed presses, the springs should not be deflected more than 25 percent of their free length. For heavy, slow-speed presses, the total deflection should not exceed 37.5 percent of the free length of the spring. The stripping force for aluminum alloys is generally taken as one-eighth the maximum blanking pressure.

## BLANKING, DRAWING, AND NECKING METALS

What is the maximum blanking force for an aluminum part if the length of the cut is 30 in (76.2 cm), the metal is 0.125 in (0.3 cm) thick, and the yield strength is 2.5 tons/in² (0.35 t/cm²)? How much force is required to draw a 12-in (30.5-cm) diameter 0.25-in (0.6-cm) thick stainless steel shell if the yield strength is 15 tons/in² (2.1 t/cm²)? What force is required to neck a 0.125-in (0.3-cm) thick aluminum shell from a 3- to a 2-in (7.6- to 5.1-cm) diameter if the necking angle is 30° and the ultimate compressive strength of the material is 14 tons/in² (1.97 t/cm²)?

### Calculation Procedure:

### 1. *Compute the maximum blanking force*

The maximum blanking force for any metal is given by $F = Lts$, where $F$ = blanking force, tons; $L$ = length of cut, in (= circumference of part, in); $t$ = metal thickness, in; $s$ = yield strength of metal, tons/in². For this part, $F = (30)(0.125)(2.5) = 0.375$ tons (0.34 t).

### 2. *Compute the maximum drawing force*

Use the same equation as in step 1, substituting the drawing-edge length or perimeter (circumference of part) for $L$. Thus, $F = (12\pi)(0.25)(15) = 141.5$ tons (128.4 t).

### 3. *Compute the required necking force*

The force required to neck a shell is $F = ts_c(d_1 - d_s)/\cos$ (necking angle), where $F$ = necking force, tons; $t$ = shell thickness, in; $s_c$ = ultimate compressive strength of the material, tons/in²;

**TABLE 10** Metal Yield Strength

| Metal | Yield strength tons/in$^2$ | Yield strength t/cm$^2$ |
|---|---|---|
| Aluminum, 2S annealed | 2.5 | 0.35 |
| Aluminum, 24S heat-treated | 23.0 | 2.23 |
| Low brass, ¼ hard | 24.5 | 3.46 |
| Yellow brass, annealed | 10.0 | 1.41 |
| Cold-rolled steel, ¼ hard | 16.0 | 2.25 |
| Stainless steel, 18-8 | 15.0 | 2.11 |

*Note:* As a general rule, the necking angle should not exceed 35°.

$d_l$ = large diameter of shell, i.e., the diameter *before* necking, in; $d_s$ = small diameter of shell, i.e., the diameter *after* necking, in. For this shell, $F$ = (0.125)(14)(3.0 − 2.0)/cos 30° = 2.02 tons (1.8 t).

**Related Calculations:** Table 10 presents typical yield strengths of various metals which are blanked or drawn in metalworking operations. Use the given strength as shown above.

## METAL PLATING TIME AND WEIGHT

How long will it take to electroplate a 0.004-in (0.1-mm) thick zinc coating on a metal plate if a current density of 25 A/ft$^2$ (269.1 A/m$^2$) is used at an 80 percent plating efficiency? How much zinc is required to produce a 0.001-in (0.03-mm) thick coating on an area of 60 ft$^2$ (5.6 m$^2$)?

### Calculation Procedure:

#### 1. *Compute the metal plating time*

The plating time $T_p$ min = $60\ An/(A_a e)$, where $A$ = A/ft$^2$ required to deposit 0.001 in (0.03 mm) of metal at 100 percent cathode efficiency; $n$ = number of thousandths of inch actually deposited; $A_a$ = current actually supplied, A/ft$^2$; $e$ = plating efficiency, expressed as a decimal. Table 11 gives typical values of $A$ for various metals used in electroplating. For plating zinc, from the value in Table 11, $T_p$ = 60(14.3)(4)/[(25)(0.80)] = 171.5 min, or 171.5/60 = 2.86 h.

**TABLE 11** Electroplating Current and Metal Weight

| Metal | Time to deposit, Ah | | Metal density | |
|---|---|---|---|---|
| | 0.001 in/ft$^2$ at 100% efficiency | 0.01 mm/m$^2$ at 100% efficiency | lb/in$^3$ | g/cm$^3$ |
| Antimony, Sb | 10.40 | 0.038 | 0.241 | 6.671 |
| Cadmium, Cd | 9.73 | 0.036 | 0.312 | 8.636 |
| Chromium, Cr | 51.80 | 0.189 | 0.256 | 7.086 |
| Cobalt(ous), Co | 19.00 | 0.069 | 0.322 | 8.913 |
| Copper(ous), Cu | 8.89 | 0.033 | 0.322 | 8.913 |
| Copper(ic), Cu | 17.80 | 0.065 | 0.322 | 8.913 |
| Gold(ous), Au | 6.20 | 0.023 | 0.697 | 19.29 |
| Gold(ic), Au | 18.60 | 0.068 | 0.697 | 19.29 |
| Nickel, Ni | 19.00 | 0.069 | 0.322 | 8.913 |
| Platinum | 27.80 | 0.102 | 0.775 | 21.45 |
| Silver, Ag | 6.20 | 0.023 | 0.380 | 10.52 |
| Tin(ous), Sn | 7.80 | 0.029 | 0.264 | 7.307 |
| Tin(ic), Sn | 15.60 | 0.057 | 0.264 | 7.307 |
| Zinc, Zn | 14.30 | 0.052 | 0.258 | 7.141 |

**2.** ***Compute the weight of metal required***

The plating metal weight = (area plated, $in^2$) (plating thickness, in) (plating metal density, lb/$in^3$). For this plating job, given the density of zinc from Table 11, the plating metal weight = $(60 \times 144)(0.004)(0.258) = 8.91$ lb (4.0 kg) of zinc. In this calculation the value 144 is used to convert 60 $ft^2$ to square inches.

**Related Calculations:** The efficiency of finishing cathodes is high, ranging from 80 to nearly 100 percent. Where the actual efficiency is unknown, assume a value of 80 percent and the results obtained will be safe for most situations.

## SHRINK- AND EXPANSION-FIT ANALYSES

To what temperature must an SAE 1010 steel ring 24 in (61.0 cm) in inside diameter be raised above a 68°F (20°C) room temperature to expand it 0.004 in (0.10 mm) if the linear coefficient of expansion of the steel is 0.0000068 in/(in·°F)[0.000012 cm/(cm·°C)]? To what temperature must a 2-in (5.08-cm) diameter SAE steel shaft be reduced to fit it into a 1.997-in (5.07-cm) diameter hole for an expansion fit? What cooling medium should be used?

### Calculation Procedure:

**1.** ***Compute the required shrink-fit temperature rise***

The temperature needed to expand a metal ring a given amount before making a shrink fit is given by $T = E/(Kd)$, where $T$ = temperature rise *above* room temperature, °F; $K$ = linear coefficient of expansion of the metal ring, in/(in·°F); $d$ = ring internal diameter, in. For this ring, $T = 0.004/[(0.0000068)(24)] = 21.5$°F (11.9°C). With a room temperature of 68°F (20.0°C), the final temperature of the ring must be $68 + 21.5 = 89.5$°F (31.9°C) or higher.

**2.** ***Compute the temperature for an expansion fit***

Nitrogen, air, and oxygen in liquid form have a low boiling point, as does dry ice (solid carbon dioxide). Nitrogen and dry ice are considered the safest cooling media for expansion fits because both are relatively inert. Liquid nitrogen boils at −320.4°F (−195.8°C) and dry ice at −109.3°F (−78.5°C). At −320°F (−195.6°C) liquid nitrogen will reduce the diameter of metal parts by the amount shown in Table 12. Dry ice will reduce the diameter by about one-third the values listed in Table 12.

With liquid nitrogen, the diameter of a 2-in (5.1-cm) round shaft will be reduced by $(2.0)(0.0022) = 0.0044$ in (0.11 mm), given the value for SAE steels from Table 12. Thus, the diameter of the shaft at −320.4°F (−195.8°C) will be $2.000 - 0.0044 = 1.9956$ in (5.069 cm). Since the hole is 1.997 in (5.072 cm) in diameter, the liquid nitrogen will reduce the shaft size sufficiently.

If dry ice were used, the shaft diameter would be reduced $0.0044/3 = 0.00146$ in (0.037 mm), giving a final shaft diameter of $2.00000 - 0.00146 = 1.99854$ in (5.076 cm). This is too large to fit into a 1.997-in (5.072-cm) hole. Thus, dry ice is unsuitable as a cooling medium.

**TABLE 12** Metal Shrinkage with Nitrogen Cooling

| Metal | Shrinkage, in/in (cm/cm) of shaft diameter |
|---|---|
| Magnesium alloys | 0.0046 |
| Aluminum alloys | 0.0042 |
| Copper alloys | 0.0033 |
| Cr-Ni alloys (18-8 to 18-12) | 0.0029 |
| Monel metals | 0.0023 |
| SAE steels | 0.0022 |
| Cr steels (5 to 27% Cr) | 0.0019 |
| Cast iron (not alloyed) | 0.0017 |

## PRESS-FIT FORCE, STRESS, AND SLIPPAGE TORQUE

What force is required to press a 4-in (10.2-cm) outside-diameter cast-iron hub on a 2-in (5.1-cm) outside-diameter steel shaft if the allowance is 0.001-in interference per inch (0.001 cm/cm) of shaft diameter, the length of fit is 6 in (15.2 cm), and the coefficient of friction is 0.15? What is the maximum tensile stress at the hub bore? What torque is required to produce complete slippage of the hub on the shaft?

### Calculation Procedure:

**1. *Determine the unit press-fit pressure***

Figure 5 shows that with an allowance of 0.001 in interference per inch (0.001 cm/cm) of shaft diameter and a shaft-to-hub diameter ratio of $2/4 = 0.5$, the unit press-fit pressure between the hub and the shaft is $p = 6800$ lb/in$^2$ (46,886.0 kPa).

**2. *Compute the press-fit force***

The press-fit force $F$ tons $= \pi f p d L/2000$, where $f$ = coefficient of friction between hub and shaft; $p$ = unit press-fit pressure, lb/in$^2$; $d$ = shaft diameter, in; $L$ = length of fit, in. For this press fit, $F = (\pi)(0.15)(6800)(2.0)(6)/2000 = 19.25$ tons (17.4 t).

**3. *Determine the hub bore stress***

Use Fig. 6 to determine the hub bore stress. Enter the bottom of Fig. 6 at 0.0010 in (0.0010 cm) interference allowance per in of shaft diameter and project vertically to $d/D = 0.5$. At the left read the hub stress as 11,600 lb/in$^2$ (79,982 kPa).

**4. *Compute the slippage torque***

The torque, in·lb, required to produce complete slippage of a press fit is $T = 0.5\ \pi f p L d^2$, or $T = 0.5(3.1416)(0.15)(6800)(6)(2)^2 = 38{,}450$ in·lb (4344.1 N·m).

**Related Calculations:** Figure 7 shows the press-fit pressures existing with a steel hub on a steel shaft. The three charts presented in this calculation procedure are useful for many different press fits, including those using a hollow shaft having an internal diameter less than 25 percent of the external diameter and for all solid steel shafts.

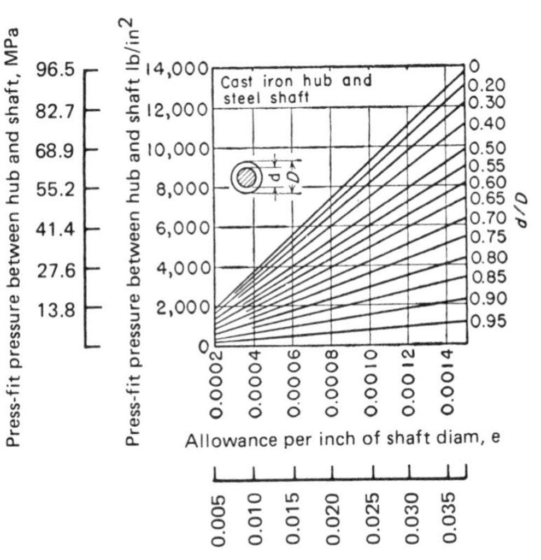


**FIG. 5** Press-fit pressures between steel hub and shaft.

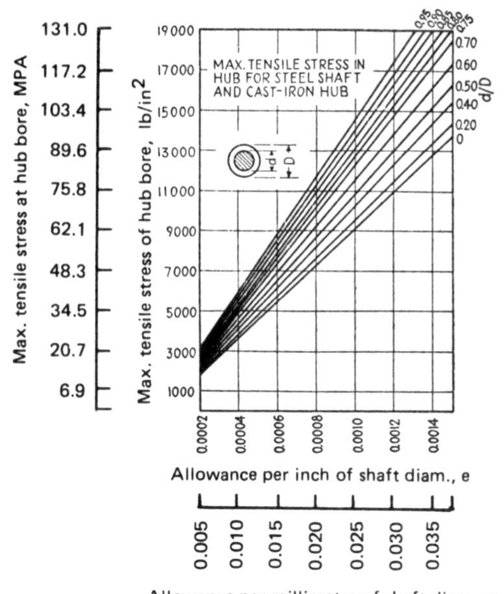


**FIG. 6** Variation in tensile stress in cast-iron hub in press-fit allowance.

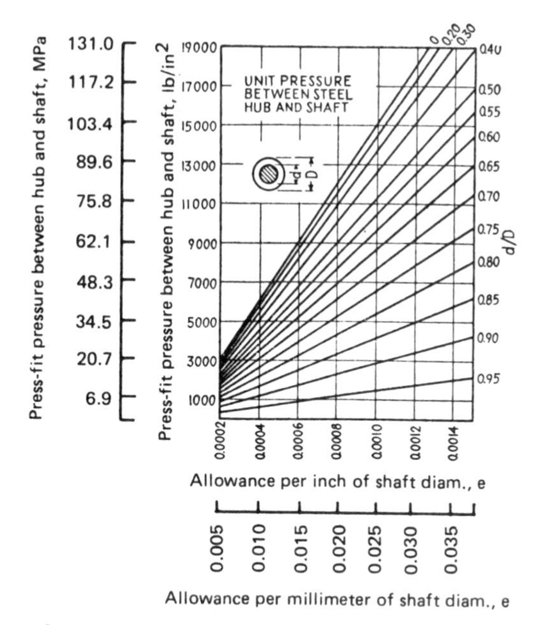


**FIG. 7** Press-fit pressures between cast-iron hub and shaft.

## LEARNING-CURVE ANALYSIS AND CONSTRUCTION

A short-run metalworking job requires five operators. The longest individual learning time for the new task is 3 days; 2 days are allowed for group familiarization with the task. If the normal output is 1000 units per 8-hr day, determine the daily allowance per operator when the standard for 100 percent performance is 0.8 worker-hour per 100 units produced.

### Calculation Procedure:

#### 1. *Plot the learning curve*

A learning curve shows the improvement that occurs with repetition of a task. Figure 8 is a typical learning curve with the learning period, days, plotted against the percent of methods time measurement (MTM) determined normal task. The shape of the curve, once determined for a given operation, does not change. The horizontal scale division is, however, changed to suit the minimum learning period for 100 percent performance. Thus, for a 3-day learning period the horizontal scale becomes 3 days. The coordinate at each of these three points (i.e., days) becomes the minimum expected task for each day. Performance above these tasks rates a bonus. The base of 60 percent of normal performance for the first day of learning for all jobs is attainable and meets management's minimum requirements.

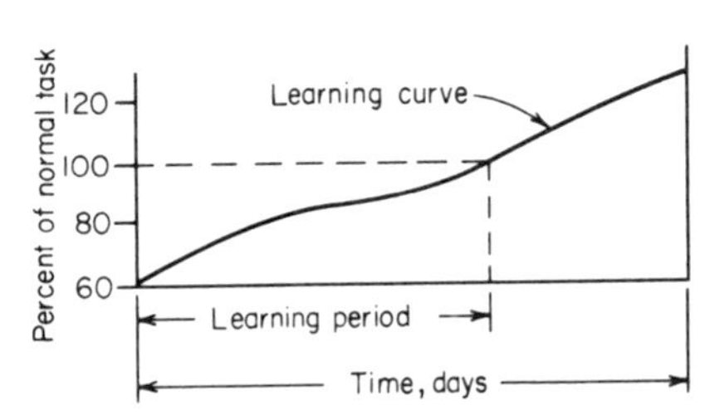


**FIG. 8** Typical learning curve for a metalworking task.

#### 2. *Determine the learning period to allow*

(*a*) Find the learning time, by test, for each work station in the group. (*b*) Select the longest individual learning time—in this instance, 3 days. (*c*) Add a group familiarization allowance when the group exceeds three operators—2 days here. (*d*) Find the sum of $b + \bar{c}$, or $3 + 2 = 5$ days. This is the learning period to allow.

#### 3. *Find the task for each day*

Divide the horizontal learning-period axis into five parts, one part for each of the 5 learning-period days allowed. Draw an ordinate for each day, and read the percentage task for that day at the intersection with the learning curve, or: 60.0; 70.5; 75.5; 80.0; 87 percent for days 1, 2, 3, 4, and 5, respectively.

#### 4. *Compute the daily task and daily allowance*

With a normal (100 percent) task of 1000 units for an 8-hr day, set up a table (like Table 13) of daily tasks and time allowance during the learning period. Begin with a column listing the number of learning days. In the next column, list the percentage learning performance read from Fig. 8. Find the daily task in units, column 3 of Table 13, by taking the product of 1000 units and the percentage learning performance, column 2, expressed as a decimal. Last, compute the daily

**TABLE 13** Learning-Curve Analysis

| Learning days | Percentage of learning performance | Daily task, units | Daily allowance per operator, in units |
|---|---|---|---|
| 1 | 60.0 | 600 | 40% × 8 = 3.20 |
| 2 | 70.5 | 705 | 29.5% × 8 = 2.36 |
| 3 | 75.5 | 755 | 24.5% × 8 = 1.96 |
| 4 | 80.0 | 800 | 20.0% × 8 = 1.60 |
| 5 | 87.0 | 870 | 13.0% × 8 = 1.04 |
| 6 | 100.0 | 1000 | 0% × 8 = 0 |

allowance per operator by finding the product of 8 h and the difference between 1.00 and the percentage of learning performance; i.e., for day 1: (1.00 − 0.60)(8) = 3.2 h. Tabulate the results in the fourth column of Table 13.

### 5. *Compute the incentive pay for the group*

In this plant the incentive pay is found by taking the product of the production in units and the standard set for 100 percent performance, or 0.80 worker-hour per 100 units produced. Thus, production of 600 units on day 1 will earn (600/100)(0.80) = 4.8-h pay for each group. Add to this the learning allowance of 3.2 h for day 1, and each group has earned 4.8 + 3.2 = 8-h pay for 8-h work.

If the group produced 700 units during day 1, it would earn (700/100)(0.80) = 5.6-h pay at this standard. With the learning allowance of 3.2 h, the daily earnings would be 5.6 + 3.2 = 8.8-h pay for 8-h work. This is exactly what is desired. The operator is rewarded for learning quickly.

**Related Calculations:** Select the length of the learning period for any new short-run task by conferring with representatives of the manufacturing, industrial engineering, and industrial relations departments. A simple operation that will be performed 1000 to 2000 times in an 8-h period would require a 3-day learning period. This is considered the minimum time for bringing such an operation up to normal speed. This is also true if a small group (three or less operators) perform equally simple operations. With larger groups (four or more operators), both simple and complex operations require an additional allowance for operators to adjust themselves to each other. Two days is a justified allowance for up to 15 operators learning to cooperate with one another under incentive conditions.

To prepare a plant-wide learning curve, keep records of the learning rates for a number of short-run tasks. Combine these data to prepare a typical learning curve for a particular plant. The method developed here was first described in *Factory*, now *Modern Manufacturing*, magazine.

## LEARNING-CURVE EVALUATION OF MANUFACTURING TIME

A metalworking process requires 1.00 h for manufacture of the first unit of a production run. If the operator has an improvement or learning rate of 90 percent, determine the time required to manufacture the 2d, 4th, 8th, and 16th units. What is the cumulative average unit time for the 16th unit? If 100 units are manufactured, what is the cumulative average time for the 100th unit? What is the unit manufacturing time for the 100th item?

### Calculation Procedure:

### 1. *Compute the unit time for the production cycle*

The learning curve relates the production time to the number of units produced. When the number of units produced doubles, the time required to produce the unit representing the doubled quantity is: (Learning rate, percent)(time, h or min, to produce the unit representing one-half the doubled quantity). Or, for the production line being considered here:

| Unit number | Production time, h |
|---|---|
| 1 | 1.00 |
| 2 | 0.90(1.00) = 0.900 |
| 4 | 0.90(0.90) = 0.810 |
| 8 | 0.90(0.81) = 0.729 |
| 16 | 0.90(0.729) = 0.656 |

### 2. *Compute the cumulative average unit time*

The cumulative average unit time for any unit in a production run = (Σ unit time for each item in the run)/(number of items in the run). Thus, computing the time for items 1 through 16 as shown in step 1, and taking the sum, we get the cumulative average unit time = 12.044 h/16 units = 0.752 h.

**TABLE 14** Learning-Curve Factors

| No. of units | Learning rate, percent | | |
|---|---|---|---|
| | 85 | 90 | 95 |
| | Time or cost, percent of unit 1 | Time or cost, percent of unit 1 | Time or cost, percent of unit 1 |
| 1 | 1.000 | 1.000 | 1.000 |
| 2 | 0.850 | 0.900 | 0.950 |
| 4 | 0.723 | 0.810 | 0.903 |
| 8 | 0.614 | 0.729 | 0.857 |
| 16 | 0.522 | 0.656 | 0.815 |
| 32 | 0.444 | 0.591 | 0.774 |
| 64 | 0.377 | 0.531 | 0.735 |
| 100 | 0.340 | 0.497 | 0.711 |

Learning-curve slopes

| Learning rate, percent | Curve slope |
|---|---|
| 70 | −0.514 |
| 75 | −0.415 |
| 80 | −0.322 |
| 85 | −0.234 |
| 90 | −0.152 |
| 95 | −0.074 |

**3. *Compute the cumulative average time for the 100th unit***

Set up a ratio of the learning factor for the 100th unit/learning factor for the 16th unit, and multiply the ratio by the cumulative average 16th unit time. Or, from the factors in Table 14, (0.497/0.656)(0.752 h) = 0.570 h.

**4. *Compute the unit time for the 100th unit***

Using the factor for the 90 percent learning curve in Table 14, the unit time for the 100th unit made = (1.00 h)(0.497) = 0.497 h.

**Related Calculations:** When using learning curves, be extremely careful to distinguish between *unit time* and *cumulative average unit time*. The unit time is the time required to make a particular unit in a production run, say the 10th, 16th, etc. Thus, a unit time of 0.5 h for the 16th unit in a production run means that the time required to make the 16th unit is 0.5 h. The 15th unit will require *more* time to make it; the 17th unit will require *less* time.

The cumulative average unit time is the *average* time to manufacture a given number of identical items. To obtain the cumulative average unit time for any given number of items, take the sum of the time required for each item up to and including that item and divide the sum by the number of items.

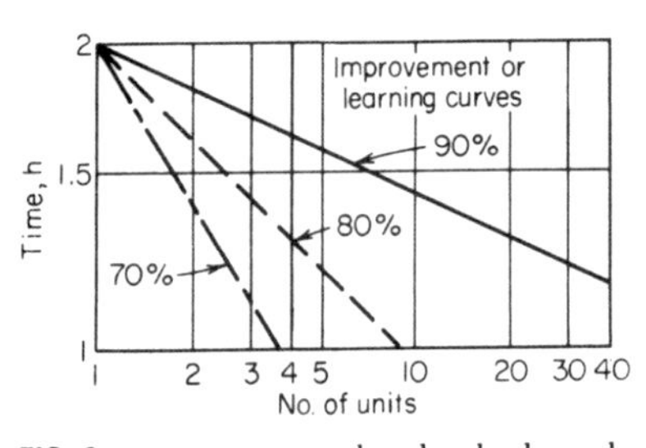


**FIG. 9** Learning curves plotted on log-log scale.

Either the *unit time or cumulative average unit time* can be used in manufacturing time or cost estimates, as long as the estimator knows which time value is being used. Failure to recognize the respective time values can result in serious errors.

A learning curve plotted on log-log coordinates is a straight line, Fig. 9. The slope of typical learning curves is listed in Table 14. Since a learning curve slopes downward—i.e., the unit manufacturing time decreases as more units are produced—the slope is expressed as a negative value.

Typical improvement or learning rates are: machining, drilling, etc., 90 to 95 percent; short-cycle bench assembly, 85 to 90 percent; equipment maintenance, 75 to 80 percent; electronics assembly and welding, 80 to 90 percent; general assembly, 70 to 80 percent. When an operation consists of several tasks having different learning rates, compute the overall learning rate for the task by taking the sum of the product of each learning rate (LR) and the percentage of the total task it represents. Thus, with $LR_1 = 0.90$ for 60 percent of the total task; $LR_2 = 0.80$ for 20 percent of the total task; $LR_3 = 0.70$ for 10 percent of the total task, the overall learning rate $LR = 0.90(0.60) + 0.80(0.20) + 0.70(0.10) = 0.77$.

Note that in machine-paced operations—i.e., those in which the speed of the machine controls the operator's activities—there is less chance for the operator to learn. Hence, the learning rate will be higher—90 to 95 percent—than in worker-paced operations that have learning rates of 70 to 80 percent. When learning or improvement ceases, the operator has reached the level-off point, and the task cannot be performed any more rapidly. The ratio set up in step 3 can use any two items in a production run, provided that the cumulative average time for the smaller item is multiplied by the ratio.

## DETERMINING BRINELL HARDNESS

A 3000-kg load is put on a 10-mm diameter ball to determine the Brinell hardness of a steel. The ball produces a 4-mm-diameter identation in 30 s. What is the Brinell hardness of the steel?

### Calculation Procedure:

**1. *Determine the Brinell hardness by using an exact equation***

The standard equation for determining the Brinell hardness is $BHN = F/(\pi d_1/2)(d_1 - \sqrt{d_1^2 - d_s^2})$, where $F$ = force on ball, kg; $d_1$ = ball diameter, mm; $d_s$ = indentation diameter, mm. For this test, $BHN = 3000/(\pi \times 10/2)(10 - \sqrt{10^2 - 4^2}) = 229$.

**2. *Compute the Brinell hardness by using an approximate equation***

One useful approximate equation for Brinell hardness is $BHN = (4F/\pi d_s^2) - 10$. For this test, $BHN = (4 \times 3000/\pi \times 4^2) - 10 = 228.5$. This compares favorably with the exact formula. For Brinell hardness exceeding 200, the approximate equation gives results that are less than 0.1 percent in error.

**Related Calculations:** Use this procedure for iron, steel, brass, bronze, and other hard or soft metals. A 500-kg test load is used for soft metals (brass, bronze, etc.). For Brinell hardness above 500, use a tungsten-carbide ball. The metal tested should be at least 10 times as thick as the indentation depth and wide enough so that no metal flows toward the edges of the specimen. The metal surface must be clean and free of defects.

## ECONOMICAL CUTTING SPEEDS AND PRODUCTION RATES

A cutting tool used to cut beryllium costs \$6 with its shank and can be reground for reuse five times. The average tool-changing time is 5 min. What is the most economical cutting speed if the machine labor rate is \$3 per hour and the overhead is 200 percent? What is the cutting speed for the maximum production rate? The cost of regrinding the tool is 35 cents per edge.

### Calculation Procedure:

**1. *Determine the tool cost factor***

The cost factor $T_c + Y/X$ for a tool is composed of $T_c$ = time to change tool, min; $Y$ = tool cost per cutting edge, including prorated initial cost plus reconditioning costs, cents; $X$ = machining rate, including labor and overhead, cents/min.

For this tool, $T_c = 5$ min. The tool can be reground five times after its original use, giving a total of $5 + 1 = 6$ cutting edges (five regrindings + the original edge) during its life. Since the tool costs \$6 new, the prorated cost per edge = \$6/6 edges = \$1, or 100 cents. The regrinding cost = 35 cents per edge; thus $Y = 100 + 35 = 135$ cents per edge.

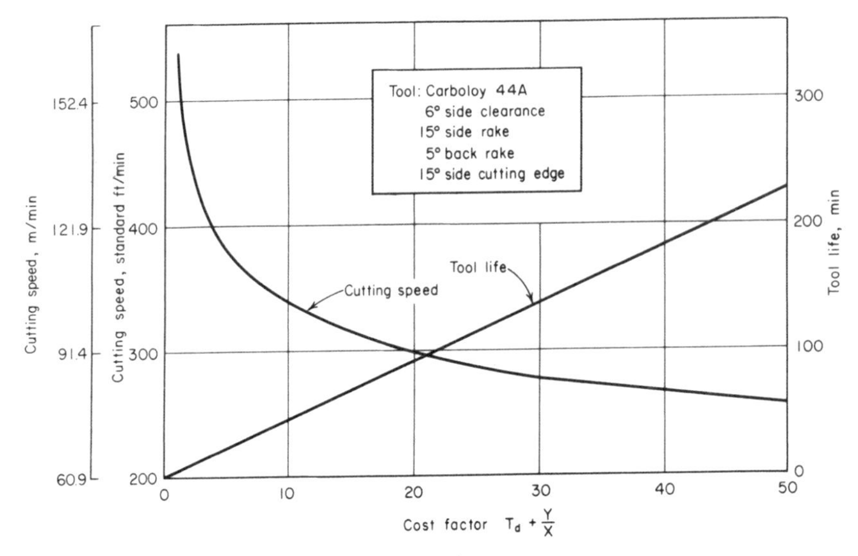


**FIG. 10** Optimum cutting-speed chart. *(American Machinist.)*

With a machine labor rate of \$3 per hour and an overhead factor of 200 percent, or 2.00(\$3) = \$6, the value of $X$ = machining rate = \$3 + \$6 = \$9, or 900 cents per hour, or 900/60 = 15 cents per minute. Then, the cost factor $T_c + Y/X = 5 + 135/15 = 14$.

**2. *Determine the cutting speed for minimum cost***

Enter Fig. 10 at a cost factor of 14 and project vertically upward until the cutting-speed curve is intersected. At the left, read the cutting speed for minimum tool cost as 320 surface ft/min.

**3. *Determine the probable tool life***

Project upward from the cost factor of 14 in Fig. 10 to the tool-life curve. At the right, read the tool life as 66 min.

**4. *Determine the speed and life for the maximum production rate***

Substitute the value of $T_c$ for the cost factor $T_c + Y/X$ on the horizontal scale of Fig. 10. As before, read the cutting speed and tool life at the intersection with the respective curves. The plotted values apply when the chip-removal suction devices will operate efficiently at the cutting speeds indicated by the curves. Thus, with $T_c = 5$, the cutting speed is 370 surface ft/min, and the tool life is 30 min.

**Related Calculations:** Figure 10, and similar optimum cutting-speed charts, is plotted for a specific land wear—in this case 0.010 in (0.03 cm). For a land wear of 0.015 in (0.04 cm), multiply the cutting speeds obtained from Fig. 10 by 1.13. However, a land wear of 0.015 in (0.04 cm) is not recommended because the wear rates are accelerated. If Carboloy 883 tools are used in place of the 44A grade plotted in Fig. 10, multiply the cutting speeds obtained from this chart by 1.12 for a 0.010-in (0.03-cm) wear land or 1.26 for a 0.015-in (0.04-cm) wear land.

Charts similar to Fig. 10 for other tool materials can be obtained from tool manufacturers. Do not use Fig. 10 for any tool material other than Carboloy 44A. The method presented here is the work of D. R. Walker and J. Gubas, as reported in *American Machinist*.

## OPTIMUM LOT SIZE IN MANUFACTURING

A manufacturing plant has a demand for 900 of its products per month on which the setup cost is \$10. The cost of each unit is \$5; the annual inventory charge is 12 percent/year of the average

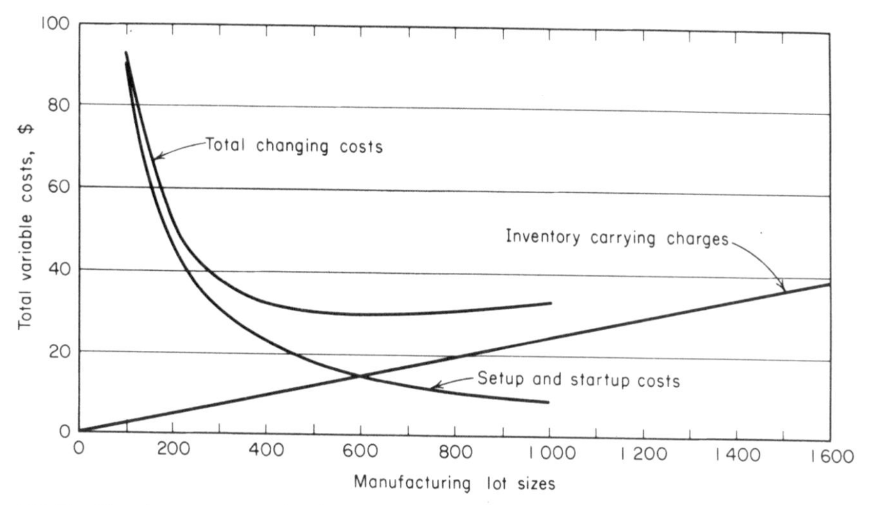


**FIG. 11** Changing costs associated with manufacturing lot sizes. *(American Machinist.)*

dollar value held in stock; the period for which the demand has occurred is 1/12 year. What is optimum manufacturing lot size? Plot a cost chart for this plant.

## Calculation Procedure:

### 1. *Determine the optimum manufacturing lot size*

Optimum manufacturing lot size can be found from: {2(demand, units, per period)(cost per setup, $)/[(demand period, fraction of a year)($ cost per unit)(annual inventory charge, percent of average $ value held in stock)]}$^{0.5}$. For this run, optimum lot size = {2(900)($10)/(1/12)($5)(0.12)]}$^{0.5}$ = 600 units.

### 2. *Plot a cost chart for this plant*

Figure 11 shows a typical cost chart. Plot each curve using production runs of 100, 200, 300, 400, . . . , 1600 units. The values for each curve are determined from: inventory carrying charges = (number of units in run)($ cost per unit)(annual inventory charge, percent)(demand period)/2; setup and startup costs = (demand during period, units)($ cost of setup)/(number of units in run); total changing costs = inventory carrying charges + setup and startup costs.

The units for these equations are the same as given in step 1. Note that the total-changing-costs curve is a minimum at the point where the inventory-carrying-charges curve and setup-and-startup-costs curve intersect. Also, the two latter curves intersect at the optimum manufacturing lot size—600 units, as computed in step 1.

**Related Calculations:** Economical lot-size relations are readily adaptable to machine-shop computations. With only slight changes, the same principles can be applied to determination of optimum-quantity purchases. The procedure described here is the work of I. Heitner, as reported in *American Machinist*.

## PRECISION DIMENSIONS AT VARIOUS TEMPERATURES

A magnesium workpiece with a dimension of 12.5000 to 12.4996 in (31.750 to 31.749 cm) is at a temperature of 85°F (29.4°C) after machining. The steel gage with which the dimensions of the workpiece will be checked is at 75°F (23.9°C). The workpiece must be gaged immediately to determine whether further grinding is necessary. Tolerance on the work is ±0.0002 in (0.005 mm). What should the dimensions of the workpiece be if there is not enough time available to allow the gage and workpiece temperatures to equalize? The standard reference temperature is 68°F (20.0°C).

### Calculation Procedure:

#### 1. *Compute the actual work dimensions*

The temperature of the workpiece is 85°F (29.4°C), or 85 − 68 = 17°F (9.4°C) above the standard reference of 68°F (20.0°C). Since the actual temperature of the workpiece is greater than the standard temperature, the dimensions of the workpiece will be larger than at the standard temperature because the part expands as its temperature increases.

To find the amount by which the workpiece will be oversize at the actual temperature, multiply the nominal dimension of the workpiece, 12.5 in (31.75 cm), by the coefficient of linear expansion of the material and by the difference between the actual and standard temperatures. For this magnesium workpiece, the average oversize amount at the actual temperature, 85°F (29.4°C), is $(12.5)(14.4 \times 10^{-6})(85 - 68) = 0.003060$ in (0.078 mm).

#### 2. *Compute the actual gage dimension*

Compute the actual gage dimension in a similar way, using the same dimension, 12.5 in (31.75 cm), but the coefficient of linear expansion of the gage material, steel, and the gage temperature, 75°F (23.9°C). Or, $(12.5)(6.4 \times 10^{-6})(75 - 68) = 0.000560$ in (0.014 mm).

#### 3. *Compute the workpiece dimension as a check*

The workpiece dimension, corrected for tolerance, plus the difference between the oversize amounts computed in steps 1 and 2 is the dimension to which the part should be checked at the existing shop and gage temperature.

Applying the tolerance, ±0.0002 in (±0.005 mm), to the drawing dimension, 12.5000 − 12.4996, gives a drawing dimension of 12.4998 ± 0.0002 in (31.7495 ± 0.0005 cm). Adding the difference between oversize dimensions, 0.003060 − 0.000560 = 0.002500 in (0.0635 mm), to 12.4998 ± 0.0002 in (31.7495 ± 0.0005 cm) gives a checking dimension of 12.5023 ± 0.0002 in (31.7558 ± 0.0005 cm). If personnel check the workpiece at this dimension, they will have full confidence that it will be the right size.

**Related Calculations:** This procedure can be used for any metal—bronze, aluminum, cast iron, etc.—for which the coefficient of linear expansion is known. Obtain the coefficient from Baumeister and Marks—*Standard Handbook for Mechanical Engineers*, or a similar reference. When a workpiece is at a temperature less than the National Bureau of Standards standard of 68°F (20.0°C), the part contracts instead of expanding. The dimension change computed in step 1 is then negative. This is also true of the gage, if it is at a temperature of less than 68°F (20.0°C). Note that the tolerance is constant regardless of the actual temperature of the part.

The procedure given here is the work of H. K. Eitelman, as reported in *American Machinist*.

## HORSEPOWER REQUIRED FOR METALWORKING

What is the input horsepower required for machining, on a geared-head lathe, a 4-in (10.2-cm) diameter piece of AISI 4140 steel having a hardness of 260 BHN if the depth of cut is 0.25 in (0.6 cm), the cutting speed is 300 ft/min (1.5 m/s), and the feed per revolution is 0.025 in (0.6 mm)?

### Calculation Procedure:

#### 1. *Determine the metal removal rate*

Compute the metal removal rate (MRR) in$^3$/min from MRR = $12fDC$, where $f$ = tool feed rate, in/r; $D$ = depth of cut, in; $C$ = cutting speed, ft/min. For this workpiece, MRR = 12(0.025)(0.25)(300) = 22.5 in$^3$/min (6.1 cm$^3$/s).

#### 2. *Determine the unit horsepower required*

Table 15 lists the average unit horsepower required for cutting various metals. The unit horsepower $hp_u$ is the power required to remove 1 in$^3$ (1 cm$^3$) of metal per minute at 100 percent efficiency of the machine. Table 15 shows that AISI 4130 to 4345 of 250 to 300 BHN, the range into which AISI 4140 260 BHN falls, has a unit hp of 0.70 (8.5 unit kW).

**TABLE 15** Average Unit hp (kW) Factors for Ferrous Metals and Alloys*

| Material classification | Brinell hardness number | | |
|---|---|---|---|
| | 201–250 | 251–300 | 301–350 |
| AISI 3160–3450 | 0.62 (7.52) | 0.75 (9.1) | 0.87 (10.6) |
| AISI 4130–4345 | 0.58 (7.04) | 0.70 (8.5) | 0.83 (10.1) |
| AISI 4615–4820 | 0.58 (7.04) | 0.70 (8.5) | 0.83 (10.1) |

*General Electric Company.

The unit horsepower must be corrected for feed. From Fig. 12 and a feed of 0.025 in/r (0.6 mm/r), the correction factor is found to be 0.90. Thus, the true unit horsepower = (0.70)(0.90) = 0.63 hp/(in$^3$·min) [28.7 W/(cm$^3$·min)].

### 3. *Compute the horsepower required at the cutter*

The horsepower required at the cutter $hp_c$ = $(hp_u)$(MRR), or $hp_c$ = (0.63)(22.5) = 14.18 hp (10.6 kW).

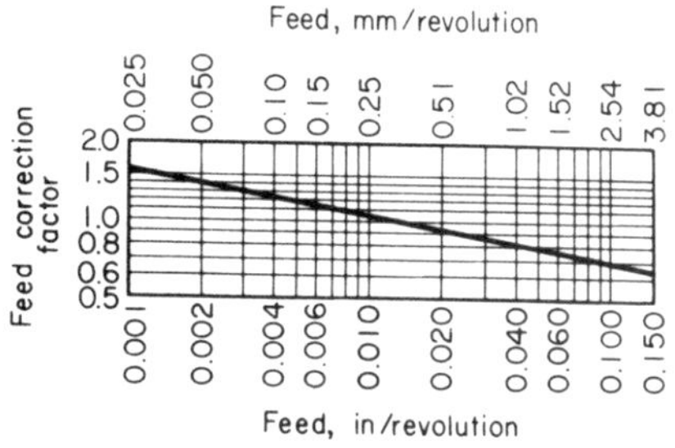

### 4. *Compute the motor horsepower required*

The power required at the cutter is the input necessary after allowing for losses in gears, bearings, and other parts of the drive. Table 16 lists typical overall machine-tool efficiencies. A gear-head lathe has an efficiency of 70 percent. Thus, $hp_m = hp_c/e$, where $e$ = machine-tool efficiency, expressed as a decimal. Or, $hp_m$ = 14.18/0.70 = 20.25 hp (15.1 kW). A 20-hp (14.9-kW) motor would be satisfactory for this machine.



**FIG. 12** Feed correction factors based on normal tool geometries. *(General Electric Co.)*

**TABLE 16** Efficiencies of Metalworking Machines*

| Typical overall machine-tool efficiency values (except milling machines), percent | Typical overall efficiencies for milling machines: Rated power of machine, hp | Rated power of machine, kW | Overall efficiency, percent |
|---|---|---|---|
| Direct spindle drive, 90 | 3 | 2.2 | 40 |
| | 5 | 3.7 | 48 |
| One-belt drive, 85 | 7.5 | 5.6 | 52 |
| Two-belt drive, 70 | 10 | 7.5 | 52 |
| | 15 | 11.2 | 52 |
| Geared head, 70 | 20 | 14.9 | 60 |
| | 25 | 18.6 | 65 |
| | 30 | 22.4 | 70 |
| | 40 | 29.8 | 75 |
| | 50 | 37.3 | 80 |

*General Electric Company.

**Related Calculations:** Use this procedure for single or multiple tools. When more than one tool is working at the same time, compute $hp_c$ for each tool and add the individual values to find the total $hp_c$. Divide the total $hp_c$ by the machine efficiency to determine the required motor horsepower $hp_m$. This procedure makes ample allowance for dulling of the tools.

Compute metal removal rates for other operations as follows. *Face milling:* MRR $= WDF_T$, where $W$ = width of cut, in; $D$ = depth of cut, in; $F_T$ = table feed, in/min. *Slot milling:* MRR $= WDF_T$, where all symbols are as before. *Planing or shaping:* MRR $= DfLS$, where $D$ = depth of cut, in; $f$ = feed, in per stroke or revolution; $L$ = length of workpiece, in; $S$ = strokes/min. *Multiple tools:* MRR $= (d_1^2 - d_s^2)\pi fR/4$, where $d_1$ = original diameter of workpiece, in, *before* cutting; $d_s$ = workpiece diameter *after* cutting, in; $R$ = rpm of workpiece; other symbols as before.

The procedure given here is the work of Robert G. Brierley and H. J. Siekmann as reported in *Machining Principles and Cost Control.*

## CUTTING SPEED FOR LOWEST-COST MACHINING

What is the optimal cutting speed for a part if the maximum feed for which an acceptable finish is obtained at 169 r/min of the workpiece is 0.011 in/r (0.3 mm/r) when the cost of labor and overhead is \$0.24 per hour, the number of pieces produced per tool change is 15, the cost per tool change is \$0.62, and the length of cut is 8 in (20.3 cm)?

### Calculation Procedure:

#### 1. *Compute the optimization factor*

When the lowest-cost machining speed for an operation is determined, one popular procedure is to choose any speed and feed at which the operation meets the finish requirements. If desired, the speed and feed at which the operation is now running might be chosen. By keeping the speed constant, the feed is increased to the maximum value for which the finish is acceptable. This is the optimal value for the feed and is called the *optimal feed.* The number of pieces produced under these conditions is measured between tool changes. Then the optimal cutting speed is computed from: optimal cutting speed, r/min = (chosen speed, r/min)(optimization factor).

For any operation, the optimization factor = {(labor and overhead cost, \$/min)(number of pieces per tool change)(length of cut, in)/[(3)(cost per tool change, \$)(chosen speed, r/min)(optimal feed, in/r)]}$^{-4}$. Substitute the given values. Thus, the optimization factor = $\{(0.24)(15)(8)/[(3)(0.62)(169)(0.011)]\}^{-4} = 1.7$.

#### 2. *Compute the optimal cutting speed*

From the relation given in step 1, optimal cutting speed = (chosen speed, r/min)(optimization factor) = (169)(1.7) = 287 r/min.

**Related Calculations:** The relation given in step 1 for the optimization factor is valid for carbide tools. It can be modified to apply to high-speed tools by changing the fourth root to an eight root and changing the 3 in the denominator to 7.

## REORDER QUANTITY FOR OUT-OF-STOCK PARTS

A metalworking process uses 10 parts during the lead time. How many parts should be reordered if an out-of-stock situation can be accepted for 10 percent of the time? For 35 percent of the time?

### Calculation Procedure:

#### 1. *Determine the out-of-stock factor*

Table 17 lists out-of-stock factors for various times during which a part might be out of stock. Thus, the acceptable out-of-stock factor for 10 percent is 1.29, and for 35 percent it is 0.39.

#### 2. *Compute the reorder quantity*

For any manufacturing process, reorder quantity = (out-of-stock factor)(usage during lead time)$^{0.5}$ + (usage during lead time). Thus, the reorder point for this process with an acceptable out-of-stock factor for 10 percent is $(1.29)(10)^{0.5} + (10) = 14.08$ parts, say 15 parts. With 35 percent, reorder point $= (0.29)(10)^{0.5} + 10 = 11.23$, or 12 parts.

**Related Calculations:** Use this procedure for any types of parts ordered from either an internal or external source. In general, reducing the allowable stock-out time will increase the time during which a process using the parts can operate.

## SAVINGS WITH MORE MACHINABLE MATERIALS

What are the gross and net savings made with a more machinable material that reduces the production time by 36 s per part when 800 lb (362.9 kg) of steel is required for 1000 parts and the total machine operating cost is $6 per hour? The more machinable material costs 4 cents per pound (8.8 cents per kilogram) more than the less machinable material, and 5000 parts are produced per day.

### Calculation Procedure:

**1.** ***Compute the gross savings possible***

The gross saving possible in a machining operation when a more machinable material is used is: gross saving, cents/lb = (machining time saved with new material, s per piece)(total cost of operating machine, cents/h)/[(3.6)(weight of material to make 1000 pieces, lb)]. For this operation, the gross saving = (36)(600)/[(3.6)(800)] = 7.5 cents per pound (16.5 cents per kilogram). With a production rate of 5000 parts per day, the gross saving is (5000 parts)(800 lb/1000 parts)(7.5 cents/lb) = 30,000 cents, or $300.

**2.** ***Compute the net savings possible***

The more machinable materials cost 4 cents more per pound than the less machinable material. Hence, the net saving is (7.5 − 4.0) = 3.5 cents per pound (7.7 cents per kilogram), or (5000 parts)(800 lb/1000 parts)(3.5 cents/lb) = 14,000 cents, or $140.

**Related Calculations:** Use this general procedure for parts made of any material—steel, brass, bronze, aluminum, plastic, etc.

**TABLE 17** Out-of-Stock Factors*

| Acceptable percentage of stock-outs | Out-of-stock factor |
|---|---|
| 50 | 0.00 |
| 45 | 0.13 |
| 40 | 0.26 |
| 35 | 0.39 |
| 25 | 0.68 |
| 15 | 1.04 |
| 10 | 1.29 |
| 5 | 1.65 |
| 4 | 1.76 |
| 3.5 | 1.82 |
| 3 | 1.89 |
| 2 | 2.06 |
| 1 | 2.33 |
| 0 | 4.0 |

*Nyles V. Reinfeld in *American Machinist*.

## TIME REQUIRED FOR THREAD MILLING

How long will it take to thread-mill a 2⅞-in (7.3-cm) diameter hard steel bolt with a 2½-in (6.4-cm) diameter 18-flute hob?

### Calculation Procedure:

**1.** ***Determine the cutting speed and feed of the hob***

Table 18 lists typical cutting speeds and feeds for various materials. For hard steel, the usual cutting speed in thread milling is 50 ft/min (0.3 m/s), and the feed per flute is 0.002 in (0.05 mm).

**2.** ***Compute the time required for thread milling***

The time required for thread milling, $T_t$ min $= \pi d/(fnR)$, where $d$ = work diameter, in; $f$ = feed, in per flute; $n$ = number of flutes on hob; $R$ = hob rpm.

From a previous calculation procedure, $R = 12C/(\pi d)$, where $C$ = hob cutting speed, ft/min; $d$ = hob diameter, in. For this hob, $R = (12)(50)/[\pi(2.5)] = 76.4$ r/min. Then $T_t = \pi(2\frac{7}{8})/[(0.002)(18)(76.4)] = 3.29$ min.

**TABLE 18** Thread-Milling Speeds and Feeds

| | Speed | | Feed per flute | |
|---|---|---|---|---|
| Material threaded | ft/min | m/s | in | mm |
| Aluminum | 500 | 2.5 | 0.0015 | 0.038 |
| Brass | 250 | 1.3 | 0.0015 | 0.038 |
| Mild steel | 100 | 0.5 | 0.0020 | 0.051 |
| Medium steel | 75 | 0.4 | 0.0020 | 0.051 |
| Hard steel | 50 | 0.3 | 0.0020 | 0.051 |

**Related Calculations:** Use this procedure for any metallic or nonmetallic material—aluminum, brass, mild steel, medium steel, hard steel, plastics, etc.

## DRILL PENETRATION RATE AND CENTERLESS GRINDER FEED RATE

What is the drill penetration rate when a drill turns at 1000 r/min and has a feed of 0.006 in/r (0.15 mm/r)? What is the feed rate of a centerless grinder having a 12-in (30.5-cm) diameter regulating wheel running at 60 r/min if the angle of inclination between the regulating and grinding wheel is 5°?

### Calculation Procedure:

**1. *Compute the rate of drill penetration***

The rate of drill penetration $P$ in/min $= fR$, where $f =$ drill feed, in/r; $R =$ drill rpm. For this drill, $P = (0.006)(1000) = 6.0$ in/min (2.5 mm/s).

**2. *Compute the grinder feed rate***

The work feed $f$ in/min in a centerless grinder is $f = \pi dR \sin a$, where $d =$ regulating-wheel diameter, in; $R =$ regulating-wheel rpm; $a =$ angle of inclination between the regulating and grinding wheel. For this grinder, $f = \pi(12)(60)(\sin 5°) = 197.6$ in/min (8.4 cm/s).

## BENDING, DIMPLING, AND DRAWING METAL PARTS

What is the minimum bend radius $R$ in for 0.02-gage Vascojet 1000 metal if it is bent transversely to an angle of 130°? What is the minimum radius $R$ in of a bend in 0.040-gage Rene 41 metal bent longitudinally at an angle of 52° at room temperature? Determine the maximum length of dimple flange $H$ in for AM-350 metal at 500°F (260°C) when the bend angle is 42° and the edge radius $R$ is 0.250 in (6.4 mm). Find the maximum blank diameter and maximum cup depth for drawing Rene 41 metal at 400°F (204°C) when using a die diameter of 10 in (25.4 cm) and 0.063-gage material. Figure 13*a*, *b*, and *c* shows the anticipated manufacturing conditions.

### Calculation Procedure:

**1. *Compute the minimum bend radius***

Table 19 shows that the critical bend angle (i.e., maximum bend angle $\alpha$ without breakage) for Vascojet 1000 metal is 118°. Hence, the required bend angle is greater than the critical bend angle. Therefore, the required bend limit equals the critical bend limit, and $R/T = 1.30$, from Table 19. Hence, the minimum radius $R_m = (R/T)(T) = (1.30)(0.02) = 0.026$ in (0.66 mm).

With Rene 41 metal, bent longitudinally at room temperature, the critical bend angle is 122°, from Table 19. Since the required bend angle of 52° is less than critical, find the $R/T$ value in the right-hand portion of Table 19. When the actual bend angle is between two tabulated angles, interpolate thus:

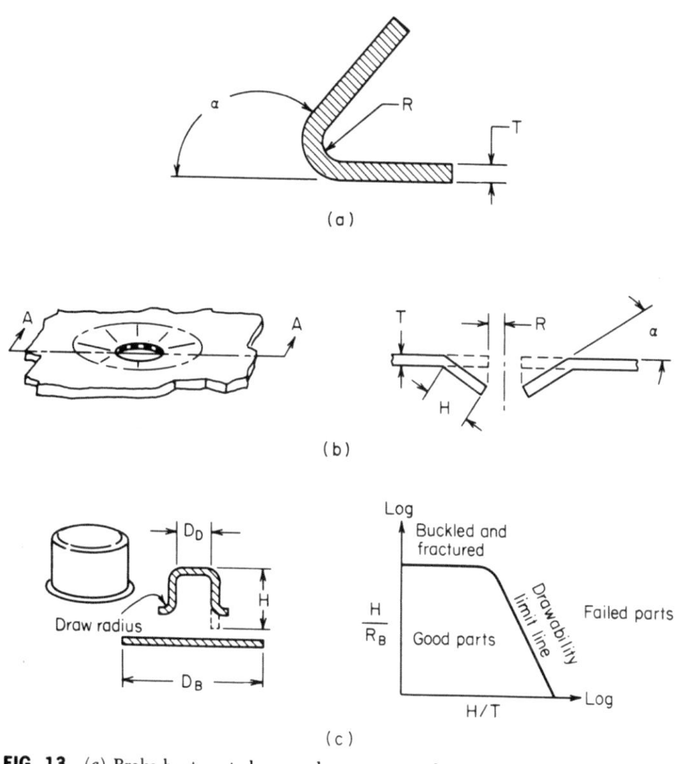


**FIG. 13** (*a*) Brake-bent part shape and parameters; (*b*) ram-coin dimpling setup; (*c*) drawing setup. (*American Machinist.*)

**TABLE 19** Brake-Bend Parts Parameters*

| Material | $L/T$ | $F$ | ∞ | $R/T$ | $R/T$ for angles ∞ below critical | | | | | | |
|---|---|---|---|---|---|---|---|---|---|---|---|
| | | | | | 30 | 45 | 60 | 75 | 90 | 105 | 120 |
| Titanium (13V-11Cr-3Al) | $L/T$ | $RT$ | 105 | 2.40 | 0.34 | 0.68 | 1.16 | 1.80 | 2.25 | 2.40 | |
| Vascojet 1000 | $L/T$ | $RT$ | 118 | 1.30 | 0.18 | 0.38 | 0.64 | 0.92 | 1.13 | 1.26 | 1.30 |
| USS 12 MoV | $L/T$ | $RT$ | 119 | 1.20 | 0.16 | 0.34 | 0.60 | 0.84 | 1.04 | 1.16 | 1.20 |
| 17-7PH | $L/T$ | $RT$ | 122 | 0.80 | 0.10 | 0.22 | 0.37 | 0.54 | 0.66 | 0.75 | 0.79 |
| AM-350 | $L/T$ | $RT$ | 122 | 0.80 | 0.10 | 0.22 | 0.37 | 0.54 | 0.66 | 0.75 | 0.79 |
| PH 15-7 Mo | $L/T$ | $RT$ | 121 | 0.86 | 0.11 | 0.23 | 0.42 | 0.60 | 0.72 | 0.80 | 0.84 |
| A-286 | $L/T$ | $RT$ | 124 | 0.66 | 0.07 | 0.15 | 0.29 | 0.43 | 0.54 | 0.62 | 0.65 |
| Hastelloy X | $L/T$ | $RT$ | 120 | 1.00 | 0.12 | 0.26 | 0.47 | 0.67 | 0.84 | 0.95 | 1.00 |
| Inconel X | $L/T$ | $RT$ | 124 | 0.64 | 0.06 | 0.14 | 0.28 | 0.41 | 0.52 | 0.60 | 0.63 |
| Rene 41 | $L$ | $RT$ | 122 | 0.80 | 0.10 | 0.22 | 0.37 | 0.54 | 0.66 | 0.75 | 0.79 |
| Rene 41 | $T$ | $RT$ | 113 | 1.64 | 0.28 | 0.53 | 0.84 | 1.16 | 1.44 | 1.58 | 1.64 |
| J-1570 | $L$ | $RT$ | 124 | 0.68 | 0.08 | 0.16 | 0.30 | 0.45 | 0.56 | 0.64 | 0.67 |

* *American Machinist*, LTV, Inc.; USAF.

*Note:* $L/T$ = grain direction, where $L$ = longitudinal and $T$ = transverse; $F$ = bending temperature; ∞ = critical bend; $R/T$ = critical bend limits.

| Angle,° | $R/T$ value |
|---|---|
| 60 | 0.37 |
| 52 | |
| 45 | 0.22 |

$[(52 - 45)/(60 - 45)](0.37 - 0.22) = 0.07$. Then $R/T$ for $52° = 0.22 + 0.07 = 0.29$. With $R/T$ known for 52°, compute the minimum radius $R_m = (R/T)(T) = (0.29)(0.040) = 0.0116$ in (0.2946 mm).

**2. *Determine the dimple-flange length***

Table 20 shows typical dimpling limits to avoid radial splitting at the edge of the hole of various modern materials. With a bend angle between the tabulated angles, interpolate thus:

| Angle, ° | $H/R$ value |
|---|---|
| 45 | 1.10 |
| 42 | |
| 40 | 1.43 |

$[(42 - 40)/(45 - 40)](1.10 - 1.43) = -0.132$, and $H/R = 1.43 + (-0.132) = 1.298$ at 42°. Then the maximum dimple-flange length $H_m = (H/R)(R) = (1.298)(0.250) = 0.325$ in (8.255 mm).

**3. *Determine the maximum blank diameter***

Table 21 lists the drawing limits for flat-bottom cups made of various modern materials. For this cup, $D_D/T = 10/0.063 = 158.6$, say 159. The corresponding $D_B/T$ and $H/D_D$ ratios are not tabulated. Therefore, interpolate between $D_D/T$ values of 150 and 200. Thus, for Rene 41:

| $D_D/T$ | $D_B/D_D$ |
|---|---|
| 200 | 1.52 |
| 159 | |
| 150 | 1.73 |

$[(159 - 150)/(200 - 150)](1.52 - 1.73) = -0.0378$, and $D_B/D_D = 1.73 + (-0.0378) = 1.692$, when $D_D/T = 159$. Then, the *maximum* value of $D_{Bm} = (D_B/D_D)(D_D) = (1.692)(10) = 16.92$ in (43.0 cm).

Interpolating as above yields $H/D_D = 0.48$ when $D_D/T = 159$. Then the maximum height $H_m = (H/D_D)(D) = (0.48)(10) = 4.8$ in (12.2 cm).

**Related Calculations:** The procedures given here are typical of those used for the newer "exotic" metals developed for use in aerospace, cryogenic, and similar advanced technologies. The three tables presented here were developed by LTV, Inc., for the U.S. Air Force, and reported in *American Machinist*.

## BLANK DIAMETERS FOR ROUND SHELLS

What blank diameter $D$ in is required for the round shells in Fig. 14*a* and *b* if $d = 12$ in (30.5 cm), $d_1 = 12$ in (30.5 cm), $d_2 = 14$ in (35.6 cm), and $h = 14$ in (35.6 cm)?

**TABLE 20** Dimpling Limits to Avoid Radial Splitting at Hole Edge*

| Material | Temperature, °F (°C) | Dimpling limit $H/R$: Standard, for various bend angles $a$; above and below standard bend angle | | | | |
|---|---|---|---|---|---|---|
| | | 30° | 35° | 40° | 45° | 50° |
| 2024-T3 | 70 (21.1) | 2.15 | 1.60 | 1.20 | 0.93 | 0.80 |
| Ti-8-1-1 | 70 (21.1) | 1.88 | 1.42 | 1.08 | 0.82 | 0.70 |
| TZM Moly | 70 (21.1) | 1.98 | 1.50 | 1.12 | 0.87 | 0.73 |
| Cb-752 | 70 (21.1) | 2.28 | 1.70 | 1.30 | 0.98 | 0.83 |
| PH 15-7 Mo | 500 (260.0) | 2.43 | 1.84 | 1.40 | 1.07 | 0.90 |
| AM-350 | 500 (260.0) | 2.46 | 1.87 | 1.43 | 1.10 | 0.93 |
| Ti-8-1-1 | 1200 (648.9) | 2.30 | 1.72 | 1.30 | 1.00 | 0.85 |
| Ti-13-11-3 | 1200 (648.9) | 2.58 | 1.95 | 1.48 | 1.15 | 0.95 |

**American Machinist*, LTV, Inc.; USAF.

**TABLE 21** Drawing Limits for Flat-Bottom Cups*

| Material | Temperature ratio, °F (°C) | Die to blank diameter ratios $D_B/D_D$; cup-depth ratios $H/D_D$: For various $D_D/T$ ratios | | | | | | | |
|---|---|---|---|---|---|---|---|---|---|
| | | 25 | 50 | 100 | 150 | 200 | 250 | 300 | 400 |
| Am-350 | 500 (260.0) $D_B/D_D$ | 2.22 | 2.18 | 2.00 | 1.71 | 1.54 | 1.42 | 1.40 | 1.30 |
| | $H/D_D$ | 0.97 | 0.95 | 0.75 | 0.50 | 0.37 | 0.30 | 0.26 | 0.20 |
| A-286 | 1000 (537.8) $D_B/D_D$ | 2.22 | 2.46 | 2.16 | 1.85 | 1.64 | 1.49 | 1.41 | 1.36 |
| | $H/D_D$ | 1.00 | 1.21 | 0.87 | 0.57 | 0.42 | 0.34 | 0.28 | 0.22 |
| Rene 41 | 400 (204.4) $D_B/D_D$ | 2.22 | 2.22 | 1.92 | 1.73 | 1.52 | 1.48 | 1.42 | 1.33 |
| | $H/D_D$ | 0.97 | 0.97 | 0.73 | 0.51 | 0.37 | 0.31 | 0.27 | 0.21 |
| L-605 | 500 (260.0) $D_B/D_D$ | 2.22 | 2.29 | 2.00 | 1.68 | 1.54 | 1.45 | 1.44 | 1.38 |
| | $H/D_D$ | 0.97 | 1.05 | 0.74 | 0.47 | 0.35 | 0.30 | 0.26 | 0.21 |
| T1-13-11-3 | 1200 (648.9) $D_B/D$ | 2.38 | 2.53 | 2.28 | 1.92 | 1.67 | 1.58 | 1.45 | 1.44 |
| | $H/D_D$ | 1.15 | 1.34 | 0.94 | 0.60 | 0.44 | 0.35 | 0.30 | 0.24 |
| Tungsten | 600 (315.6) $D_B/D_D$ | 2.08 | 2.11 | 1.98 | 1.66 | 1.53 | 1.46 | 1.38 | 1.34 |
| | $H/D_D$ | 0.83 | 0.87 | 0.69 | 0.45 | 0.35 | 0.29 | 0.24 | 0.20 |

**American Machinist*, LTV, Inc.; USAF.

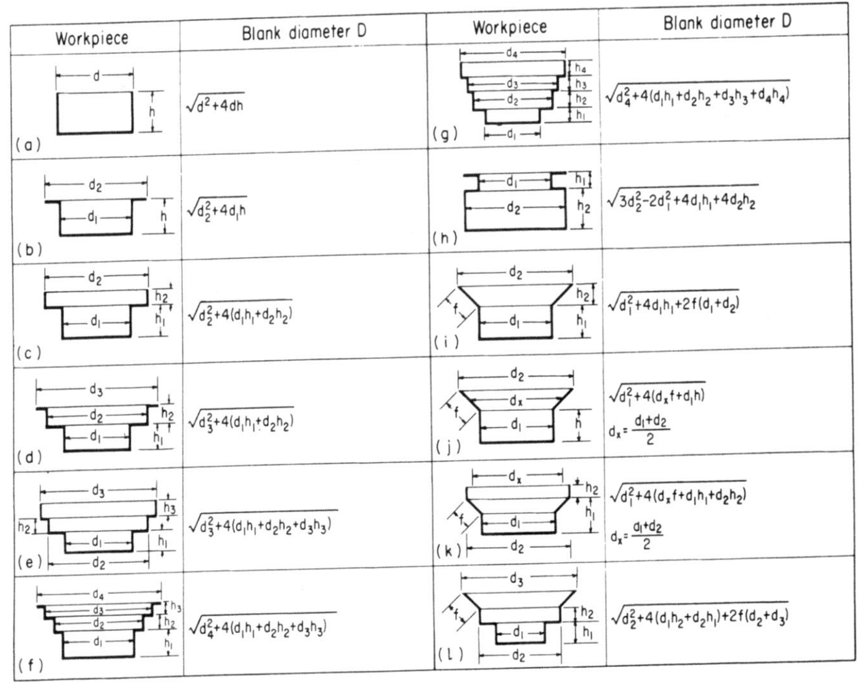


**FIG. 14** Blank diameters for round shells. *(American Machinist.)*

## Calculation Procedure:

### 1. *Compute the plain-cup blank diameter*

Figure 14*a* shows the plain cup. Compute the required blank diameter from $D = (d^2 + 4dh)^{0.5} = (12^2 + 4 \times 12 \times 4)^{0.5} = 18.33$ in (46.6 cm).

### 2. *Compute the flanged-cup blank diameter*

Figure 14*b* shows the flanged cup. Compute the required blank diameter from $D = (d_2^2 + fd_1h)^{0.5} = (14^2 + 4 \times 12 \times 4)^{0.5} = 19.7$ in (49.3 cm).

**Related Calculations:** Figure 14 gives the equations for computing 12 different round-shell blank diameters. Use the same general procedures as in steps 1 and 2 above. These equations were derived by Ferene Kuchta, Mechanical Engineer, J. Wiss & Sons Co., and reported in *American Machinist*.

## BREAKEVEN CONSIDERATIONS IN MANUFACTURING OPERATIONS

A manufacturing plant has the net sales and fixed and variable expenses shown in Table 22. What is the breakeven point for this plant in units and sales? Plot a conventional and an alternative breakeven chart for this plant.

## Calculation Procedure:

### 1. *Compute the breakeven units*

Use the relation $BE_u$ = fixed expenses, \$/[(sales income per unit, \$ − variable costs per unit, \$)], where $BE_u$ = breakeven in units. By substituting, $BE_u$ = \$400,000/(\$20 − \$12) = 50,000 units.

**TABLE 22** Manufacturing Business Income and Expenses

Condensed Income Statement
*For year ending Dec. 31, 19—*

| | | | |
|---|---|---|---|
| Net sales (60,000 units @ $20 per unit) | | | $1,200,000 |
| Less costs and expenses: | *Variable* | *Fixed* | |
| Direct material | $195,000 | . . . | |
| Direct labor | 215,000 | . . . | |
| Manufacturing expenses | 100,000 | $200,000 | |
| Selling expenses | 50,000 | 150,000 | |
| General and administrative expenses | 160,000 | 50,000 | |
| Total | 720,000 | 400,000 | 1,120,000 |
| Net profit before federal income taxes | | | 80,000 |

### 2. *Compute the breakeven sales*

Two methods can be used to compute the breakeven sales. With the breakeven units known from step 1, $BE_s$ = $BE_u$(unit sales price, \$), where $BE_s$ = breakeven sales, \$. By substituting, $BE_s$ = (50,000)(\$20) = \$1,000,000.

Alternatively, compute the profit-volume (PV) ratio: PV = (sales, \$ − variable costs, \$)/sales, \$ = (\$1,200,000 − \$720,000)/\$1,200,000) = 0.40. Then $BE_s$ = fixed costs, \$/PV = \$400,000/0.40 = \$1,000,000. This is identical to the breakeven sales computed in the previous paragraph.

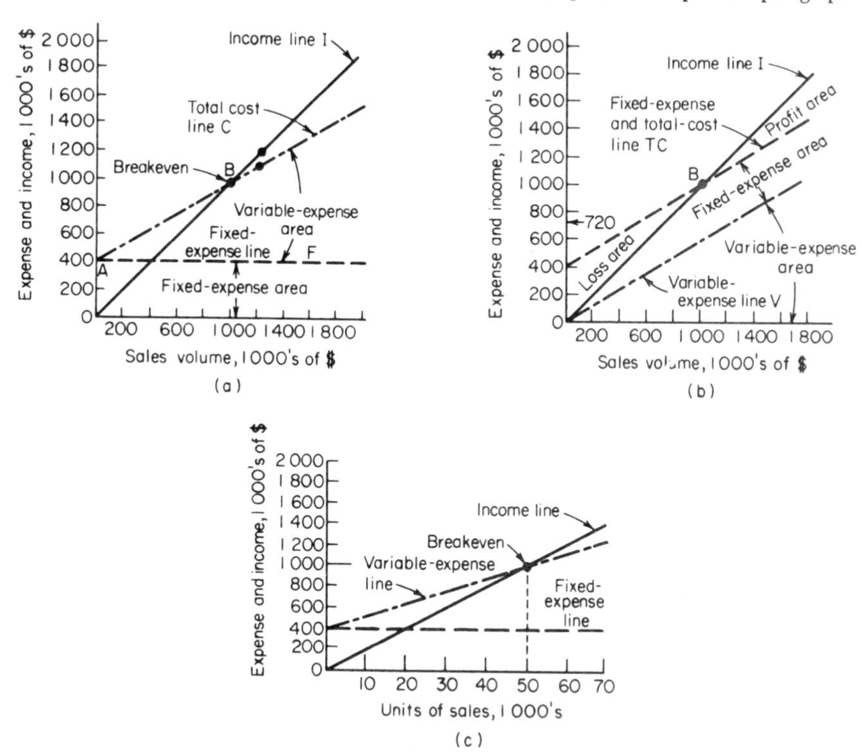


**FIG. 15** Three forms of the breakeven chart as used in metalworking activities.

### 3. *Draw the conventional and alternative breakeven charts*

Figure 15*a* shows the conventional breakeven chart for this plant. Construct this chart by drawing the horizontal line $F$ for the fixed expenses, the solid sloping line $I$ for the income or sales, and the dotted sloping line $C$ for the total costs. Note that the vertical axis is for the expenses and income and the horizontal axis is for the sales, all measured in monetary units.

The breakeven point is at the intersection of the income and total-cost curves, point $B$, Fig. 15*a*. Projecting vertically downward shows that point $B$ corresponds to a sales volume of $1,000,000, as computed in step 2.

Alternative breakeven charts are shown in Fig. 15*b* and *c*. Both charts are constructed in a manner similar to Fig. 15*a*.

**Related Calculations:** Breakeven computations are valuable tools for analyzing any manufacturing operation. The concepts are also applicable to other business activities. Thus, typical PV values for various types of businesses are:

| Business | Typical activity | Typical PV |
|---|---|---|
| Consumer appliances | Fully automated; high-volume output | 0.15–0.25 |
| Standard centrifugal pumps | Batch output in large volume | 0.20–0.30 |
| Acid-handling centrifugal pumps | Batch output in small volume | 0.25–0.35 |
| Standard prototype, one-of-a-kind | Ships, machine tools | 0.30–0.40 |
| Special-design one-of-a-kind prototype | Buildings, factories | 0.35–0.50 |

# INDEX